T0328268

The Atlantic Walrus

Multidisciplinary Insights into Human-Animal Interactions

The Atlantic Walrus

Multidisciplinary Insights into Human-Animal Interactions

Edited by

Xénia Keighley

Globe Institute, University of Copenhagen, Copenhagen, Denmark;
Arctic Centre, University of Groningen, Groningen, The Netherlands

Morten Tange Olsen

Globe Institute, University of Copenhagen, Copenhagen, Denmark

Peter Jordan

Department of Archaeology and Ancient History, Lund University, Lund, Sweden

Sean Desjardins

Arctic Centre/Groningen Institute of Archaeology, University of Groningen, Groningen, The Netherlands;
Canadian Museum of Nature, Ottawa, Ontario, Canada

Academic Press is an imprint of Elsevier
125 London Wall, London EC2Y 5AS, United Kingdom
525 B Street, Suite 1650, San Diego, CA 92101, United States
50 Hampshire Street, 5th Floor, Cambridge, MA 02139, United States
The Boulevard, Langford Lane, Kidlington, Oxford OX5 1GB, United Kingdom

British Library Cataloguing-in-Publication Data
A catalogue record for this book is available from the British Library

Library of Congress Cataloging-in-Publication Data
A catalog record for this book is available from the Library of Congress

ISBN: 978-0-12-817430-2

Original cover art and design and scientific illustrations created by Elena Kakoshina.
To view more of her work, please visit: www.artkakos.com.

Publisher: Charlotte Cockle
Acquisitions Editor: Anna Valutkevich
Editorial Project Manager: Lindsay Lawrence
Production Project Manager: Maria Bernard
Cover Typesetting: Greg Harris

Typeset by MPS Limited, Chennai, India

Contents

Section II Walruses and Indigenous peoples

Section III European walrus use from the Norse to present

Section IV Future directions and innovations in Atlantic walrus research

List of contributors

Liselotte W. Andersen
Department of Bioscience, Kalø, Aarhus University, Rønde, Denmark

Jette Arneborg
National Museum of Denmark, Middle Ages, Renaissance and Numismatics, Copenhagen, Denmark

James H. Barrett
McDonald Institute for Archaeological Research, Department of Archaeology, University of Cambridge, Downing Street, Cambridge, United Kingdom

Department of Archaeology and Cultural History, NTNU University Museum, Trondheim, Norway

Trinity Centre for Environmental Humanities, Arts Block, Trinity College Dublin, College Green, Dublin, Ireland

Robert W. Boessenecker
Department of Geology and Environmental Geosciences, College of Charleston, Charleston, South Carolina, United States

Mace Brown Museum of Natural History, College of Charleston, Charleston, South Carolina, United States

University of California Museum of Paleontology, University of California, Berkeley, California, United States

Erik W. Born
Greenland Institute of Natural Resources, Nuuk, Greenland

Morgan Churchill
Department of Biology, University of Wisconsin Oshkosh, Oshkosh, Wisconsin, United States

Christyann M. Darwent
Department of Anthropology, University of California, Davis, Calirfornia, United States

Sean Desjardins
Arctic Centre/Groningen Institute of Archaeology, University of Groningen, Groningen, The Netherlands

Canadian Museum of Nature, Gatineau, Quebec, Canada

Canadian Museum of Nature, Ottawa, Ontario, Canada

Eva Garde
Greenland Institute of Natural Resources, Nuuk, Greenland

Ian Gjertz
Department for Oceans, Research Council of Norway, Oslo, Norway

Anne B. Gotfredsen
Globe Institute, University of Copenhagen, Copenhagen, Denmark

Rikke G. Hansen
Greenland Institute of Natural Resources, Nuuk, Greenland

Magnus W. Jacobsen
Section for Marine Living Resources, National Institute of Aquatic Resources, Technical University of Denmark, Silkeborg, Denmark

Peter Jordan
Department of Archaeology and Ancient History, Lund University, Lund, Sweden

Xénia Keighley
Globe Institute, University of Copenhagen, Copenhagen, Denmark

Arctic Centre, University of Groningen, Groningen, The Netherlands

Genevieve M. LeMoine
Peary–MacMillan Arctic Museum, Bowdoin College, Brunswick, Maine, United States

Morten Tange Olsen
Globe Institute, University of Copenhagen, Copenhagen, Denmark

Paul Szpak
Department of Anthropology, Trent University, Peterborough, Ontario, Canada

Fern Wickson
North Atlantic Marine Mammal Commission (NAMMCO), Tromsø, Norway

Øystein Wiig
Natural History Museum, University of Oslo, Oslo, Norway

Biographies

Xénia Keighley completed her PhD as part of the Marie Curie Horizon 2020 ArchSci2020 network, investigating ancient genomics of the Atlantic walrus. She has been based at the University of Copenhagen (Denmark) and the University of Groningen (the Netherlands) and defended her PhD thesis in January 2021. Xénia's academic background is in Biology, with a Bachelor of Philosophy (Science) (Honours) awarded by the Australian National University. Her undergraduate studies focused on taxonomy, phylogeography, botany and environmental sciences.

Morten Tange Olsen is an associate professor at the Globe Institute, University of Copenhagen, Denmark. His research seeks to document and understand the effects of environmental change, pathogens and human activities on marine mammals and ecosystems, using ancient, environmental and modern DNA analyses, museum collections and observations in the wild. He is a member of multiple marine mammal expert groups, including the Helsinki Commission (HELCOM) and the North Atlantic Marine Mammal Commission (NAMMCO), as well as the Danish Ministry of Environment and Environmental Protection Agency.

Peter Jordan is currently a professor of Archaeology at the Department of Archaeology and Ancient History, Lund University, Sweden. His research examines long-term human–animal–environment interactions in the circumpolar regions, including northern Europe and Northeast Asia. He was previously Director of the Arctic Centre at the University of Groningen and the Netherlands Scientist-in-Charge in the ArchSci2020 Network. He also served as the Netherlands representative in the Council of the International Arctic Science Committee (IASC) and was also Chair of IASC's Polar Archaeology Network (PAN).

Sean Desjardins is a postdoctoral researcher at the Arctic Centre and the Groningen Institute of Archaeology, University of Groningen (the Netherlands) and a research associate at the Canadian Museum of Nature (Canada). An anthropological archaeologist with more than a decade of experience investigating human–walrus interactions in Inuit Nunangat (the traditional Inuit territories of Arctic Canada), his research interests include Indigenous knowledge systems, food security/food safety among hunting–gathering–fishing populations and long-term human responses to climate change in the Arctic. He represents the Netherlands in the Sustainable Development Working Group (SDWG) of the Arctic Council.

Acknowledgements

The genesis of this book was a series of three annual Walrus Research Day workshops (2016–18) organised and hosted by editors Keighley and Olsen at the University of Copenhagen; all of the editors would like to sincerely thank the participants of these workshops – many of whom contributed to this volume – for their ideas, enthusiasm and support. This includes Astrid Andersen, Liselotte Andersen, Martin Appelt, Jette Arneborg, Erik Born, Andrea Cabrera, Matthew Collins, Janne Flora, Andy Foote, Eva Garde, Tom Gilbert, Anne Birgitte Gotfredsen, Kristian Gregersen, Rikke Guldborg, Snæbjörn Pálsson, Eirik H. Røe and Bastiaan Star. Additionally, we thank those collaborators who were not able to attend the workshops, namely James Barrett, Sanne Boessenkool, Robert Boessnecker, Morgan Churchill, Christyann Darwent, Ian Gjertz, Mads Peter Heide-Jørgensen, Lesley Howse, Lars Jacobsen, Genevieve LeMoine, Hilmar Malmquist, Vicki Szabo, Paul Szpak and Fern Wickson.

This edited book emerged out of the PhD thesis research which was funded from the European Union's EU Framework Programme for Research and Innovation Horizon 2020 under Marie Curie Actions Grant Agreement No. 676154 (Keighley, Olsen and Jordan) and a Dutch Research Council (Dutch: *Nederlandse Organisatie voor Wetenschappelijk Onderzoek*; *NWO*) Veni grant (Desjardins). Finally, we would like to thank Elena Kakoshina for all her beautiful and illuminating scientific illustrations throughout the book.

CHAPTER

Introduction

1

Xénia Keighley[1,2], Morten Tange Olsen[1], Peter Jordan[4] and Sean Desjardins[3,5]

[1]*Globe Institute, University of Copenhagen, Copenhagen, Denmark*
[2]*Arctic Centre, University of Groningen, Groningen, The Netherlands*
[3]*Arctic Centre/Groningen Institute of Archaeology, University of Groningen, Groningen, The Netherlands*
[4]*Department of Archaeology and Ancient History, Lund University, Lund, Sweden*
[5]*Canadian Museum of Nature, Ottawa, Ontario, Canada*

For people around the world, the walrus is an iconic symbol of the Arctic. Its massive body, rough and thick with blubber, its imposing tusks and its tenacity are all features that seem uniquely suited to the cold waters of the North. Far from resource-poor, Arctic regions feature highly productive marine ecosystems, which have supported many species – including humans – for millennia. The walrus is one example of a species well-adapted to the freezing temperatures, dark winters and endless summer days of the world's most northerly latitudes. The walrus is also rapidly joining the polar bear as a popular symbol for the wide-ranging impacts of human activities on the Arctic. The whole-ecosystem effects of recent anthropogenic climate change are already being felt as the steadily increasing temperatures reduce the extent and duration of sea ice, which forms such a crucial part of the Arctic marine ecosystem. With these fast-changing human-environment dynamics in mind, we have organised a timely synthesis of long-term human–walrus relationships to frame our current knowledge within a historical and holistic perspective.

The subspecies of walrus most familiar to western Europeans and Indigenous peoples east of Alaska is the Atlantic walrus (*Odobenus rosmarus rosmarus*), which inhabits the waters of the subarctic and Arctic regions of the North Atlantic Ocean. Subsistence hunting of this Atlantic subspecies has sustained Indigenous cultures for thousands of years. More recently, Atlantic walruses have been subjected to several centuries of intense commercial harvesting for ivory, skin, meat and blubber. An up-to-date and integrated compendium of the depth and breadth of such relationships in the North Atlantic and Arctic has been lacking in the literature; this is largely because the wealth of important and widely-applicable research is scattered across a range of discipline-specific academic literature, or has been published separately within national and international management or policy documents.

The Atlantic Walrus. DOI: https://doi.org/10.1016/B978-0-12-817430-2.00010-8

Through this volume, we aim to bridge academic divides and consolidate the existing, extensive knowledge of human–walrus interactions for both general and specialist audiences; including those interested in purely biological topics (e.g., how Atlantic walruses forage for food), purely (pre)historic or cultural aspects (e.g., the practical and cosmological importance of walruses to ancient Arctic cultures, and to modern Inuit), as well as the intersection of the two. The book brings together contributors with a wide variety of academic backgrounds to provide in-depth thematic review chapters covering exciting new methodological developments and offering fresh insights into the future of walrus research.

The first set of chapters provides an essential biological and ecological basis for understanding Atlantic walruses. In **Chapter Two**, "*Blubber and tusks: the surprising evolutionary heritage of the modern walrus as chronicled by the fossil record*", Boessnecker and Churchill explore the rich fossil record of the family *Odobenidae*, of which modern walruses are the only remaining living species. The authors describe the geological and geographic contexts of known walrus fossils, as well as the current state of our knowledge about the phylogenetic relationships of ancient walrus species to one another and to other species. The authors also explore how the morphology of walruses has evolved to adapt to a range of ecological niches. While our knowledge of walrus evolution is drawn mainly from fossil specimens from the North Pacific, it also stretches to subtropical latitudes, which are well outside the range of modern Atlantic *or* Pacific subspecies. Boessnecker and Churchill argue that that the evolution of walruses began with the origin of all Odobenids in the North Pacific in the early to mid-Miocene (approx. 16–17 million years ago) and explore Atlantic walrus biogeography up until the present day.

In **Chapter Three**, "*Ecology and behaviour of Atlantic walruses*", Born and Wiig describe the biology and ecology of Atlantic walruses. They begin by outlining the major differences between the two subspecies before providing an overview of the current and historic distribution of walruses across the Atlantic. They detail the habitat requirements of the species, specifically sea ice cover, haul out sites (either terrestrial or on sea ice) as well as choice and abundance of prey. Key morphological traits in walruses are discussed, with reference to scientific illustrations presented here for the first time. Differences between male and female walruses are also outlined, with respect to age of sexual maturity, appearance and behaviour, both during and outside of the breeding season. Additional life history characteristics are presented, including life expectancy, fecundity, lactation, sex ratio and mortality rates. Throughout the chapter, the authors emphasise the social nature of walruses, including their gregariousness and communication. They conclude by describing the predators and pathogens threatening Atlantic walruses today, leaving the impact of human activities and climate change for chapters by Wickson (Chapter 11) and Born, Wiig and Olsen (Chapters 12 and 13).

In **Chapter Four**, "*Stocks, distribution and abundance*", Garde and Hansen examine the distribution and relative abundance of modern Atlantic walruses across Canadian, Greenlandic, northern European and Russian waters. They present the seasonal variation in walrus distribution, as well as the major annual migration patterns. The authors discuss distribution and abundance for Atlantic walruses overall by combining results from a multitude of population surveys, and also provide fine-scale details for individual populations region-by-region. The various methods used to monitor walrus populations are described, in particular how tagging and aerial surveys have been used to obtain population estimates, as well as an understanding of behaviour, population structure and dispersal. The authors also briefly touch upon the degree to which walrus population size is likely to have changed over time. The detailed analysis of human impacts included in this chapter and those that follow highlight the often exploitative nature of human–walrus interactions. The long-term legacies of these interactions on current walrus population size and distribution cannot be understated.

The authors of Chapters 5 through 9 provide a deep cultural and historical context of past exploitation practices. They also examine – as best as can be determined from historical, archaeological or ethnographic evidence – hunting methods, as well as the potential ecological impacts of human exploitation on Atlantic walrus populations. Modern walrus management and conservation measures focus on the concept of 'sustainability' – aiming to ensure hunting practices will not jeopardise ecological, economic or cultural futures. However, accurately defining sustainable practices requires detailed knowledge of historic and ancient 'baselines', as well as a detailed understanding of how walruses have responded to past human activities. Chapters 5 and 6 draw upon archaeological data to describe the earliest phases of walrus hunting after people first settled the Canadian Arctic Archipelago and Greenland approximately 6000 years ago. The two subsequent chapters focus on early commercial European hunting during the medieval period (Chapter 8), and the most recent and intense period of industrialised walrus hunting (Chapter 9). In this book we make a clear distinction between Indigenous and early 'commercial' or later large-scale industrial practices.

In **Chapter Five**, "*Pre-Inuit walrus use in Arctic Canada and Greenland, c. 2500 BCE to 1250 CE*", Darwent and LeMoine provide a highly detailed archaeological reconstruction of Pre-Inuit (Paleo-Inuit) walrus use, from around 2500 BCE until the arrival of Inuit in Arctic Canada (Inuit Nunangat) and Greenland (Kalaallit Nunaat), sometime in the 13th century CE. The chapter also includes a description of walrus ivory, highlighting how it differs in appearance, properties and use to other materials available to Arctic Indigenous peoples. The synthesis and analysis presented by the authors shows the various Pre-Inuit cultures varied considerably in their economic reliance on walruses, ranging from scavenging for ivory to organised group subsistence hunting. Details regarding the abundance of walrus finds, as well as the kinds of tools and techniques with which the materials were processed, reveal significant variability across time and space. These changes relate to technological innovations and changing social organisation, in turn likely affected by climatic conditions and the associated abundance of walruses.

The interrelation between changing climates, walrus populations and the ability for humans to effectively adapt is also key to understanding human–animal interactions among later Indigenous populations. In **Chapter Six**, "*Subsistence walrus hunting in Inuit Nunangat (Arctic Canada) and Kalaallit Nunaat (Greenland) from the 13th century CE to present*", authors Desjardins and Gotfredsen draw upon archaeological, ethnographic and oral-historical information to explore premodern and contemporary Inuit walrus use from approximately 1300 CE to present. Upon entering former Pre-Inuit lands, Thule Inuit brought with them an already well-developed tradition of hunting Pacific walruses (as well as many other species of marine mammal). This chapter focuses on walrus hunting and use by Thule and historic Inuit in two main geographic regions: northern Foxe Basin, central Nunavut and Avanersuq in North-West Greenland. In addition to contextualising our past knowledge of Inuit walrus exploitation, the authors explore topics as diverse as (a) butchery, storage and consumption of walrus meat; (b) the use of bone and ivory for tools; (c) the variation in hunting practices; (d) factors influencing annual catch size and (e) changes in technology for transport and capture. The chapter concludes with an exploration of the significance of the walrus in Inuit cosmology.

While Pre-Inuit and Inuit peoples were hunting walruses primarily for subsistence, a new kind of relationship emerged between humans and walruses during medieval times. Walrus hunting became increasingly 'commercialised' by early European settlers moving into the North Atlantic. **Chapter Seven**, "*Early European and Greenlandic walrus hunting: Motivations, techniques and practices*", by Arneborg documents these early European interests in walrus products (most notably, ivory, which was used primarily for trade and exchange) from the earliest Norwegian written accounts of the 9th century CE until the collapse of the Norse Greenlandic settlements in the 13th–14th centuries. This chapter focuses on how the demand for ivory encouraged early trade between what is now Scandinavia and Russia, and then likely motivated, at least in part, the exploration and settlement of Iceland and Greenland by the Norse during the Viking Age. Arneborg draws upon historical records and an analysis of medieval ivory artefacts to discuss how walruses were likely hunted to extinction in Iceland, and how the annual walrus hunt became central to the remote Norse settlements in Greenland in later periods. The author also demonstrates how Norse efforts to meet the demand for walrus ivory in European markets fundamentally shaped the social structure, wealth, connections and, ultimately, survival, of remote human settlement across the wider North Atlantic.

In **Chapter Eight**, "*The exploitation of walrus ivory in medieval Europe*", Barrett builds on the topics introduced in Chapter 7 by tracing the wider trade, exchange, consumption and production networks of Europe, and by linking these to remote Arctic hunting grounds. He begins by outlining the uses of ivory around 1000 CE, both in trading hubs and further afield, and then charts the changing demand for walrus ivory in Romanesque and Gothic art, highlighting the uses and diminishing value of walrus ivory until its sharp decline in the mid-15th century CE.

Early European and Greenlandic walrus hunting motivations, techniques and practices explored in Chapter 7 are revisited but with a different focus – the demand arising from European markets, as well as the trade routes that ivory followed. The ways in which walrus ivory was transported, worked, redistributed and traded are pieced together, with a detailed description of various workshops and suggestions of likely trading routes. Barrett concludes the chapter by discussing the extent of knowledge of walruses among Europeans during the medieval period.

In **Chapter Nine**, "*Modern European commercial walrus exploitation, 1700 to 1960 CE*", Gjertz examines the next stage in human exploitation of walruses: the industrial-scale hunts, processing and production undertaken by Europeans in the 18th–20th centuries. Gjertz demonstrates that the intensive hunting during this period was driven largely by the international market in a wide range of walrus products that had significant impacts on Atlantic walrus populations across their range. This industrialised phase of hunting saw increased supply and affordability of walrus products compared with earlier medieval trade, which had largely focused on a smaller quantity of material for the production of prestige items. This difference emerges as Gjertz reveals the value and uses of walrus products, from skins essential to various manufacturing equipment, through to ivory for dentures and blubber for soap. Gjertz analyses the methods, motivations, geographic distribution, pace, approach and technology used. The chapter concludes with an analysis of hunting intensity within distinct geographic regions and its impact on specific walrus populations.

The final four chapters of the book examine the methods, technologies, regulation and management approaches that are being applied to answer questions regarding past, current and future human–walrus interactions, and provides an updated assessment of the effects of human activities and climate change on the Atlantic walrus. In **Chapter 10**, "*Molecular advances in achaeological and biological research on Atlantic walrus*", Andersen, Szpak and Jacobsen explore the major recent advances in genetic, stable isotope and lipid analyses on modern, historical, archaeological and ancient walrus samples. These new approaches offer unique and exciting opportunities to not only reconstruct the past, but also predict future impacts on walrus populations to inform management and conservation efforts. The authors also present an in-depth, critical review of the various biomolecular methods employed in these fields, evaluating the applicability of each to the study of long-term population structure delineation, demography, life history, diet and the ecology of Atlantic walruses. Importantly, the authors conclude that biomolecular analyses are set to play a major role in improving our understanding of current Atlantic walrus biology, as well as of the consequences of long-term interactions between humans and walruses throughout recent millennia. These insights will have direct relevance to the design of modern management strategies and future conservation aims.

Current Atlantic walrus management, conservation and regulation is explored in detail in **Chapter 11**, "*Atlantic walrus management, regulation, and conservation*," by Wickson, who provides summaries for several regions (including Norway, Russia, Greenland and Canada), outlining the timing and approaches used in managing walrus populations across the North Atlantic. The author also discusses various efforts towards collaborative or co-management of subsistence hunting, international agreements relevant to management, and the role of regional management institutions (e.g., the North Atlantic Marine Mammal Commission, NAMMCO). The chapter concludes with a discussion of the current and future challenges for management, hunting and conservation in a changing Arctic.

In **Chapter 12**, "*Anthropogenic impacts on the Atlantic walrus*", Born, Wiig and Olsen highlight past and present human impacts on walruses, and discuss how walruses may be impacted in the future as the projected climate-induced reduction in sea ice is anticipated to increase the level of human activities in the Arctic. The chapter begins by summarizing responses and effects of hunting, and proceed by discussing effects of shipping, fishery, aircraft traffic, non-renewable natural resource extraction, pollution and tourism. When applicable, the activities are discussed in terms of their putative direct effects on walrus behaviour and health, as well as their indirect effects on prey and habitat. The chapter concludes by providing an overall assessment and recommendations for mitigating human impacts on walruses.

Finally, in **Chapter 13**, "*The future of Atlantic walrus in a rapidly warming Arctic*", Born, Wiig and Olsen discuss the challenges faced by walruses in a rapidly warming Arctic. The chapter discusses the anticipated effects of diminishing sea ice and increased temperatures on walrus abundance, distribution, behaviour and health. It concludes by providing a set of recommendations for research initiatives to increase our understanding of how walruses may respond to future global warming.

The dynamic nature of academic research, generally, means that the rich, multidisciplinary information we have brought together in this volume is continually expanding and developing. As a result, the book highlights that many questions remain unanswered (and many have yet to be formulated), and that important themes and questions still await further research to improve our understanding of long-term human interactions with Atlantic walruses.

We hope this book will act as a catalyst to stimulate further research into the shared past, present and future of humans and walruses. In a changing world, there is much to be gained from developing our understanding of the interconnectedness of climate, animals and human cultures across the Arctic. Furthermore, we hope that our holistic approach to the cultural and biological history of walruses will inspire a new generation of students and researchers, and inform new inter-, multi- and transdisciplinary projects and collaborations. This book has been designed in part for those with professional and personal experience across the Arctic, particularly those involved with Atlantic walrus management, including policymakers, wildlife conservation authorities and northern Indigenous peoples. The future of Atlantic walruses and many of these local Arctic communities is closely intertwined. The path forward requires careful consideration, informed by knowledge gained across nations, academic fields and human cultures. As such, we hope to inspire others to explore the histories and modern legacies of human interactions with the natural world for a more sustainable future in the Arctic and beyond.

SECTION

I

Atlantic walrus evolution, ecology and behaviour

CHAPTER

The surprising evolutionary heritage of the Atlantic walrus as chronicled by the fossil record

Robert W. Boessenecker[1,2,3] and Morgan Churchill[4]

[1]*Department of Geology and Environmental Geosciences, College of Charleston, Charleston, South Carolina, United States*

[2]*Mace Brown Museum of Natural History, College of Charleston, Charleston, South Carolina, United States*

[3]*University of California Museum of Paleontology, University of California, Berkeley, California, United States*

[4]*Department of Biology, University of Wisconsin Oshkosh, Oshkosh, Wisconsin, United States*

Chapter Outline

The Atlantic Walrus. DOI: https://doi.org/10.1016/B978-0-12-817430-2.00006-6

Introduction

The walrus (*Odobenus rosmarus*) is one of the most iconic mammals of the Arctic, significantly differing from all other pinnipeds by its enormous size and prominent tusks (Fay, 1982). Walruses are uniquely specialized for benthic suction feeding, allowing a single walrus to consume up to 6000 clams per foraging session (Fay, 1985).

Odobenus rosmarus is the sole surviving member of the family Odobenidae, a family with much higher diversity in the past as demonstrated across the rich

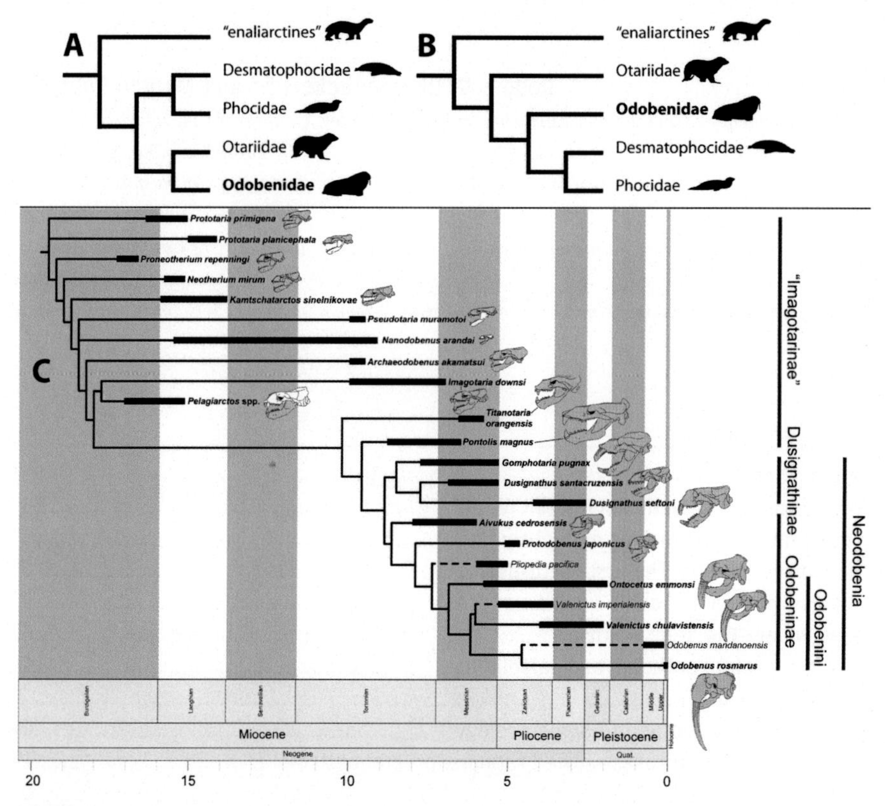

FIGURE 2.1

Phylogenetic relationships of the walruses (Odobenidae). (A) Phylogenetic hypothesis showing walruses as otarioids (after Furbish, 2015 and Boessenecker and Churchill, 2018). (B) Phylogenetic hypothesis showing walruses as phocomorphs (after Berta and Wyss, 1994). (C) Time-calibrated phylogenetic tree of Odobenidae with skulls and mandibles to scale; grey indicates preserved elements, white indicates missing elements; composite tree topology based on Boessenecker and Churchill (2013), Magallanes et al. (2018), and Velez-Juarbe and Salinas-Márquez (2018).

Modified from Berta, A., Churchill, M., Boessenecker, R.W., 2018. The origin and evolutionary biology of pinnipeds: seals, sea lions, and walruses. Annual Review of Earth and Planetary Sciences 46, 203–228.

fossil record. In fact, walruses may have one of the most complete fossil records of any pinniped group, with fossil from Odobenidae demonstrating the gradual evolution of their iconic traits, and showing them to be one of the most diverse and successful pinniped clades of the past 17 million years (Fig. 2.1).

Two subspecies are currently recognised (Berta and Churchill, 2012). The Atlantic walrus (*O. rosmarus rosmarus*) is today only found in the Canadian Arctic, Greenland, Svalbard and the western portions of the Russian Arctic, although its range historically extended south to Sable Island, Nova Scotia. A second subspecies, the Pacific walrus (*O. rosmarus divergens*), is found in the Bering, Chukchi and Beaufort Seas, from Alaska to Siberia. A third proposed subspecies, the Laptev walrus (*O. rosmarus laptevi*), has previously been found to represent an isolated population of the Pacific walrus in the Laptev Sea (Lindqvist et al., 2009) (see Chapter 4, this volume).

Occurrence, preservation and fossil record

Like other pinnipeds, most walrus fossils are preserved in shallow marine rocks deposited in continental shelf settings (Berta et al., 2018). Walrus fossils typically occur as partial skeletons (e.g., see Repenning and Tedford, 1977: plate 11), isolated bones or, rarely, as articulated skeletons (Magallanes et al., 2018). Curiously, complete and partial skeletons are known only from the North Pacific. In contrast, Pliocene and Pleistocene specimens from the western North Atlantic consist solely of isolated remains (Kohno and Ray, 2008). Complete and partial skeletons of several North Pacific Miocene and Pliocene odobenids have been recovered, and include *Proneotherium repenningi*, *Archaeodobenus akamatsui*, *Imagotaria downsi*, *Titanotaria orangensis*, *Aivukus cedrosensis*, *Gomphotaria pugnax*, *Dusignathus seftoni*, and *Valenictus chulavistensis* (Mitchell, 1968; Repenning and Tedford, 1977; Barnes and Raschke, 1991; Deméré, 1994a,b; Deméré and Berta, 2001; Tanaka and Kohno, 2015; Magallanes et al., 2018). These remains are preserved across several different stratigraphic horizons and shelf settings of the Asian and North American continental margins. However, few western North Atlantic walruses originate from outer-shelf deposits, probably due to the lower gradient of Atlantic continental shelves. The broad difference in preservation between the North Atlantic and North Pacific is probably related to substantially slower sedimentation rates along the Atlantic coast of North America – as lower sedimentation rates impart a greater chance for skeletal scattering (e.g., see Berta et al., 2018). Differences in mollusc and shark preservation between fossil assemblages of the same age in both California and Maryland have been attributed to these differences in deposition (Kidwell, 1993; Boessenecker et al., 2019).

Walrus fossils are typically found in inner-middle shelf sandstones and siltstones, in phosphatic bonebeds (Mitchell, 1961; Deméré, 1994a; Tanaka and Kohno, 2015; Boessenecker, 2017) and, occasionally, within outer-shelf mudrocks and diatomites (Barnes and Raschke, 1991; Mitchell, 1968). Preservation of some walruses (e.g., *G. pugnax*) in outer-shelf settings may reflect 'bloat and float' (Schäfer, 1972), rather than true pelagic habits. Dead walruses can bloat from decay gases for 3–4 weeks (Espinoza et al., 1997), allowing their bodies to drift for hundreds of kilometres, depending upon the currents (e.g., Blanco and Rodriguez, 2001).

Perhaps unique amongst pinnipeds is the large number of Pliocene–Pleistocene odobenid specimens which continue to be dredged from the seafloor (e.g., Miyazaki et al., 1992; Erdbrink and Van Bree, 1999a,b,c; Gallagher et al., 1989). These specimens are often skulls, mandibles, and tusks, which are somewhat larger and denser than other pinniped material and, thus, have higher preservation potential. Most of the recovered specimens are presumably from now-flooded marine or coastal deposits which would have been further out on continental shelves during the Pleistocene.

No mass walrus mortality assemblages are known, but one locality (Taylor Quarry, Santa Margarita Sandstone in Santa Cruz County, California, United States) may represent a deposit adjacent to an *I. downsi* rookery based on the presence of juvenile remains (Repenning and Tedford, 1977). However, initial field examination by author Boessenecker found little taphonomic evidence to support this interpretation. The fossil assemblage from Taylor Quarry also preserves sharks, sea birds, baleen whales and fur seals (Domning, 1978), suggesting a gradual accumulation of skeletons, bones and teeth over a long period of time. Formal taphonomic evaluation of this locality is still needed.

No published examples are known of walrus fossils with gut contents, bite marks from sharks or fish, or acid etching from partial digestion from a shark, despite such features having been documented in other pinniped fossils (Cozzuol, 2001; Boessenecker et al., 2014). Walrus carcasses and skeletal material can be inhabited by other species postmortem, including borings in a dusignathine tooth by the bone-eating worm *Osedax* (Boessenecker et al., 2014) and postmortem colonisation of a skull by barnacles and oysters (Kohno and Ray, 2008).

Unlike all other marine mammals, walruses appear to have produced trace fossils. Trace fossils identified as feeding pits formed during hydraulic jetting by *Odobenus* have been reported from upper-middle Pleistocene (200–120 ka) terrace deposits near Willapa Bay, Washington, United States (Gingras et al., 2007). Careful field examination of well-exposed, walrus-bearing Pliocene strata (e.g., the Purisima and San Diego formations, California; the Yorktown Formation, North Carolina; and the Tamiami Formation, Florida, United States) may one day reveal older feeding traces, as earlier odobenines (e.g. *Ontocetus*, *Valenictus*) appear to have been suction feeders like *Odobenus*.

Phylogenetic relationships of Odobenidae

Walruses belong to the Carnivoran clade Arctoidea, along with bears, raccoons, and weasels, as well as all other pinnipeds (seals and sea lions). However, the genetic interrelationships within Pinnipedia, and more specifically of walruses to other pinnipeds, have been debated. Traditionally, Pinnipedia (the clade including walruses, sea lions, and seals) was considered diphyletic (not sharing a common ancestor; Tedford, 1976;

Repenning et al., 1979; Muizon, 1982; Barnes, 1989; Koretsky and Barnes, 2006). Under this paradigm, walruses were considered to be most closely related to Otariidae (fur seals and sea lions), and were included within the clade Otarioidea, understood to have evolved from a bear-like basal arctoid ancestor. Earless seals (Phocidae) were believed to represent a separate adaptation to marine environments more closely related to musteloids, a clade that includes otters, martens, raccoons, skunks, and weasels. Rigorous cladistic studies of morphology began to cast doubt on this diphyletic hypothesis and strongly suggested Pinnipedia was a monophyletic (a clade sharing a common ancestor) group within Pinnipedimorpha [which includes a variety of extinct stem taxa including *Enaliarctos*, *Pteronarctos* and *Pacificotaria* (Wyss, 1987; Berta and Wyss, 1994)]. These 'enaliarctines' are a paraphyletic (a group lacking all the descendants of a common ancestor) early radiation of pinnipedimorphs, which are directly ancestral to the various clades of modern pinnipeds. The diphyletic hypothesis of pinniped origins has been further discredited by every molecular study to date that has sought to address carnivoran or pinniped relationships, with consistent findings of a monophlyetic Pinnipedia (Bininda-Emonds et al., 1999; Arnason et al., 2006; Higdon et al., 2007; Fulton and Strobeck, 2010; Nyakatura and Bininda-Emonds, 2012).

Within Pinnipedia, the relationships of walruses to other seals has been contentious (Fig. 1A and B). Early cladistic studies that supported pinniped monophyly also found strong evidence that walruses were closely related to earless seals (phocids; Fig. 1B), thereby forming the clade Phocomorpha along with the extinct Desmatophocidae (Berta and Wyss, 1994). This clade was, in turn, the sister group to Otariidae. More recent molecular studies in contrast, found support for a clade (Otarioidea) containing otariids and odobenids, with phocids being the sister group to this clade (Bininda-Emonds et al., 1999; Arnason et al., 2006; Higdon et al., 2007; Fulton and Strobeck, 2010; Nyakatura and Bininda-Emonds, 2012). More recent morphologic phylogenetic analyses have identified odobenids as more closely related to phocids than otariids (Boessenecker and Churchill, 2013), albeit with lower statistical support, or unclear relationships with other pinniped families (Kohno, 2006; Tanaka and Kohno, 2015; Velez-Juarbe, 2018). One recent study combining genetic and morphological evidence found otarioids to be a monophyletic group (Fig. 1A), and that Phocidae were most closely related to the extinct Desmatophocidae (Furbish, 2015).

Lastly, one morphology-only cladistic analysis recovered (though with weak support) otarioid monophyly and a desmatophocid + phocid (Phocoidea) clade (Fig. 1A; Boessenecker and Churchill, 2018).

The 'Imagotariines': The first walruses

Walruses have a relatively complete fossil record, which illuminates a gradual evolution from animals little different from their 'enaliarctine' ancestors (Kohno et al., 1995a) to the massive tusked animals of today. This rich record is in contrast to phocids and otariids, which have considerable gaps within their fossil record obscuring their early evolution (Koretsky and Holec 2002; Boessenecker and Churchill, 2015). The wealth of walrus fossils makes them one of the best examples of macroevolution in the mammal fossil record. This fossil record is almost entirely based in the North Pacific, the centre of origin for the family (Fig. 2.2) (Deméré et al., 2003).

The earliest walruses date to the early-middle Miocene (17–16 Ma) and belong to a group often referred to informally as the 'Imagotariinae' (Fig. 2.3). Originally considered a grouping of fossil walruses closely related to the other walrus subfamilies (Mitchell, 1968; Barnes, 1989), phylogenetic analyses have consistently found 'Imagotariinae' (*sensu* Barnes, 1989) to be paraphyletic (Deméré, 1994a; Kohno, 1994; Boessenecker and Churchill, 2013; Tanaka and Kohno, 2015; Velez-Juarbe, 2018; Magallanes et al., 2018), with many 'imagotariine' taxa representing successive sister taxa to Odobeninae (true walruses) and Dusignathinae (double-tusked walruses).

The earliest diverging 'imagotariine' walruses date to the late early and early-middle Miocene, and include the Japanese species *Prototaria primigenia* (Takeyama and Ozawa, 1984) and *Prototaria planicephala* (Kohno, 1994), the Oregonian *P. repenningi* (Fig. 2.3D) (Kohno et al., 1995a; Deméré and Berta 2001), and the Californian *Neotherium mirum* (Fig. 2.3A, G–H, J) (Kellogg, 1931; Kohno et al., 1995a,b; Velez-Juarbe, 2018). When alive, a casual observer would have found it difficult to distinguish these taxa from their 'enaliarctine' forbearers; they were comparable in body size to a small fur seal (Churchill et al., 2014), lacked tusks, and were probably piscivorous (Churchill and Clementz, 2016). There is also evidence that these early 'imagotariine' walruses were sexually dimorphic to a degree comparable to modern sea lions (Velez-Juarbe, 2018). The differences in size between male and female 'imagotariine' walruses indicates some form of polygnous mating system, likely inherited from their enaliarctine ancestors (Cullen et al., 2014). The 'Imagotariinae' possessed several unique traits that indicate their affinities towards

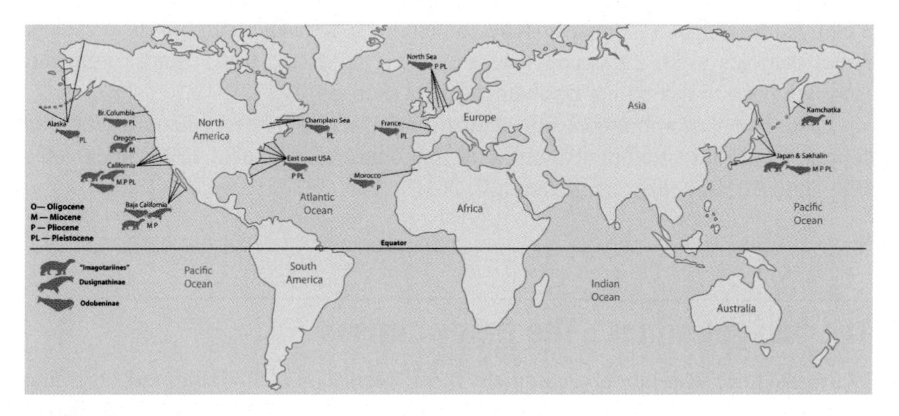

FIGURE 2.2

Map of fossil occurrences of Odobenidae.

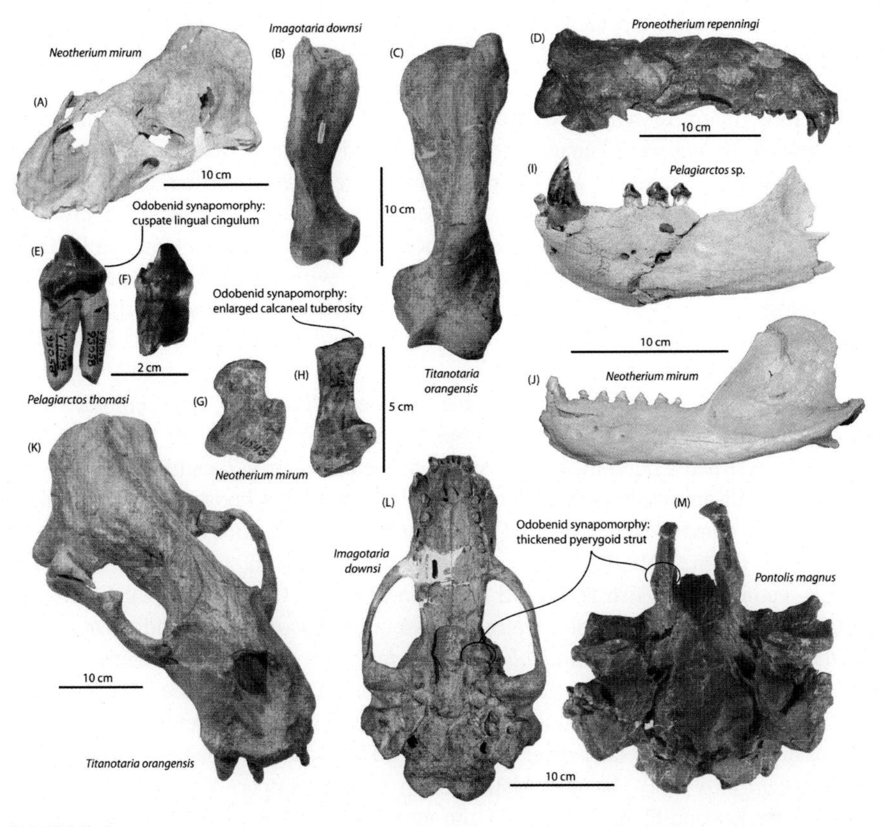

FIGURE 2.3

Representative fossils of the 'Imagotariinae' – the earliest walruses. (A) Referred skull of *Neotherium mirum* (LACM 131950); (B) referred humerus of *Imagotaria downsi* (USNM 23870); (C) holotype humerus of *Titanotaria orangensis* (OCPC 11141); (D) holotype skull of *Proneotherium repenningi* (USNM 205334); (E and F) referred teeth of *Pelagiarctos thomasi* (UCMP 93058, TATE 2694); (G and H) paralectotype astragalus and lectotype calcaneum of *N. mirum* (USNM 11543 and 11542); (I) referred mandible of *P. thomasi* (SDNHM 131041); (J) referred mandible of *N. mirum* (UCMP 81665); (K) holotype skull of *T. orangensis* (OCPC 11141); (L) referred skull of *I. downsi* (USNM 23858); (M) holotype braincase of *Pontolis magnus* (USNM 3792).

All photographs taken by the authors.

other walruses, including a thickened pterygoid strut, cuspate lingual cingulum on the upper premolar teeth, a bony tentorium closely appressed to the petrosal, and expanded calcaneal tuberosity (Kohno, 1994). The distribution of these taxa shows that even early on, Odobenidae had already achieved a wide distribution across the North Pacific. Later diverging and geologically younger 'imagotariines' continued to maintain the small body size of more basal forms. These include *Pseudotaria muramotoi* from the upper Miocene of Japan (Kohno, 2006), and *Kamtschatarctos sinelnikovae* from the middle Miocene of Kamchatka, Russia (Dubrovo, 1981).

Increased body size is seen during the middle-late Miocene in later diverging odobenids, many of which were contemporary with other, smaller walruses. For example *A. akamatsui* (Tanaka and Kohno, 2015), known from a partial skeleton and incomplete cranium, was approximately three meters in length (comparable in size to modern sea lions) and is known from the same deposits that produced the smaller *Pseudotaria*. Similar occurrences of multiple walrus taxa can be found in the Sharktooth Hill bonebed of California, with *Neotherium* and the later diverging *Pelagiarctos*. This suggests rapid speciation of walruses during the middle Miocene, possibly driven by climate change and rapid fluctuations in sea level (Tanaka and Kohno, 2015).

Pelagiarctos, known from the middle Miocene 'Topanga' and Temblor formations, remains one of the more mysterious walruses (Fig. 2.3E and F) (Barnes, 1988; Boessenecker and Churchill, 2013). Although identified from localities with an abundance of well-preserved pinniped material, *Pelagiarctos* is known only from unusually robust mandibles and teeth. This walrus has been inferred to have fed on vertebrates (Barnes, 1988), although evidence for this is slim, and it likely was a generalist, eating mostly fish and squid (Boessenecker and Churchill, 2013; Loch et al., 2016). Another enigmatic early walrus known only from isolated mandibular material is the recently described *Nanodobenus arandai*, from the upper Miocene Tortugas Formation of Baja California Sur (Velez-Juarbe and Salinas-Márquez, 2018). At only 1.65 m in length, this diminutive walrus contrasts with the overall trend towards increasing body size in later diverging walrus taxa, and is the smallest known walrus.

Pelagiarctos may be the sister taxon to the upper Miocene *Imagotaria* (Boessenecker and Churchill, 2013), the namesake of the 'imagotariines', and perhaps one of the best known early walruses. These walruses form a clade with several other later diverging 'imagotariine' taxa, as well as with members of the clade Neodobenia, comprising dusignathine and odobenine walruses (Magallanes et al., 2018). The clade including *Imagotaria* and all later diverging walruses is united by a suite of characters, including lateral lower incisors greater than medial lower incisors in size, bulbous postcanine tooth crowns, single and bilobed P^3 roots, a shallow glenoid fossa of the squamosal, a large mastoid process, and a broad, pentagonal basioccipital (Boessenecker and Churchill, 2018). *Imagotaria* is known from multiple complete specimens and skulls. The type species *I. downsi* was described from the Sisquoc Formation of central California (Mitchell, 1968). Additional specimens of *Imagotaria* are also known from the Santa Margarita Formation of California (Fig. 2.3B and L) (Repenning and Tedford, 1977), Empire Formation of Oregon (Deméré, 1994a; Deméré et al., 2003) and the Aoso and Koetoi Formations of Japan (Miyazaki et al., 1995). This walrus was around the size of a sea lion (Churchill et al., 2014), and may have filled a similar ecological niche. Material from the Santa Margarita formation shows evidence of two distinct size classes, suggesting sexual dimorphism (Repenning and Tedford, 1977) and indicating a polygynous mating system as found in many extant seals.

During much of the early-mid Miocene, walruses were just one component of a diverse North Pacific pinniped fauna that included 'enaliarctines', rare otariids, and extinct desmatophocids. The latter group was the dominant pinniped clade during the middle Miocene and included the first truly giant pinnipeds (Churchill et al., 2014; Velez-Juarbe, 2017; Boessenecker and Churchill, 2018). After the extinction of desmatophocids, walruses begin to attain similar, and eventually larger, body sizes (Churchill et al., 2014; Velez-Juarbe, 2017; Boessenecker and Churchill, 2018).

These giant walruses included *T. orangensis* (Fig. 2.3C and K) (Magallanes et al., 2018) and *Pontolis magnus* (Fig. 2.3M) (True, 1905a,b), which were also the youngest 'imagotariines'. *Titanotaria* was recently described from the upper Miocene Capistrano Formation of southern California on the basis of a nearly complete skeleton. *Titanotaria* generally resembles *Imagotaria* in cranial, dental, and postcranial features, but is larger and more robust (Magallanes et al., 2018). *Pontolis magnus* from the Empire Formation of Oregon was initially described from only a basicranium and occiput; more complete material, including intact skulls, have since been identified as belonging to this taxon (Deméré, 1994a). *P. magnus* was enormous, with a potential total body length of more than four meters (Churchill et al., 2014), making it not only the largest walrus that ever lived, but one of the largest pinnipeds to have ever evolved, surpassed only by male elephant seals (*Mirounga*). Some phylogenetic analyses have recovered *Pontolis* as a dusignathine (Deméré, 1994a), but more recent studies generally support *Pontolis* as being the sister taxon to the clade Neodobenia (Boessenecker and Churchill, 2013; Tanaka and Kohno, 2015; Magallanes et al., 2018).

The Dusignathinae: Double-tusked walruses

The Dusignathinae are the sister taxon to the Odobeninae and include the earliest diverging lineages of Neodobenia (Fig. 2.4) (Boessenecker and Churchill, 2013; Tanaka and Kohno, 2015; Magallanes et al., 2018). Dusignathines were a successful radiation of walruses known from the late Miocene and Pliocene of California and Mexico. Often referred to as 'double-tusked walruses', this clade is united by the possession of tusk-like canines in both the upper and lower dentition, as well as possession of single-rooted molars M^1 and M_2 (Boessenecker and Churchill, 2013). Known taxa include the gigantic *G. pugnax* from the upper Miocene Capistrano Formation of southern California (Fig. 2.4D, E and G) (Barnes and Raschke, 1991), *Dusignathus santacruzensis* from the upper Miocene Purisima and Almejas Formations of Central California and Mexico (Fig. 2.4B, C and H, I) (Repenning and Tedford, 1977; Deméré, 1994a), and *D. seftoni* (the youngest dusignathine) from the upper Pliocene San Diego and Purisima Formations of California (Fig. 2.4A and F) (Deméré, 1994b; Boessenecker, 2013a).

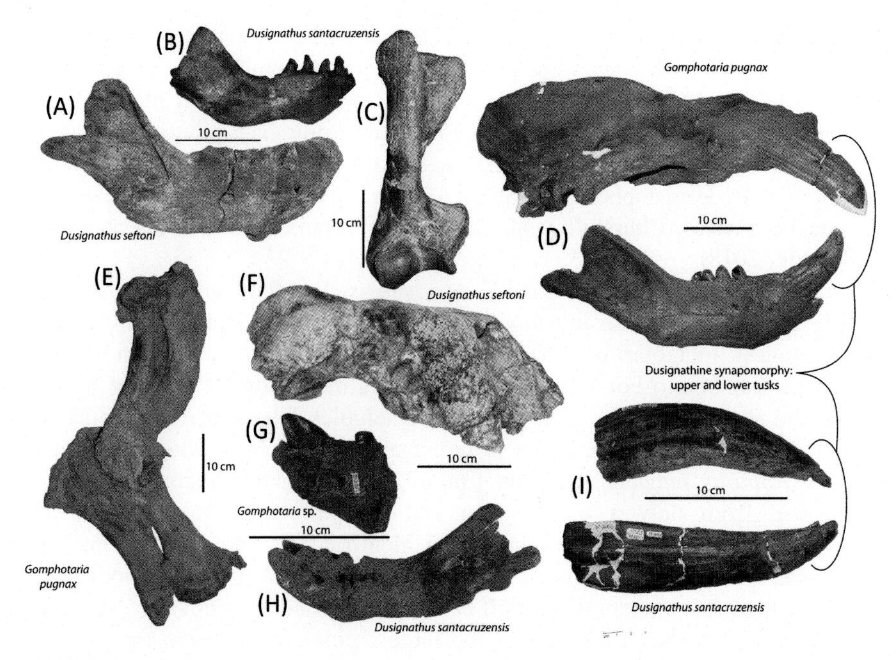

FIGURE 2.4

Representative fossils of the Dusignathinae, the 'double-tusked' walruses. (A) Referred mandible of *Dusignathus seftoni* (SDNHM 20801); (B) holotype mandible of *Dusignathus santacruzensis* (UCMP 27121); (C) referred humerus of *D. santacruzensis* (UCMP 65318); (D and E) holotype skull, mandible, and fused forelimb of *Gomphotaria pugnax* (LACM 121508); (F) holotype skull of *D. seftoni* (SDNHM 38342); (G and H), mandibles of *Gomphotaria* and *D. santacruzensis* from the Wilson Grove Formation (UCMP 125570 and 125567, respectively); (I) associated upper and lower tusks of an undetermined dusignathine (UCMP 219501).

All photographs taken by the authors.

Although represented by only three species, a great deal of dietary diversity may have been present within Dusignathinae. Wear patterns and dental features have suggested a molluscivorous diet for *Gomphotaria* and *D. seftoni* (Barnes and Raschke 1991; Berta et al., 2018) and piscivory for *D. santacruzensis*. *Gomphotaria* possesses a greatly enlarged sagittal crest similar to California sea lion males (*Zalophus californianus*) (Barnes and Raschke, 1991), and although female specimens of *Gomphotaria* are not yet known, it could suggest extreme sexual dimorphism, or perhaps high bite force associated with male–male combat. Extreme dental wear seen in *G. pugnax* is suggestive of sediment abrasion, benthic feeding or crushing of molluscs (Barnes and Raschke, 1991). Adam and Berta (2002) identified a number of suction-feeding adaptations in *Gomphotaria* and *Dusignathus*, including a transversely vaulted and elongated palate and reduced incisors.

Unusually robust and short forelimb bones suggest that dusignathines were likely forelimb-dominant swimmers (Barnes and Raschke, 1991), similar to present-day otarids. One specimen of *G. pugnax* has a pathologically-fused elbow that evidently did not hinder its ability to forage, providing further evidence for dusignathines preying upon benthic/epibenthic mollusks (rather than agile fish or squid) in nearshore environments (Barnes and Raschke, 1991).

Dusignathines appear to have been restricted to the eastern North Pacific (Fig. 2.2), though large isolated teeth from the Pliocene of Japan, currently attributed to large sea lions (Kohno and Tomida, 1993), hint that there may be future discoveries of dusignathines from the western North Pacific.

Odobeninae: The 'true' walruses

The odobenines comprise the group that gave rise to, and includes, the modern species *O. rosmarus* (Fig. 2.5). Many extinct odobenines would be recognisable as walruses to the casual observer today. The modern walrus, *O. rosmarus*, is neither the largest nor the most derived member of this clade. Like earlier diverging walruses, odobenine fossils are typically found at low latitudes, with the North Pacific being the relative centre of their diversity. A variety of morphological features characterise odobenine walruses, including the possession of a short ascending process of the maxilla, a dorsally projected postorbital process of the jugal, foreshortened braincase, reduced squamosal fossa on the external auditory meatus, premolariform incisor I^3, loss of tooth enamel, and molar M^2 (Boessenecker and Churchill, 2013; Tanaka and Kohno, 2015).

The earliest diverging and oldest odobenine, *A. cedrosensis*, was a small sea lion-like walrus with a proportionally-elongate rostrum and unenlarged canines, known from the upper Miocene Almejas Formation of Cedros Island, Baja California Sur, Mexico (Repenning and Tedford, 1977). Forelimb elements are similar to *Odobenus*, but notably more gracile. Extreme attritional tooth wear in *Aivukus* is likely caused by molluscivory (as in *Gomphotaria*), or perhaps sediment ingestion as in *Odobenus*. Another primitive odobenine, *Protodobenus japonicus* from the lower Pliocene Tamugigawa Formation of Japan (Horikawa, 1995), is similarly quite small, lacks elongate canines, and bears a moderately vaulted palate like *Aivukus*. It shares with more derived odobenids a greatly inflated and anteriorly truncated rostrum as well as an inflated, foreshortened braincase. This unusual morphology appears to be caused by massively-enlarged canine roots, despite the canine crowns being quite short. *Protodobenus* does not have tooth wear consistent with molluscivory and likely fed on fish.

A more exclusive clade, the Odobenini, consists of only the tusk-bearing walruses (*Ontocetus*, *Valenictus*, *Odobenus*), and is characterised by features including elevated bony nares, enlarged canine C^1, and mastoid processes (Deméré, 1994b). Within Odobenini is the most geographically-widespread and most

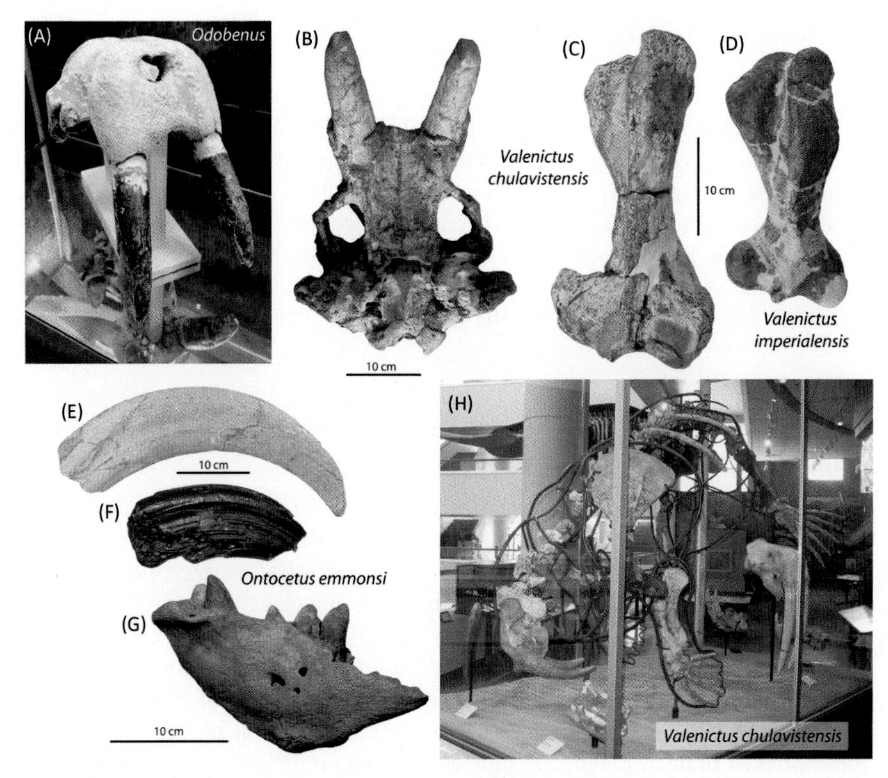

FIGURE 2.5

Representative fossils of the Odobeninae, the 'true' walruses: (A) Trawled Pleistocene skull of *Odobenus rosmarus* from the North Carolina continental shelf, specimen on display at the Aurora Fossil Museum, Aurora, North Carolina; (B) paratype skull of *Valenictus chulavistensis* (SDNHM 38227) highlighting toothless palate; (C) holotype humerus of *V. chulavistensis* (SDNHM 36786); (D) cast of the holotype humerus of *Valenictus imperialensis* (LACM 3926); (E) youngest known tusk of *Ontocetus emmonsi*, early Pleistocene, South Carolina (CCNHM 1144); (F) holotype tusk of *O. emmonsi* (329064); (G) referred mandible of *O. emmonsi* (USNM 475482); and (H) composite skeleton of *V. chulavistensis*, on display at the San Diego Natural History Museum.

All photographs taken by the authors.

commonly-discovered fossil walrus: *Ontocetus*. *Ontocetus* is also noteworthy for having the most confusing taxonomic history (Fig. 2.5E–G) (Kohno and Ray, 2008). The oldest known records of *Ontocetus* (unnamed *Ontocetus* sp. from the early Pliocene of Japan) indicate that this genus originated in the western North Pacific and quickly dispersed through the Arctic to the eastern United States, completely bypassing the eastern North Pacific (Fig. 2.6; Kohno et al., 1995b, 1998; Kohno and Ray, 2008).

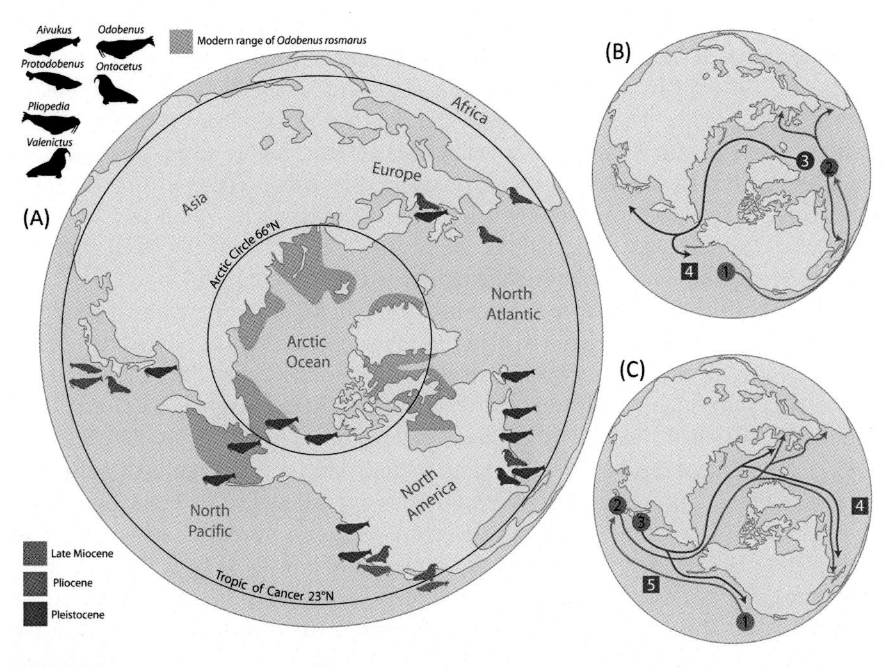

FIGURE 2.6

Fossil record and biogeography of the Odobeninae. (A) Polar projection map showing distribution of Odobenine fossil records from the Miocene, Pliocene, and Pleistocene, and the distribution of modern *Odobenus rosmarus*; (B) Odobenine biogeographic hypothesis proposed by Repenning et al. (1979): (1) late Miocene origin of Odobeninae (*Aivukus*) in eastern North Pacific followed by dispersal through the Central American Seaway to North Atlantic, (2) evolution of *Ontocetus emmonsi* and dispersal along western and eastern North Atlantic during Pliocene; (3) direct evolution of *Odobenus* from *O. emmonsi* followed by Pleistocene dispersal to the North Pacific, and followed by (4) Pleistocene extinction of remaining non-*Odobenus* walruses in temperate North Pacific; (C) updated Odobenine biogeographic hypothesis as detailed by Kohno et al. (1995a,b), Kohno and Ray (2008), and Boessenecker et al. (2018): (1) late Miocene origin of Odobeninae in eastern North Pacific followed by dispersal to western North Pacific, (2) Pliocene evolution of *Ontocetus* followed by dispersal through the Arctic to the North Atlantic, (3) Pliocene evolution of *Odobenus* followed by (4) Pleistocene extinction of *Ontocetus* and subsequent dispersal of *Odobenus* to North Atlantic and eastern North Pacific, and followed by (5) Pleistocene extinction of non-*Odobenus* walruses in temperate North Pacific.

Ontocetus emmonsi is known from lower Pliocene through to lower Pleistocene strata of the east coast of the United States, England, the Netherlands, Belgium, and Morocco; at least seven previously-described species from four different genera (mostly based on fragmentary or undiagnostic remains) are now recognised as a single species (Kohno and Ray, 2008). The holotype tusk was originally identified as a sperm whale tooth, explaining the genus name, as the root 'cetus' means 'whale'. *Ontocetus* had a skull approximately 15% larger than *O. rosmarus*, and the two taxa shared elongated tusks, a vaulted palate, and foreshortened braincase (Deméré, 1994b; Kohno and Ray, 2008). However, *Ontocetus* tusks were notably procumbent, shorter, and more strongly curved than in *Odobenus* (Deméré, 1994b; Kohno and Ray, 2008). A deeply-vaulted palate and highly-worn, peg-like teeth indicate a similar benthic mollusk-specialized feeding ecology to *Odobenus*. *Ontocetus* is notable for inhabiting subtropical latitudes as far south as Florida and Morocco (Kohno and Ray, 2008), with the species surviving well into the Pleistocene in the western North Atlantic (Boessenecker et al., 2018).

Arguably the most bizarre odobenine, *Valenictus*, was originally reported on the basis of a single, yet distinctive and pachyosteosclerotic (thickened and dense) humerus (Fig. 2.5D; *Valenictus imperialensis*) from the lower Pliocene Deguynos Formation of southern California (Mitchell, 1961). Repenning and Tedford (1977) originally assigned *Valenictus* to the Dusignathinae (Repenning and Tedford, 1977). Subsequent discoveries, including partial skeletons and skulls of a different species (Fig. 2.5B, C and H; *V. chulavistensis*) from the Pliocene San Diego Formation of southern California (Deméré, 1994a), revealed that *Valenictus* completely lacked teeth aside from the upper canine tusks (Fig. 2.5b), and had a skeleton with widespread pachyosteosclerosis (resulting in especially thick and dense bones). In addition, the skull possessed a foreshortened braincase, like *Odobenus*, but shared the longer, more basal rostrum found in *Ontocetus*. *Valenictus* is known from several basins in central and southern California (Mitchell, 1961; Repenning and Tedford, 1977; Deméré, 1994a,b; Boessenecker, 2017). *Valenictus* inhabited now-vanished inland seas and hypersaline lagoons, particularly *V. imperialensis* and a dwarf species from the proto-Gulf of California (Atterholt et al., 2007; Barnes, 2005).

The lack of postcanine teeth in *Valenictus* parallels the condition evident in suction-feeding odontocetes that have undergone tooth loss (Ziphiidae, *Grampus, Monodon*). Modern *Odobenus* does not use its teeth to chew or pierce molluscan prey, and therefore *Valenictus* is, surprisingly, a more specialized suction feeder than the modern walrus. The fragmentary remains of the walrus *Pliopedia pacifica* consists of postcranial bones and a *Valenictus*-like braincase from the lower Pliocene of central California (Kellogg, 1921; Repenning and Tedford, 1977).

The current Arctic distribution of *O. rosmarus* is considerably more restricted than the fossil records of other *Odobenus*, as well as odobenines and odobenids more generally. The oldest known specimens of *Odobenus* are an isolated tusk from the Pliocene–Pleistocene Shiranuka Formation of Japan (Kohno et al., 1995a,b) and a nearly complete Plio-Pleistocene skull from the bottom of the Sea of Okhotsk (Miyazaki et al., 1992). The now extinct species *Odobenus mandanoensis*, is based on a partial jaw from the middle Pleistocene Mandano Formation of Japan (Tomida, 1989). Tusks and skulls of *Odobenus* are well represented in upper Pleistocene deposits stretching south along the east coast of North America from the Canadian high Arctic and Maritimes provinces (Dyke et al., 1999; McLeod et al., 2014) to South Carolina (Fig. 2.5A; Sanders, 2002) and along the western coastlines of Europe until southern France (Kardas, 1965). Specimens are accidentally discovered by fishing trawlers on the east coast of the United States (New Jersey to North Carolina) and on the Dutch and Belgian margins of the North Sea quite frequently (Erdbrink and Van Bree, 1999a,b,c).

Until recently, a lack of well-dated fossil occurrences meant that there was little consensus as to the biogeographic origin of *Odobenus*, or the timing of transarctic dispersal. Misinterpretation of fossils now understood to be *Ontocetus* led to an original hypothesised North Atlantic origin for modern walruses. A western North Pacific origin (Fig 2.6) (Kohno et al., 1995a,b) is supported by new evidence, including the discovery of late Pliocene–early Pleistocene *Odobenus* fossils from Japan and the Sea of Okhotsk (Miyazaki et al., 1992; Kohno et al., 1995b), reevaluation of *Ontocetus* (Deméré, 1994b; Kohno and Ray, 2008), and clarification of the geochronology of walruses in the North Atlantic (Boessenecker et al., 2018). Subsequently, during the mid Pleistocene, walruses originating from the North Pacific dispersed into the North Atlantic through the Arctic (Fig. 2.6) (Boessenecker et al., 2018).

The fossil record of *Odobenus* requires further clarification in three important areas. Firstly, despite the existence of diagnosable extinct species of *Odobenus*, many Pleistocene walrus fossils have been poorly described and/or uncritically identified as *O. rosmarus*. Therefore, critical reassessment of specimens identified as *O. rosmarus* (e.g. as in Kohno et al., 1995b) will improve our knowledge of the evolution and biogeography of the modern walrus. Secondly, occasional Pleistocene occurrences of *Odobenus* have yielded radiocarbon dates approaching the analytical limit of the isotope (40,000 years; e.g. Harington and Beard, 1992), and these specimens may in fact be much older. Improved stratigraphic and biostratigraphic control or the use of other absolute dating methods (e.g. Uranium–Thorium dating) is recommended to reassess these dates. Lastly, paleoclimatic proxies (e.g. associated invertebrates, pollen, microfossils) associated with fossil walruses remain virtually unstudied, and further study of this material may allow reexamination of the climatic tolerances for *Odobenus* in the past.

Trends in walrus evolution and paleobiology

Over the course of walrus evolution, several major trends in morphological evolution can be observed, most notably changes in locomotion, overall increases in body size, and changes in cranial and dental morphology associated with the evolution of suction feeding and tusks.

Locomotion

Modern pinniped families exhibit different swimming styles. Otariids 'fly' underwater using forelimb-dominated pectoral rowing and possess enlarged foreflippers and smaller hindlimbs used mostly for steering (English, 1976; Gordon, 1983). Phocid seals are hindlimb-dominated swimmers, swimming by lateral undulation of the hindflippers and steering with their foreflippers (Fish et al., 1988). As a result, the ankle bones of phocids are locked so the hindflippers cannot be rotated, preventing phocid seals from 'walking' on land like otariids. The modern walrus, on the other hand, is unique in that it uses both modes of swimming: hindlimb-dominated swimming at faster speeds and forelimb paddling at slower speeds (Gordon, 1981). Walruses also show both patterns of terrestrial locomotion: Juveniles can walk like a sea lion owing to their small size, while adults lunge forward with their forelimbs and pull their bodies during each lunge in a way that more closely resembles a phocid (Gordon, 1981). Based on the earliest-known pinniped, *Enaliarctos mealsi*, pinnipeds were initially capable of a combination fore- and hindlimb-dominated swimming (Berta and Ray, 1990; Bebej, 2009). The early 'imagotariine' walrus *P. repenningi* has similar vertebrae and hindlimb elements suggesting similar locomotion to *Enaliarctos*; a stiffened ankle joint in *Proneotherium* additionally indicates otariid-like walking as in *Enaliarctos* (Deméré and Berta, 2001). The unusually-robust forelimb bones of *Gomphotaria* led to the hypothesis that it was a forelimb-dominated swimmer (Barnes and Raschke, 1991), therefore implying a reversal within Odobenidae (Berta and Adam, 2001). This change in locomotion is further supported by comparison of forelimb elements attributed to *Dusignathus* (Repenning and Tedford, 1977; Barnes and Raschke, 1991).

Analyses of neural canal diameter can differentiate between the three styles of swimming in modern pinnipeds, and have been applied to extinct pinnipeds (Giffin, 1992). Unsurprisingly, given its sister taxon relationship with *Odobenus*, the extinct toothless walrus *Valenictus* has neural canals indicating a similar mode of forelimb/hindlimb combined locomotion (Giffin, 1992). A unique swimming style has been suggested based upon an undescribed postcranial skeleton of *Pontolis* (Giffin, 1992); however, this remains unclear and requires detailed examination. Future study of odobenid locomotion should examine existing dusignathine skeletal remains in more detail, new (unpublished) skeletons of *Valenictus*, and the virtually complete skeleton of *Titanotaria*.

Dental evolution and foraging ecology

Walruses show the most extreme examples of dental modification of any pinniped group with adaptations to suction feeding within odobenine walruses. Despite this, fossil walrus diet, functional anatomy, and ecology still remain poorly studied. There is also substantial disagreement among researchers on how to describe and categorise methods of prey capture and stages of prey processing (Hocking et al., 2017b; Kienle et al., 2017), as well as a wide variety of feeding behaviours present in extant taxa, using cranial and dental modifications (Churchill and Clementz, 2015).

Initial adaptations seen in walrus dentition are associated with the transition towards tooth usage focused on raptorial biting of fish prey (pierce-feeding) rather than mastication which had already been lost in their 'enaliarctine' ancestors (Churchill and Clementz, 2016). Parallel trends in dental simplification also occurred independently in other pinniped clades. The earliest odobenids retained the protocone shelf, metaconid and hypoconid cusps of earlier diverging arctoids, while also displaying more derived characters such as reduced and simplified upper molars, increased homodonty, reduction in molar and premolar P^4 roots, and loss of the P^4-M^1 embrasure pit (Deméré and Berta, 2001; Churchill and Clementz, 2016). As walruses became further adapted to aquatic feeding, the aforementioned cusps were reduced and ultimately lost, resulting in a simple peg-like dentition as seen in both Dusignathinae and Odobeninae. Alongside crown simplification, the tooth roots also gradually coalesced; from the double-rooted condition known in the earliest walruses, to the partially fused bilobate roots of *Imagotaria*, to the single-rooted condition seen in Neodobenia (Boessenecker and Churchill, 2013). Thinning of the enamel also became widespread among Neodobenia (Boessenecker and Churchill, 2013), perhaps corresponding to increasing use of suction, rather than teeth, in prey capture.

While simplification of dentition, including reduction in cusps and fusion of roots, is prevalent in Pinnipedia (Churchill and Clementz, 2016), odobenine walruses took tooth simplification to the extreme, particularly within Odobenini – the clade including *Ontocetus*, *Odobenus*, and *Valenictus*. Odobenini show enamel thinning (present in Dusignathinae as well) and eventual complete enamel loss, molarification of the canine C_1 and incisor I^3, and complete loss of molars and the fourth premolar (Boessenecker and Churchill, 2013, Churchill and Clementz, 2016). At its most extreme, *Valenictus* have a complete loss of all dentition except for the canine C1, which are retained as prominent tusks (Deméré, 1994b).

This simplification of teeth in walruses was likely due to suction-feeding specialisation, a method of prey capture reconstructed by Adam and Berta (2002) to have evolved at the base of Neodobenia. Other features of the mandible and skull, which support specialisation towards suction feeding, include fusion of the mandibular symphysis (Adam and Berta, 2002), enlargement of the pterygoid hamulus (reinforcing the soft palate while generating negative intraoral pressures needed for suction feeding; Kastelein and Gerrits, 1990), and elongation and vaulting of the palate (increasing suction pressure; Kastelein and Gerrits, 1990; Deméré, 1994a). Modern walruses are specialised suction feeders of bivalves, and it is reasonable to assume that closely-related taxa, such as *Ontocetus* and *Valenictus*, had a similar diet given their similar morphology. However, it is difficult to identify when molluscivorous specialisations first evolved as the earliest walruses were likely unspecialised piscivores. Suction feeding likely preceded specialisation in diet. Wear patterns in dusignathines have been used to suggest molluscivory in *Gomphotaria* and *D. seftoni* (Barnes and Raschke 1991; Berta et al., 2018), but the pattern of wear is different from that observed in extant *Odobenus*. Extreme wear observed in *Gomphotaria* and *D. seftoni* may simply be a result of feeding in environments with high sediment loads, rather than true molluscivory. Detailed analysis of tooth wear, as well as morphometric analysis of skull shape, may help in determining the diet of these extinct walruses and provide further information on the transition from pierce-feeding to specialized suction feeding.

The evolution of tusks – 'sled runners' or 'oral antlers?'

The modern walrus is arguably most famous for sporting a pair of greatly-elongated canines, or tusks. In *O. rosmarus*, these canines are often twice as long as the length of the skull and oriented perpendicular to the palate, or basal plane of the skull. Initially assumed to be used in feeding, Fay (1982) demonstrated that walruses do not use their tusks during benthic feeding and instead the tusks drag passively behind the skull. Sexual dimorphism and a mating behaviour described as lekking and/or female-defence polygyny in Pacific walruses led Deméré (1994a) to interpret tusks as an analogue for deer antlers: secondary sex characteristics used for display and/or combat. In contrast, tusks of the bizarre, extinct 'walrus-faced whale' *Odobenocetops*, found in Peru, along with those of *Odobenus*, have also been interpreted as a tool for stabilising the head during benthic suction feeding (referred to as 'sled runners') (Muizon et al., 2002), although a social role was also acknowledged.

Tusks of several forms are now known in multiple genera within Neodobenia, including the short, procumbent upper and lower tusks of *Gomphotaria* and *Dusignathus*; the greatly-enlarged, tusk-like roots of *Protodobenus*; and the short, somewhat procumbent and highly-curved tusks of *Ontocetus* and *Valenictus* (Barnes and Raschke, 1991; Deméré, 1994b; Horikawa, 1995; Kohno and Ray, 2008). None of these tusks are long enough or oriented perpendicular to the palate to fulfil a 'sled runner' role (Boessenecker and Churchill, 2013). Instead, the orientation of these tusks and variety of tusk forms is evocative of antler disparity as envisioned by Deméré (1994a). While modern *Odobenus* may have elongated and appropriately oriented tusks to aid in benthic feeding, the earliest tusks found in dusignathines and extinct odobenines were not suited to such tasks, indicating that walrus tusks initially evolved purely as secondary sexual characteristics (Boessenecker and Churchill, 2013).

Body size

Another obvious trend that can be observed in walrus evolution is the increase in body size over time. The earliest walruses were small in size, approximately the same as their 'enaliarctine' ancestors (a little larger than a modern fur seal) (Churchill et al., 2014). Walruses would remain relatively modest in size until the Tortonian–Messinian, when an increase in diversity coincided with an increase in body size within walruses (Fig. 2.1; Velez-Juarbe, 2017; Boessenecker and Churchill, 2018). This increase in body size was probably spurred on by cooling of the climate and an increase in coastal upwelling (Boessenecker and Churchill, 2018), and the small pinniped niche was seemingly taken over by otariid seals.

Sexual dimorphism

The modern walrus is sexually dimorphic, with differences between males and females such as body mass, tusk length, curvature, and divergence (Fay, 1982). However, it is important to note that differences in walrus body size and skull morphology between the sexes are not as extreme as in some phocids (e.g. *Mirounga*) and otariids (e.g. *Callorhinus ursinus*). Pinnipeds likely evolved sexual dimorphism early in their evolution, as evidenced by 'enaliarctine' fossils (Cullen et al., 2014). Sexual dimorphism is also inferred for some imagotariines, such as *Neotherium* and *Imagotaria*, based upon large sample sizes of mandibles and skulls (respectively, Repenning and Tedford, 1977; Barnes, 1988; Kohno et al., 1995a,b; Velez-Juarbe, 2018). Tusks of *O. emmonsi* show evidence of sexual dimorphism (Kohno and Ray, 2008), and further study of new material of *Valenictus*, *Dusignathus*, *Pontolis*, *Titanotaria*, and others will reveal whether sexual dimorphism is a clade-wide phenomenon amongst walruses.

Baculum

Walruses have the largest baculum (penis bone) of any mammal in both absolute terms and relative to body mass. Bacula aid in prolonged intromission. Amongst pinnipeds, baculum size is correlated with aquatic versus terrestrial copulation; aquatic copulators possess the largest bacula (Scheffer and Kenyon, 1962). The baculum is preserved for several extinct odobenids (*Gomphotaria*, *Titanotaria*, *Valenictus*) and may shed light on the evolution of copulatory behaviour in walruses. For example, modestly-sized otariid-like bacula in *Gomphotaria* and *Titanotaria* may suggest terrestrial copulation and a more recent evolution of aquatic copulation. Of note, a gigantic baculum without an associated skeleton has been found in middle Miocene rocks in Japan (Hasegawa, 2007), comparing best with extant *Odobenus*, and suggesting the existence of a truly enormous or particularly well-endowed aquatic copulating walrus awaiting discovery of more complete remains. Further study of bacula could provide important information on the evolution of social behaviour in walruses and other pinnipeds.

Walrus biogeography and diversity

The walrus is one of the most iconic marine mammals of the Arctic Ocean, alongside the polar bear (*Ursus maritimus*) and narwhal (*Monodon monoceros*). However, as should be apparent from our review of the fossil record, their occupation of Arctic environments represents a relatively recent adaptation. Walruses formed a common and important component of ancient marine mammal faunas in the subtropical and temperate North Pacific and North Atlantic. Phylogenetic analyses and the fossil record agree on a North Pacific origin for Odobenidae in the Miocene (Fig. 2.2; Deméré, 1994b; Kohno et al., 1995a,b; Miyazaki et al., 1995), with dispersal outside of this ocean basin occurring during the Pliocene (Fig. 2.6). Walruses were also sufficiently diverse in the Miocene and Pliocene to have been preserved alongside one another in multispecies assemblages in the North Pacific (Boessenecker, 2013a; Velez-Juarbe, 2017; Magallanes et al., 2018). Multispecies walrus assemblages have been recorded from the Sharktooth hill bonebed ($n = 3$; Velez-Juarbe, 2018) and the Capistrano ($n = 3$; Velez-Juarbe, 2017), Purisima ($n = 3-4$; Boessenecker, 2013a), Empire ($n = 4$; Deméré et al., 2003), Wilson Grove ($n = 5$; Powell et al., 2019), and San Diego Formations ($n = 3$; Deméré, 1994a; Boessenecker, 2013a).

The earliest walruses were morphologically similar to their 'enaliarctine' ancestors, and although morphological variety was limited, taxonomic diversity was high. When examining only raw numbers of taxa, Boessenecker and Churchill (2018) show a peak in diversity of four taxa in the Burdigalian (20.44–15.97 Ma), followed by a decline in diversity into the Tortonian (11.63–7.246 Ma), although this may represent a sampling bias. Magallanes et al. (2018) found a similar overall trend, with higher levels of diversity particularly during the Burdigalian when accounting for the presence of ghost lineages and stratigraphic uncertainty.

As walrus diversity either declined (Magallanes et al., 2018) or plateaued (Boessenecker and Churchill, 2018) the diversity of the extinct Desmatophocidae increased, with as many as six different taxa in the Langhian (15.97–13.82 Ma; Boessenecker and Churchill, 2018). This suggests that the success of Desmatophocidae, possibly as a result of unusually warm sea surface temperatures (Boessenecker and Churchill, 2018), may have constrained walrus diversity at this time. Only with the extinction of desmatophocids at the end of the Tortonian did walrus diversity expand, with eight taxa known from the late Tortonian through Messinian (Boessenecker and Churchill, 2018; Magallanes et al., 2018). This period of time also sees the extinction of smaller-sized walruses (Boessenecker and Churchill, 2018; Velez-Juarbe and Salinas-Márquez, 2018). The small-bodied pinniped niche would be filled by the diversifying otariids during the Pliocene and by new phocid immigrants in the Pleistocene (Miyazaki et al., 1995; Boessenecker, 2013a).

While there is little doubt that walruses originated in the North Pacific, the number of dispersals and route of dispersal into the Atlantic has been controversial (Fig. 2.6). Originally, *Ontocetus* was considered directly ancestral to modern *Odobenus*. Under this hypothesis, the ancestor of *Ontocetus* dispersed along a southern route through the tropical Central American Seaway. *Ontocetus* then quickly spread across the North Atlantic during the latest Miocene or earliest Pliocene (Repenning and Tedford, 1977). Once *Odobenus* had evolved from *Ontocetus*, they then dispersed back through the Arctic Ocean to the North Pacific (Repenning and Tedford, 1977; Repenning et al., 1979; Tomida, 1989; Miyazaki et al., 1995).

However, new fossil evidence suggests that two separate dispersals out of the North Pacific occurred (Fig. 2.6). The first dispersal, through the Arctic Ocean, allowed *Ontocetus* to expand into the Atlantic Ocean (Kohno et al., 1995a,b; Kohno and Ray, 2008). While *Ontocetus emmonsi* was a successful odobenine, it was ultimately an evolutionary dead-end with no living descendants

(Deméré, 1994a,b; Kohno and Ray, 2008). Instead, *Odobenus* likely evolved in the North Pacific (Kohno et al., 1995a,b, 1998; Kohno and Ray, 2008) and dispersed independently of *Ontocetus* into the North Atlantic region at a later date, most likely through the Arctic. Further support for this hypothesis can be seen in the description of new *Ontocetus* material from South Carolina, which provides evidence that *Ontocetus* went extinct in the North Atlantic one million years before *Odobenus* appears in the fossil record (Boessenecker et al., 2018).

Although formerly the centre of their distribution, walruses are currently absent from most of the North Pacific (Fig. 2.6). Causes for the Pliocene decline and subsequent this extirpation are not yet understood (Boessenecker and Churchill, 2018; Magallanes et al., 2018). Climate change, with its associated effects on productivity and sea level, may have played an important role. The loss of abundant shallow embayments and seaways along the California coastline likely led to the extinction of benthic-feeding odontocete and mysticete specialists, and may also be responsible for the loss of temperate odobenines like *Valenictus* (Boessenecker, 2013a,b, 2017; Boessenecker and Poust, 2015). Broad-scale analyses have identified a period of faunal turnover at the end of the Pliocene that seems to have included the extinction of low latitude walruses and many extinct marine mammals with highly specialized feeding adaptations (Boessenecker, 2013a; Pimiento et al., 2017). However, the chronology and pattern of faunal change in pinnipeds and other marine mammals during the Pleistocene is poorly known for most ocean basins owing to a limited fossil record. Did the extinction of endemic walrus taxa allow for otariid diversification and the dispersal of phocids into the North Pacific, or did competition between these different pinniped groups play a significant role in the loss of walrus diversity? If competition did not play a role, did walrus diversity decline due to other ecologic or environmental drivers? Further study is still sorely needed, including the prospecting and describing of fossils from late Pliocene and early-middle Pleistocene fossil localities (Boessenecker, 2013a; *contra* Pimiento et al., 2017:1104).

Conservation paleobiology of *Odobenus rosmarus*

The modern walrus lives chiefly within the Arctic Circle but extends as far south as 53°N in the Bering Sea. This is in clear contrast with ancient walruses as many occurrences of *Odobenus* have been noted as far south as Shimane and Chiba prefectures of Japan in the western North Pacific (35°N; Kohno et al., 1995a,b), San Francisco (37°N; California, United States; Harington, 1984) in the eastern North Pacific, South Carolina (United States) in the western North Atlantic (32°N; Sanders, 2002) and the North Sea in the eastern North Atlantic (51°N; United Kingdom and the Netherlands; Erdbrink and Van Bree, 1999a,b,c, and references therein). Few fossils are associated with rigorous assessments of paleotemperature, however it is broadly inferred that *Odobenus* tolerated warmer water than today, as many finds are reported from Sangamonian interglacial deposits only slightly cooler than present conditions (Ray et al., 1968). At least one middle Pleistocene occurrence of *Odobenus* sp. from Hokkaido, Japan, is associated with molluscs indicating annual average temperatures of 7°C–12°C (Akamatsu and Suzuki, 1990), similar to current temperatures (±3°C) and thus indicating warmer climatic tolerance for *Odobenus* in the past. However, in some cases, molluscs associated with fossils of *O. rosmarus* do show evidence of substantially colder temperatures (e.g. Kempsville Member of the Acredale Formation, Virginia; Spencer and Campbell, 1987). Detailed assessment of associated molluscan assemblages has the potential to illuminate the recent (e.g. Pliocene–Pleistocene) adoption of cold waters by odobenines, as recently applied to belugas (Ichishima et al., 2019), but not yet studied in the context of walrus evolution.

In this regard, the Maritimes walrus is of particular interest. The Maritimes walrus formed a morphologically and genetically distinct southern population of *O. rosmarus* endemic to the Maritimes region of Canada (New Brunswick, Nova Scotia, Sable Island), before being extirpated due to overhunting in the 16th–18th centuries (McLeod et al., 2014). The Maritimes walrus was somewhat larger than the Atlantic walrus, with larger tusks and greater tusk asymmetry (McLeod et al., 2014). Though modern walruses typically breed on ice (Fay, 1982), the Maritimes walrus appears to have inhabited the region since the end of the Pleistocene, inhabiting ice-free coastlines for up to 12,800 years (McLeod et al., 2014). The Maritimes walrus may indicate that extant *Odobenus* populations may be tolerant of warming seas, or perhaps was physiologically distinct and extant Pacific and Atlantic populations represent a geologically recent cold-specialized clade without the tolerance for a warmer climate. Evidence of an extirpated Medieval period walrus population in Iceland (ice free at the time) may suggest that other populations can tolerate warmer, ice-free coastlines (Keighley et al., 2019).

The modern walrus is perhaps more capable of tolerating a warmer climate than typically assumed given the survival until recent times of a cold-temperate, ice-free breeding population (Maritimes walrus) and broader climatic tolerance in the Pliocene and Pleistocene for odobenines and *Odobenus.* Further study of the Maritimes walrus and the paleoclimatic distribution of fossil *Odobenus* has the potential to provide an evolutionary perspective that could inform conservation decisions or predict how *Odobenus* will respond to forecasted ice-free conditions in the future.

Acknowledgements

We thank the editors X. Keighley, M. T. Olsen, P. Jordan, and S. Desjardins for inviting us to contribute to this volume. We are indebted to the many staff of institutions housing walrus fossils who have permitted us to study fossils under their care, chiefly including Aurora Fossil Museum, California Academy of Sciences, Natural History Museum of Los Angeles County, National Museum of Nature and Science (Tokyo), San Diego Natural History Museum, University of California Museum of Paleontology, US National Museum of

Natural History. Thanks to S.J. Boessenecker for critical comments which improved an earlier draft of this chapter. Thanks to L.G. Barnes, B.L. Beatty, A. Berta, S.J. Boessenecker, T.A. Deméré, N. Kohno, W. McLaughlin, A. Poust, Y. Tanaka and J. Velez-Juarbe for countless discussions of walrus palaeontology, anatomy, and evolution; without these continuing discussions, such a review would not have been possible. Thanks to the reviewers whose constructive criticism greatly benefited this chapter.

References

Adam, P.J., Berta, A., 2002. Evolution of prey capture strategies and diet in the Pinnipedimorpha (Mammalia, Carnivora). Oryctos 4, 83–107.

Akamatsu, M., Suzuki, A., 1990. Pleistocene molluscan faunas in central and southwestern Hokkaido. Hokkaido University, Series IV. Journal of the Faculty of Science 22, 529–592.

Arnason, U., Gullberg, A., Janke, A., Kullberg, M., Lehman, N., Petrov, E.A., et al., 2006. Pinniped phylogeny and a new hypothesis for their origin and dispersal. Molecular Phylogenetics and Evolution 41, 345–354.

Atterholt, J., Jefferson, G.T., Schachner, E., 2007. New marine vertebrates from the Yuha Member of the Deguynos Formation of Anza-Borrego Desert state Park. Journal of Vertebrate Paleontology 27, 42A.

Barnes, L.G., 1988. A new fossil pinniped (Mammalia: Otariidae) from the middle Miocene Sharktooth Hill Bonebed, California. Contributions in Science 396, 1–11.

Barnes, L.G., 1989. A new enaliarctine pinniped from the Astoria Formation, Oregon, and a classification of the Otariidae (Mammalia: Carnivora). Contributions in Science, Natural History Museum of Los Angeles County 403, 1–28.

Barnes, L.G., 2005. Dense-boned late Miocene and Pliocene fossil walruses of the Imperial Desert and Baja California: possible buoyancy-control mechanisms for feeding on benthic marine invertebrates in the Proto-Gulf of California. In: Reynolds R.E., (Ed.), *The 2005 Desert Symposium*, California State University, Fullerton, 87.

Barnes, L.G., Raschke, R.E., 1991. *Gomphotaria pugnax*, a new genus and species of late Miocene dusignathine otariid pinniped (Mammalia: Carnivora) from California. Natural History Museum of Los Angeles County Contributions in Science 426, 1–27.

Bebej, R.M., 2009. Swimming Mode Inferred from Skeletal Proportions in the Fossil Pinnipeds *Enaliarctos* and *Allodesmus* (Mammalia, Carnivora). Journal of Mammalian Evolution 16, 77–97.

Berta, A., Adam, P.J., 2001. Evolutionary biology of Pinnipeds. In: Mazin J.-M., de Buffrénil V., (Eds), *Secondary Adaptation of Tetrapods to Life in Water*, Verlag Dr. Friedrich Pfeil, Munich, Germany, 235–260.

Berta, A., Ray, C.E., 1990. Skeletal morphology and locomotor capabilities of the archaic pinniped *Enaliarctos mealsi*. Journal of Vertebrate Paleontology 10, 141–157.

Berta, A., Wyss, A.R., 1994. Pinniped phylogeny. Proceedings of the San Diego Society of Natural History 29, 33–56.

Berta, A., Churchill, M., 2012. Pinniped taxonomy: review of currently recognized species and subspecies, and evidence used for their description. Mammal Review 42, 207–234.

Berta, A., Churchill, M., Boessenecker, R.W., 2018. The origin and evolutionary biology of pinnipeds: seals, sea lions, and walruses. Annual Review of Earth and Planetary Sciences 46, 203–228.

Bininda-Emonds, O.R.P., Gittleman, J.L., Purvis, A., 1999. Building large trees by combining phylogenetic information: a complete phylogeny of the extant Carnivora (Mammalia). Biological Reviews 74, 143–175.

Blanco, P.J., Rodriguez, L.A., 2001. Surprising drifting of bodies along the coast of Portugal and Spain. Legal Medicine 3, 177–182.

Boessenecker, R.W., 2013a. A new marine vertebrate assemblage from the Late Neogene Purisima Formation in Central California, Part II: Pinnipeds and cetaceans. Geodiversitas 35, 815–940.

Boessenecker, R.W., 2013b. Pleistocene survival of an archaic dwarf baleen whale (Mysticeti: Cetotheriidae). Naturwissenschaften 100, 365–371.

Boessenecker, R.W., 2017. A new early Pliocene record of the toothless walrus *Valenictus* (Carnivora, Odobenidae) from the Purisima Formation of Northern California. Paleobios 34, 1–6.

Boessenecker, R.W., Churchill, M., 2013. A reevaluation of the morphology, paleoecology, and phylogenetic relationships of the enigmatic walrus *Pelagiarctos*. PLoS One 8, e54311.

Boessenecker, R.W., Churchill, M., 2015. The oldest known fur seal. Biology Letters 11. Available from: https://doi.org/10.1098/rsbl.2014.0835.

Boessenecker, R.W., Poust, A.W., 2015. Freshwater occurrence of the extinct dolphin *Parapontoporia* (Cetacea: Lipotidae) from the upper Pliocene nonmarine Tulare Formation of California. Palaeontology 58, 489–496.

Boessenecker, R.W., Churchill, M., 2018. The last of the desmatophocid seals: a new species of *Allodesmus* from the upper Miocene of Washington, USA. Zoological Journal of the Linnaean Society 184, 211–235.

Boessenecker, R.W., Perry, F.A., Schmitt, J.G., 2014. Comparative taphonomy, taphofacies, and bonebeds of the Mio-Pliocene Purisima Formation, Central California: strong physical control on marine vertebrate preservation in shallow marine settings. PLoS One 9, e91419.

Boessenecker, S.J., Boessenecker, R.W., Geisler, J.H., 2018. Youngest record of the extinct walrus *Ontocetus emmonsi* from the early Pleistocene of South Carolina and a review of North Atlantic walrus biochronology. Acta Palaeontologica Polonica 63, 279–286.

Boessenecker, R.W., Ehret, D.J., Long, D.J., Churchill, M., Martin, E., Boessenecker, S.J., 2019. The early Pliocene extinction of the mega-toothed shark *Otodus megalodon*: a view from the eastern North Pacific. PeerJ 7, e6088.

Churchill, M., Clementz, M.T., 2015. The evolution of aquatic feeding in seals: insights from *Enaliarctos* (Carnivora: Pinnipedimorpha), the oldest known seal. Journal of Evolutionary Biology 29, 319–334.

Churchill, M., Clementz, M.T., 2016. Functional implications of variation in tooth spacing and crown size in Pinnipedimorpha (Mammalia: Carnivora). The Anatomical Record 298, 878–902.

Churchill, M., Clementz, M.T., Kohno, N., 2014. Cope's rule and the evolution of body size in Pinnipedimorpha (Mammalia: Carnivora). Evolution 69, 201–215.

Cozzuol, M.A., 2001. A “northern” seal from the Miocene of Argentina: implications for phocid phylogeny and biogeography. Journal of Vertebrate Paleontology 21, 415–421.

Cullen, T.M., Fraser, D., Rybczynski, N., Schröder-Adam, C.J., 2014. Early Evolution of sexual dimorphism and polygyny in Pinnipedia. Evolution 68, 1469–1484.

Deméré, T.A., 1994a. The family Odobenidae: a phylogenetic analysis of fossil and living taxa. Proceedings of the San Diego Society of Natural History 29, 99–123.

Deméré, T.A., 1994b. Two new species of fossil walruses (Pinnipedia: Odobenidae) from the Upper Pliocene San Diego Formation, California. Proceedings of the San Diego Society of Natural History 29, 77–98.

Deméré, T.A., Berta, A., 2001. A reevaluation of *Proneotherium repenningi* from the Miocene Astoria Formation of Oregon and its position as a basal odobenid (Pinnipedia: Mammalia). Journal of Vertebrate Paleontolog 21, 279–310.

Deméré, T.A., Berta, A., Adam, P.J., 2003. Pinnipedimorph evolutionary biogeography. Bulletin of the American Museum of Natural History 279, 32–76.

Domning, D.P., 1978. Sirenian Evolution in the North Pacific Ocean, vol. 18. University of California Publications in Geological Sciences, pp. 1–176.

Dubrovo, I.A., 1981. A new subfamily of fossil seals (Pinnipedia: Kamtschatarctinae subfam. nov.). Proceedings of the Academy of Sciences of the USSR 256, 970–974 (in Russian).

Dyke, A.S., Hooper, J., Harington, C.R., Savelle, J.M., 1999. The late Wisconsinan and Holocene record of walrus (*Odobenus rosmarus*) from North America: a review with new data from Arctic and Atlantic Canada. Arctic 52, 160–181.

English, A.W.M., 1976. Functional anatomy of the hands of fur seals and sea lions. American Journal of Anatomy 147, 1–18.

Erdbrink, D.P., Van Bree, P.J.H., 1999a. Fossil appendicular skeletal walrus material from the North Sea and the estuary of the Schede (Mammalia, Carnivora). Beaufortia 49, 63–81.

Erdbrink, D.P., Van Bree, P.J.H., 1999b. Fossil axial skeletal walrus material from the North Sea and the estuary of the Schelde, and a fossil sirenian rib (Mammalia, Carnivora; Sirenia). Beaufortia 49, 11–20.

Erdbrink, D.P., van Bree, P.J.H., 1999c. Fossil cranial walrus material from the North Sea and the estuary of the Schelde (Mammalia, Carnivora). Beaufortia 49, 1–9.

Espinoza, E.O., Yates, B.C., Mann, M.J., Crane, A.R., Goddard, K.W., LeMay, J.P., et al., 1997. Taphonomic indicators used to infer wasteful subsistence hunting in Northwest Alaska. Anthropozoologica 25, 103–112.

Fay, F.H., 1982. Ecology and biology of the Pacific Walrus, *Odobenus rosmarus divergens* Illiger. North American Fauna 74, 1–279.

Fay, F.H., 1985. *Odobenus rosmarus*. Mammalian Species 238, 1–7.

Fish, F.E., Innes, S., Ronald, K., 1988. Kinematics and estimated thrust production of swimming harp and ringed seals. Journal of Experimental Biology 137, 157–173.

Fulton, T.L., Strobeck, C., 2010. Multiple markers and multiple individuals refine seal phylogeny and bring molecules and morphology back in line. Proceedings of the Royal Society B: Biological Sciences 277, 1065–1070.

Furbish, R., 2015. Something Old, Something New, Something Swimming in the Blue: An Analysis of the Pinniped Family Desmatophocidae, Its Phylogenetic Position, and Swimming Mode. San Diego State University, p. 75.

Gallagher, W.B., Parris, D.C., Grandstaff, B.S., DeTample, C., 1989. Quaternary mammals from the continental shelf off New Jersey. Mosasaur 4, 101–110.

Giffin, E.B., 1992. Functional implications of neural canal anatomy in recent and fossil marine carnivores. Journal of Morphology 214.

Gingras, M.K., Armitage, I.A., Pemberton, S.G., Clifton, H.E., 2007. Pleistocene walrus herds in the Olympic Peninsula area: trace-fossil evidence of predation by hydraulic jetting. Palaios 22, 539–545.

Gordon, K., 1981. The locomotor behavior of the walrus. Journal of the Zoological Society of London 195, 349–367.

Gordon, K., 1983. Mechanics of the limbs of the walrus (*Odobenus rosmarus*) and the California sea lion (*Zalophus californianus*). Journal of Morphology 175, 73–90.

Harington, C.R., 1984. Quaternary marine and land mammals and their paleoenvironmental implications – some examples from northern North America. Carnegie Museum of Natural History Special Publication 8, 511–525.

Harington, C.R., Beard, G., 1992. The Qualicum walrus: a Late Pleistocene walrus (*Odobenus rosmarus*) skeleton from Vancouver Island, British Columbia, Canada. Annales Zoologica Fennici 28, 311–319.

Hasegawa, M., 2007. Note on a large baculum from the Aoki Formation of middle Miocene age, Matsumoto, central Japan. Bulletin of the Gunma Museum of Natural History 11, 37–42.

Higdon, J.W., Bininda-Emonds, O.R.P., Beck, R.M.D., Ferguson, S.H., 2007. Phylogeny and divergence of the pinnipeds (Carnivora: Mammalia) assessed using a multigene dataset. BMC Evolutionary Biology 7, 216.

Hocking, D.P., Marx, F.G., Park, T., Fitzgerald, E.M.G., Evans, A.R., 2017b. Reply to comment by Kienle *et al.* 2017. Proceedings of the Royal Society B 284. Available from: https://doi.org/10.1098/rspb.2017.1836.

Horikawa, H., 1995. A primitive odobenine walrus of Early Pliocene age from Japan. The Island Arc 3, 309–328.

Ichishima, H., Furusawa, H., Tachibana, M., Kimura, M., 2019. First monodontid cetacean (Odontoceti: Delphinoidea) from the early Pliocene of the north-western Pacific Ocean. Papers in Palaeontology 5, 323–342.

Kardas, S., 1965. Notas sobre el genero *Odobenus* (Mammalia, Pinnipedia). 1. Una nueva suespecie fosil del Pleistoceno superior-Holoceno. Boletin de la Real Sociedad Espanola de Historia Natural. Organo del Instituto de Ciencias Naturales Joseph de Acosta Seccion. Geologica 63, 363–380.

Kastelein, R.A., Gerrits, N.M., 1990. The anatomy of the walrus head (*Odobenus rosmarus*). Part 1: The skull. Aquatic Mammals 16, 101–119.

Keighley, X., Pálsson, S., Einarsson, B.F., Petersen, A., Fernández-Coll, M., Jordan, P., et al., 2019. Disappearance of Icelandic walruses coincided with Norse settlement. Molecular Biology and Evoluition 36, 2656–2667.

Kellogg, R., 1921. A new pinniped from the Upper Pliocene of California. Journal of Mammalogy 2, 212–226.

Kellogg, R., 1931. Pelagic mammals from the Temblor Formation of the Kern River region, California. Proceedings of the California Academy of Sciences 19, 217–397.

Kidwell, S.M., 1993. Taphonomic expressions of sedimentary hiatuses: field observations on bioclastic concentrations and sequence anatomy in low, moderate, and high subsidence settings. Geologische Rundschau 82, 189–202.

Kienle, S.S., Law, C.J., Costa, D.P., Berta, A., Mehta, R.S., 2017. Revisiting the behavioral framework of feeding in predatory aquatic mammals. Proceedings of the Royal Society B 20171035.

Kohno, N., 1994. A new Miocene pinniped in the genus *Prototaria* (Carnivora: Odobenidae) from the Moniwa Formation, Miyagi, Japan. Journal of Vertebrate Paleontology 14, 414–426.

Kohno, N., 2006. A new Miocene odobenid (Mammalia: Carnivora) from Hokkaido, Japan, and its implications for odobenid phylogeny. Journal of Vertebrate Paleontology 26, 411–421.

Kohno, N., Tomida, Y., 1993. Marine mammal teeth (Otariidae and Delphinidae) from the early Pleistocene Setana Formation, Hokkaido, Japan. Bulletin of the National Science Museum, Tokyo 19, 139–146.

Kohno, N., Ray, C.E., 2008. Pliocene Walruses From the Yorktown Formation of Virginia and North Carolina, and a Systematic Revision of the North Atlantic Pliocene Walruses, vol. 14. Virginia Museum of Natural History Special Publication, pp. 39–80.

Kohno, N., Barnes, L.G., Hirota, K., 1995a. Miocene fossil pinnipeds of the genera *Prototaria* and *Neotherium* (Carnivora; Otariidae; Imagotariinae) in the North Pacific Ocean: Evolution, relationships and distribution. The Island Arc 3, 285–308.

Kohno, N., Tomida, Y., Hasegawa, Y., Furusawa, H., 1995b. Pliocene tusked odobenids (Mammalia: Carnivora) in the Western North Pacific, and their paleobiogeography. Bulletin of the National Science Museum, Tokyo Series C (Geology & Paleontology) 21, 111–131.

Kohno, N., Narita, K., Hakuichi, K., 1998. An early Pliocene odobenid (Mammalia: Carnivora) from the Joshita Formation, Nagano Prefecture, central Japan. Research Reports of the Shinshushinmachi Fossil Museum 1, 1–7.

Koretsky, I.A., Holec, P., 2002. A primitive seal (Mammalia: Phocidae) from the early middle Miocene of central Paratethys. Smithsonian Contributions to Paleobiology 93, 163–178.

Koretsky, I.A., Barnes, L.G., 2006. Pinniped evolutionary history and paleobiogeography. In: Csiki, Z. (Ed.), Mesozoic and Cenozoic Vertebrates and Paleoenvironments: Tributes to the Career of Prof. Dan Grigorescu. Ars Docendi, Bucharest, pp. 143–153.

Lindqvist, C., Bachmann, L., Andersen, L.W., Born, E.W., Arnason, U., Kovacs, K.M., et al., 2009. The Laptev Sea walrus *Odobenus rosmarus laptevi*: an enigma revisited. Zoologica Scripta 38, 113–127.

Loch, C., Boessenecker, R.W., Churchill, M., Kieser, J.A., 2016. Enamel ultrastructure of fossil and modern pinnipeds: evaluating hypotheses of feeding adaptations in the extinct walrus *Pelagiarctos*. The Science of Nature 103, 44.

Magallanes, I., Parham, J.F., Santos, G.P., Velez-Juarbe, J., 2018. A new tuskless walrus from the Miocene of Orange County, California, with comments on the diversity and taxonomy of odobenids. PeerJ 6, e5708.

McLeod, B.A., Frasier, T.R., Lucas, Z., 2014. Assessment of the extirpated maritimes walrus using morphological and ancient DNA analysis. PLoS One 9, e99569.

Mitchell, E.D., 1961. A new walrus from the imperial Pliocene of Southern California: with notes on odobenid and otariid humeri. Contributions in Science 44, 1–28.

Mitchell, E.D., 1968. The Mio-Pliocene pinniped *Imagotaria*. Journal of the Fisheries Research Board of Canada 25, 1843–1900.

Miyazaki, S., Kimura, M., Ishiguri, H., 1992. On a Pliocene walrus (*Odobenus* sp.) discovered in the Northern Pacific Ocean. The Journal of the Geological Society of Japan 98, 723–740.

Miyazaki, S., Horikawa, H., Kohno, N., Hirota, K., Kimura, M., Hasegawa, Y., et al., 1995. Summary of the fossil record of pinnipeds of Japan, and comparisons with that from the eastern North Pacific. The Island Arc 361–372.

Muizon, C.D., 1982. Phocid phylogeny and dispersal. The Annals of the South African Museum 89, 175–213.

Muizon, C.D., Domning, D.P., 2002. *Odobenocetops* peruvianus, the walrus-convergent delphinoid (Mammalia: Cetacea) from the early Pliocene of Peru. Smithsonian Contributions to Paleobiology 93, 223–261.

Nyakatura, K., Bininda-Emonds, O.R.P., 2012. Updating the evolutionary history of Carnivora (Mammalia): a new species-level supertree complete with divergence time estimates. BMC Evolutionary Biology 10, 1–31.

Pimiento, C., Griffin, J.N., Clements, C.F., Silvestro, D., Varela, S., Uhen, M.D., et al., 2017. The Pliocene marine megafauna extinction and its impact on functional diversity. Nature Ecology & Evolution 1, 1100–1106.

Powell II, C.L., Boessenecker, R.W., Smith, N.A., Fleck, R.J., Carlson, S.J., Allen, J.R., et al., 2019. Geology and Paleontology of the Late Miocene Wilson Grove Formation at Bloomfield Quarry, Sonoma County, California. In: US Geological Survey Scientific Investigations Report 2019-5021, pp. 1–77.

Ray, C.E., Wetmore, A., Dunkle, D.H., 1968. Fossil vertebrates from the marine Pleistocene of southeastern Virginia. Smithsonian Miscellaneous Collections 153, 1–25.

Repenning, C.A., Tedford, R.H., 1977. Otarioid seals of the Neogene. In: US Geological Survey Professional Paper 992, pp. 1–87.

Repenning, C.A., Ray, C.E., Grigorescu, D., 1979. Pinniped biogeography. In: Gray, J., Boucot, A.J. (Eds.), Historical Biogeography, Plate Tectonics, and the Changing Environment. Oregon State University Press, Corvallis, OR.

Sanders, A.E., 2002. Additions to the Pleistocene mammal faunas of South Carolina, North Carolina, and Georgia. Transactions of the American Philosophical Society 92, 1–152.

Schäfer, W., 1972. Ecology and Paleoecology of Marine Environments. University of Chicago Press, Chicago, IL.

Scheffer, V.B., Kenyon, K.W., 1962. Baculum size in pinnipeds. Zeitschrift für Säugetierkunde 28, 38–41.

Spencer, R.S., Campbell, L.D., 1987. The fauna and paleoecology of the Late Pleistocene marine sediments of southeastern Virginia. Bulletins of American Paleontology 92, 1–124.

Takeyama, K.-I., Ozawa, T., 1984. A new Miocene otarioid seal from Japan. Proceedings of the Japan Academy, Series B 60, 36–39.

Tanaka, Y., Kohno, N., 2015. A new late Miocene odobenid (Mammalia: Carnivora) from Hokkaido, Japan suggests rapid diversification of basal Miocene odobenids. PLoS One 10, e0131856.

Tedford, R.H., 1976. Relationships of pinnipeds to other carnivores (Mammalia). Systematic Zoology 25, 363–374.

Tomida, Y., 1989. A new walrus (Carnivora, Odobenidae) from the Middle Pleistocene of the Boso Peninsula, Japan, and its implication on odobenid paleobiogeography. Bulletin of the National Science Museum, Tokyo, Series C 15, 109–119.

True, F.W., 1905a. Diagnosis of a new genus and species of fossil sea-lion from the Miocene of Oregon. Smithsonian Miscellaneous Collections 48, 47–49.

True, F.W., 1905b. New name for *Pontoleon*. Proceedings of the Biological Society of Washington 17, 253.

Velez-Juarbe, J., 2017. *Eotaria citrica*, sp. nov., a new stem otariid from the "Topanga" formation of Southern California. PeerJ 5, 3022.

Velez-Juarbe, J., 2018. New data on the early odobenid *Neotherium mirum* Kellogg, 1931, and other pinniped remains from the Sharktooth Hill Bonebed, California. Journal of Vertebrate Paleontology 38, 1–14.

Velez-Juarbe, J., Salinas-Márquez, F.M., 2018. A dwarf walrus from the Miocene of Baja California. Royal Society Open Science 5. Available from: https://doi.org/10.1098/rsos.180423.

Wyss, A.R., 1987. The walrus auditory region and the monophyly of pinnipeds. American Museum Novitates 2871, 1–31.

CHAPTER

Ecology and behavior of Atlantic walruses

3

Erik W. Born[1] and Øystein Wiig[2]

[1]*Greenland Institute of Natural Resources, Nuuk, Greenland*

[2]*Natural History Museum, University of Oslo, Oslo, Norway*

Chapter Outline

The Atlantic Walrus. DOI: https://doi.org/10.1016/B978-0-12-817430-2.00001-7

Introduction

This chapter presents a summary of the biology and ecology of Atlantic walruses (*Odobenus rosmarus rosmarus* Linnaeus, 1758) and emphasises the distinctive traits that make walruses unique among Arctic pinnipeds. When specific information about Atlantic walruses is insufficient or lacking in the literature, information is drawn from existing knowledge of the closely-related Pacific walrus subspecies (*O. r. divergens* Illiger, 1811).

The biology and ecology of walruses in general have previously been reviewed by Allen (1880), Fay (1981, 1982, 1985), Born (1992), Kastelein (2002), Garlich-Miller et al. (2011), Kasser and Wiedmer (2012), MacCracken et al. (2017), Higdon and Stewart (2018) and Lydersen (2018). Information on biology and ecology of the Atlantic subspecies in particular was summarised by Reeves (1978), Born et al. (1995), Stewart (2002), Born (2005a,b), COSEWIC (2006, 2017) and Stewart et al. (2014a).

Systematics and general appearance

Two extant subspecies of walruses are now recognised based upon their distribution, morphology and genetics: Atlantic walrus (*O. r. rosmarus*), which is distributed in the Atlantic Arctic, and Pacific walrus (*O. r. divergens*), which is distributed in the Bering Sea–Bering Strait–Chuckchi Sea–eastern Siberia–western Alaska region (Fay, 1985; Cronin et al., 1994). A third subspecies (*O. r. laptevi*) had earlier been proposed by Chapskii (1940) based on morphological characterisation of a small number of samples. However, more recent genetic and morphometric studies have confirmed that walruses in the Laptev Sea actually represent the westernmost extension of Pacific walrus populations (Lindqvist et al., 2008, 2016).

Pacific and Atlantic walruses differ in facial outline, dimensions of the tusks, certain skull characters and body size (Allen, 1880). Generally, the tusks of the Pacific walrus are larger and thicker, and consequently, its snout is broader than that of the Atlantic subspecies (Allen, 1880; Pedersen, 1962a; Fay, 1982). Furthermore, adult male Pacific walruses generally have more fibrous tubercles ('bosses') on the skin on the neck and shoulders than do Atlantic males (Allen, 1880; Fay, 1982).

Distribution and numbers

Atlantic walruses are widely distributed across the North Atlantic Arctic and occur in at least nine subpopulations or management units (Born et al., 1995; Stewart, 2008; NAMMCO, 2006, 2018, 2020; Stewart et al., 2014a). The discreteness of some of the putative subpopulations or 'stocks' is still debated (NAMMCO, 2018), and therefore, the exact number is not clear (see Chapter 4, this volume).

Atlantic walruses exist as far west as the Bathurst Island–Penny Strait region (ca. 103° W) in the Canadian High Arctic (Stewart, 2008) and as far east as the Novaya Zemlya–Yamal Peninsula region (ca. 70° E) in the western Russian Arctic (Born, 2005b; NAMMCO, 2006; Spiridonov et al., 2014; Higdon and Stewart, 2018). Latitudinally, they are distributed from northern James Bay in Canada (ca. 54° N) north to Ellesmere Island, Canada (ca. 81°33′N), North-East Greenland (ca. 81°40′N) (Stewart, 2008; Born et al., 1997; Yurkowski et al., 2019) and Franz Joseph Land, Russia (ca. 81°49′N) (Gjertz et al., 1992; Born et al., 1995). There is a geographic gap of at least 1500 km in the Canadian Arctic between the westernmost distribution of Atlantic walruses and the eastern limit of Pacific walruses in the Bering Strait region. There is also a 500–800 km gap in distribution between Atlantic walruses in the Kara Sea area and Pacific walruses in the Laptev Sea.

Extralimital occurrences of individual walruses have been recorded along the European coasts as far south as northern Spain (Born, 1988, 1992; Gjertz et al., 1993; Born et al., 2014). In the western Atlantic, stragglers have been reported from west of the Bathurst Island–Penny Strait region in the Canadian High Arctic (Harington, 1966) and along the coasts of Labrador and Nova Scotia (i.e., between ca. 45° N and ca. 59° N) (COSEWIC, 2017). Most vagrant walruses are presumably males (Gjertz et al., 1993; Born et al., 2014 and references therein; COSEWIC, 2017). In 1991, an adult male walrus with a unique 'Pacific walrus haplotype' was shot at the entrance to Scoresby Sound (ca. 70° 22′N to ca. 21° 57′W) in central East Greenland, demonstrating that stragglers occasionally may move over very great distances (Andersen et al., 1998).

In historical times, the distribution of Atlantic walruses extended further south across the North Atlantic than today. In the 17th Century, walruses were reported to be abundant in the Maritimes (eastern Canada) as far south as approximately 44° N (i.e., at Sable Island; Born et al., 1995 and references therein). However, walruses in this area were heavily hunted from the 16th to 18th centuries (particularly Gulf of St. Lawrence and Sable Island), and were consequently extirpated (Richard and Campbell, 1988; Born et al., 1995; McLeod et al., 2014). There was also a local population of walruses in Iceland still present at the arrival of Norse settlers during the late 800s to early 900s CE. However, due to overexploitation by the Norse, this subpopulation became extinct around 1100 CE (Frei et al., 2015; Star et al., 2018; Keighley et al., 2019).

According to Magnus (1555), walruses were considered to be resident and hunted in northern Norway, and continued to be regular visitors to this region until as late as the 17th to 18th centuries CE (Collett, 1912; Lund, 1954). According to Ritchie (1921) and Ray (1960), walruses were resident in northern Scotland in the late 16th century and during the 17th century CE. The last 'breeding' herd of Atlantic walruses (a bull and three females) were shot off Shetland in 1846 CE (Clarke, 1999). Generally, the historical reports of local walruses herds in the North Atlantic (except Svalbard) are not detailed enough to determine how permanent or large these groups were.

The total number of Atlantic walruses today is not known with certainty, but has recently been estimated to number around 30,000 animals (Laidre et al., 2015 and references therein; Higdon and Stewart, 2018 and references therein). Although overall, Atlantic walruses are less abundant than before (Born et al., 1995; Witting and Born, 2005, 2014; Kovacs et al., 2014; Kovacs, 2016), some subpopulations have shown signs of recovery (Stewart et al., 2014a and references therein; Kovacs et al., 2014; Witting and Born, 2014; NAMMCO, 2018, 2019), some appear to be stable and for others, their population trend is unknown due to insufficient data. The situation is further complicated as some estimates of numbers represent only a portion of a subpopulation and the discreteness of some subpopulations is uncertain (Laidre et al., 2015; Higdon and Stewart, 2018).

Despite ongoing climate change and the increasing loss of sea ice (Lind et al., 2018; MOSJ, 2020), it is worth noting that after decades of commercial overexploitation, the number of walruses in Svalbard is increasing exponentially, at approximately 8% per year between 2006 and 2012 CE (Kovacs et al., 2014), and a further 48% in total between 2012 and 2019 (NAMMCO, 2019). Likewise, the reduction of harvest levels in Canada and Greenland, and new management measures recently introduced in Greenland (Wiig et al., 2014), are also likely to allow a number of other subpopulations to recover or stabilise (Kovacs, 2016).

Habitat requirements

Walruses occupy a relatively narrow ecological niche and feed mainly on bottom-dwelling molluscs. Generally the distribution of walruses is determined by three basic ecological requirements: (1) access to areas where the sea ice is moving and/or is not too thick or dense, (2) suitable haul-out sites (either on sea ice or on land) and (3) shallow waters supporting molluscs, especially bivalves (Vibe, 1950; Fay, 1982; Gjertz and Wiig, 1992; Born et al., 1995; Born et al., 2003).

Ice conditions

During winter, walruses are found in areas where there is suitable food and moving pack ice or little ice. Such conditions occur in and around major recurrent polynyas (Vibe, 1950; Fay, 1982; Stirling, 1997; Garde et al., 2018). Walruses can break through solid ice up to approximately 20 cm thick (Fay, 1982), however, if the ice becomes thicker, walruses retreat to areas with lighter ice conditions, or to moving pack ice (Fay, 1982; Laidre and Regehr, 2017 and references therein).

Suitable ice conditions allow Atlantic walruses to winter in the large polynyas in Foxe Basin of Arctic Canada, the North Water of northern Baffin Bay-Smith Sound, the Northeast Water off the northeastern coast of Greenland, and near the polynyas off southern Svalbard, around Franz Josef Land, and at the Yamal Peninsula in the western Russian Arctic (Vibe, 1950; Born et al., 1995, 1997; Stirling, 1997; Vorontsov et al., 2007; Heide-Jørgensen et al., 2013). Walruses can also winter in smaller polynyas in the Canadian High Arctic (Kiliaan and Stirling, 1978; Sjare and Stirling, 1996; Born et al., 1995) and along the eastern coast of Greenland (Born et al., 1997, 2005). However, in both Svalbard and Franz Josef Land, Atlantic walruses have also been observed during winter in areas with dense pack ice, far from their coastal summering areas (Freitas et al., 2009; Hamilton et al., 2015). In general, many individual walruses tend to show a high degree of perennial fidelity to certain wintering areas, returning year after year to the same terrestrial haul-out site (Sjare, 1993; Born et al., 1995, 2005; Freitas et al., 2009).

Haul-out substrate

During summer, walruses can be found in shallow coastal waters, often in areas with floes of drifting ice that they use for hauling-out (Fay, 1982; Born and Knutsen, 1997). According to Fay (1982), females and their young prefer such areas, and it is generally assumed that most walruses prefer ice when given the choice of ice or land as a resting place (Laidre et al., 2008). This can be seen in Svalbard, where male walruses haul out on land for most of the summer and switch to hauling-out on ice when suitable sea ice becomes available in late autumn (Hamilton et al., 2015). However, this preference is not absolute, as even when suitable ice pans are available near land, Atlantic walruses may be found on land only, or on ice only or a combination of the two (Salter, 1979, 1980; Stewart et al., 2014b). Additionally, in several places within their range (e.g., Southampton Island in northern Hudson Bay and Young Sound in East Greenland), Atlantic walruses of both sexes and all age-classes haul out on land irrespective of whether or not there is ice in the area (Salter, 1979; Born et al., 1995; Born and Knutsen, 1997). In the Pacific, several thousand Pacific walruses are known to haul out on land in the Bristol Bay area (Alaska) in relatively warm and ice-free latitudes (Fay, 1982). The relationship between sea ice and walruses is therefore complicated, and as Stewart et al. (2014b) noted, it is not clear whether the presence of sea ice is a factor that greatly influences hauling-out behaviour.

Irrespective of whether sea ice is present or not, Atlantic walruses in all parts of their range use terrestrial haul-outs (Inuktitut: *uglit*) during the period from break-up of the land-fast ice in late spring and early summer, to formation of fast ice in autumn (Born et al., 1995; Hamilton et al., 2015). These *uglit* are usually situated on low, rocky shores or sandy beaches with relatively steep or stepped subtidal zones (Loughrey, 1959; Mansfield, 1959; Salter, 1979; Miller and Boness, 1983; Born, 2005b). Tracking studies have shown that Atlantic walruses may exhibit a high degree of fidelity to specific terrestrial haul-outs (Born et al., 1997; Born and Acquarone, 2007; Acquarone and Born, 2007; Born et al., 2005; Freitas et al., 2009). Consequently, in several areas subjected to prolonged or repeated disturbances (most particularly, hunting on land) walruses abandon specific *uglit* (Born et al., 1994, 1995 and referenced therein; Gjertz and Wiig, 1994; COSEWIC, 2006, 2017). However, there are encouraging signs that walruses will return to former habitat and colonise new terrestrial haul-outs following reduced hunting or the introduction of protection measures. Since complete protection in 1952, walruses have increased in range and number in Svalbard (Gjertz and Wiig, 1994; Kovacs et al., 2014). Additionally, in certain parts of Greenland, walruses have reoccupied former summering habitats. This has occurred in North-East Greenland after walruses became protected in the 1950s (Born et al., 1997), and more recently in North-West Greenland following a reduction in local hunting pressure (Andersen et al., 2018; Orbicon, 2020).

Access to prey

Walruses are highly specialised feeders that mainly consume bivalves taken at shallow depths <100 m (Vibe, 1950; Fay, 1982). Walruses can also feed on a wide variety of bottom-dwelling invertebrates (species from across nine different phyla) (Fay, 1982; Maniscalco et al., 2020). They may also predate on vertebrates (see section: Food preferences). Despite this diversity, the bulk of their diet is comprised of only a few species of bivalves (Vibe, 1950; Fay, 1982; Fay et al., 1984a; Gjertz and Wiig, 1992; Fisher and Stewart, 1997; Sheffield et al., 2001; Born et al., 2003).

Morphology

Walruses are well adapted to living in icy Arctic seas, with morphological traits including (Born, 2005b): (1) a large, rotund body to conserve energy; (2) tusks that can aid in mobility and be used for protection or chopping sea ice, (3) thick, tough skin that functions as a form of protective armour and is particularly rough and highly cornified on palmar and plantar surfaces of the flippers to aid hauling-out or walking on ice and (4) a thick blubber layer (Fay, 1982, 1985).

Body size

Walruses are the largest pinnipeds in the Arctic (King, 1964). Large body size is, in terms of energy, advantageous in cold environments (Costa, 1993). A massive body also helps in smashing through solid ice from beneath (Fay, 1982) when walruses surface to breathe. Walruses are sexually dimorphic with the total body mass (TBM) of adult males is about 1.5 times that of adult females (Born, 2005b). Adult male Atlantic walruses reach an average asymptotic standard body length (SL) of about 315 cm and an asymptotic TBM of around 1100 kg. Adult females reach a SL of around 270 cm and a TBM of about 700 kg (Knutsen and Born, 1994). There are, however, slight regional differences in body size within Atlantic walruses (Knutsen and Born, 1994; Garlich-Miller and Stewart, 1998; Born, 2005b). It is generally assumed that Pacific walruses reach a larger body size than Atlantic walruses (see Kastelein, 2002). However, measurements (Born, 2005b) and estimates (Wiig and Gjertz, 1996) of TBM taken from individuals in North-East Greenland and Svalbard respectively, indicate that some male Atlantic walruses may reach a TBM similar to Pacific walruses, at around 1800–1900 kg.

Male walruses may continue to grow in TBM for most of their lives, even after termination of growth in body length (Knutsen and Born, 1994; Born, 2005b); this is indicative of strong sexual selection on body size. A 'secondary growth spurt' associated with attainment of social maturity has been noted in male Pacific walruses (Fay, 1982), but has not been observed in Atlantic walruses (Mansfield, 1958; Knutsen and Born, 1994; Garlich-Miller and Stewart, 1998; Born, 2005b). Notably, the penis bone (baculum) of adult male Pacific walruses is about 55 cm long on average (Fay, 1982), and may reach the same length in the Atlantic subspecies (Ferguson and Lariviere, 2004; Born, unpublished data). Females have a 1–2 cm long clitoris bone (Fay, 1982).

Skull and tusks

The walrus skull has some features that relate to the animal's behaviour. The characteristic tusks are actually elongated upper canines (Fig. 3.1). To accommodate the tusks, the maxillary bone is greatly enlarged and the skull has become almost rectangular in shape (Kastelein and Gerrits, 1990). Generally, males have straighter and stouter tusks than females. Tusks grow throughout life, continuing to grow in thickness with age, whereas growth in length is usually balanced by tooth wear (Fay, 1985; Kastelein, 2002; Born, unpublished data). The tusks serve several purposes: (1) they may be anchored in the ice to assist when a walrus is pulling its heavy body up onto sea ice; (2) together with body size tusks signal the rank and dominance order in a herd, and are used by males for fighting with other males during the mating season and (3) they are used for protection against natural enemies (Fig. 3.2) (Fay, 1982, 1985; Kastelein, 2002). The zygomatic arch below the orbital cavity contains a strip of cartilage that probably serves to dampen shocks to the brain case from any impact to the tusks. In contrast to most

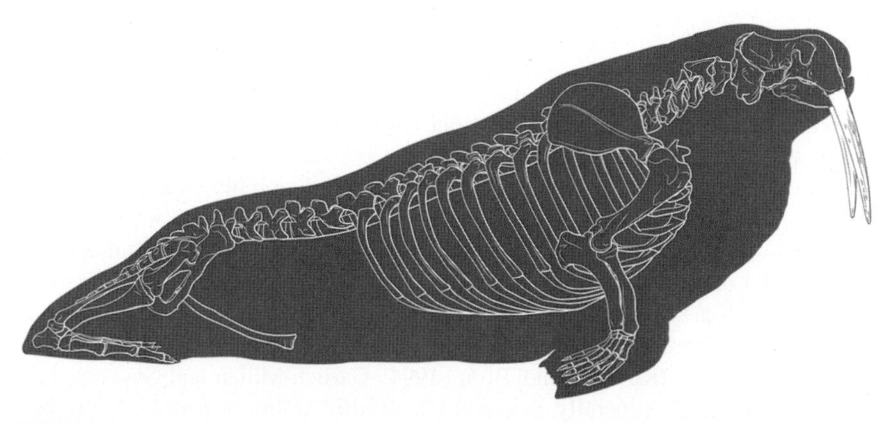

FIGURE 3.1

The skeleton of an adult male walrus. The bones of the forelimbs are large and well-developed structures allowing walruses to move their heavy body on land and sea ice. *Illustration: Elena Kakoshina.*

FIGURE 3.2

A threat display by two adult male Atlantic walruses in Young Sound, North-East Greenland (August 1994).

Photo: E.W. Born.

pinnipeds, the orbital cavity is not closed on the dorsal side of the skull, allowing walruses to look upwards when ploughing through the substrate (Kastelein and Gerrits, 1990; Kastelein, 2002). Walrus eyes are small relative to body size when compared to other pinniped species. The eyes are positioned high on the head, can protrude when looking for prey or enemies and retracted as protection from the cold (Kastelein et al., 1993; Kastelein, 2002). Walrus cheek muscles are strong, allowing the animal to produce powerful water jets from the mouth to wash sediments away from their prey. Their tongue muscles are also very strong, and are used to create negative pressure in the mouth that suck the soft parts from bivalves and snails (Fay, 1982; Kastelein and Mosterd, 1989; Kastelein et al., 1994). The mystacial pad ('upper lip') has numerous stiff vibrissae (whiskers, ranging in number from around 400 to 700) that are densely packed together and are mainly positioned forward (Fay, 1982 and references therein; Milne et al., 2020). Walruses have no external ear flaps (pinnae) (Berta et al., 2015), a trait shared with true seals (Phocidae).

Skin and blubber

Walrus skin is approximately 2–4 cm thick. It is very tough, and is thickest around the neck. The tough skin protects against attacks from polar bears (*Ursus maritimus*) and killer whales (*Orcinus orca*), as well as against abrasion from sea ice (Fay, 1982; Brodie, 2000). The skin on the neck of adult males is thicker than that of females, and is covered with fibrous tubercles ('bosses') that protect the underlying tissues during combat with other males. These bosses are also an important secondary sexual characteristic (Fay, 1982, 1985) (Fig. 3.3).

Across a walrus' body, the blubber layer can reach thicknesses of 15 cm (Fay, 1985). On average, blubber constitutes only approximately 18% of TBM in Atlantic walruses (Knutsen and Born, 1994). In this respect, walruses resemble otariids more than phocids, in which blubber generally makes up a greater proportion of TBM (Boness and Bowen, 1996). Walruses generally

FIGURE 3.3

An adult male Atlantic walrus. Adult male walruses can generally be distinguished from adult females by the numerous tubercles on their thorax and a more massive head. *Illustration: Elena Kakoshina.*

have a negative buoyancy (i.e., they are heavier than sea water). However, by inflating their pharyngeal pouches (Fay, 1960a), they can rest in water (a behaviour known as 'bottling') when there is no land or ice to rest on. This practice has been observed in both males and females (Freuchen and Salomonsen, 1959).

Reproduction and life history

Generally, the life history of Atlantic and Pacific walruses appears to be very similar (Witting and Born, 2014). However, the segregation of different sex and age classes for most of the year, and the selective hunting pattern of subsistence hunters from which biological material is obtained for determining biological parameters (e.g. Born et al., 2017 and references therein), make it difficult to obtain unbiased samples. Parameters of interest include age at first reproduction, fecundity (calf per year, per fecund female), population sex and age structure and mortality rates (Fay, 1982; Witting and Born, 2014).

Life-history

Life expectancy of walruses has been estimated at between 30 and 40 years (Fay, 1982). In captivity, this may reach 40 years of age (Zech, 2014). According to Taylor et al. (2018), longevity is at least 44 years for some Pacific walruses. However, samples obtained from the subsistence hunt, which is typically directed at adults (e.g. Garlich-Miller et al., 2006; Witting and Born, 2014; Born et al., 2017) indicate that in the wild, male walruses only rarely surpass 30 years of age and females rarely more than around 25 years of age (Fay, 1982; Knutsen and Born, 1994; Stewart and Stewart, 2005; Garlich-Miller et al., 2006). To our knowledge, the oldest Atlantic walruses documented in the wild were males of approximately 35 years (Mansfield, 1958; Garlich-Miller and Stewart, 1998).

The majority of female Atlantic walruses become sexually mature (first ovulation) when they are 6 (Born, 2001) or 7 years old (Mansfield, 1958). However, first ovulation may occur between 4 and 11 years of age (Born, 2001). Changes over time in the age of first reproduction and reproductive capacity have been observed in female Pacific walruses (Garlich-Miller et al., 2006; Tempel and Atkinson, 2020). It is, however, not clear to what extent these changes are related to changes in abundance, carrying capacity or environmental changes (Allen and Angliss, 2015; Tempel and Atkinson, 2020). Male Atlantic walruses become sexually mature on average at 11 years of age (range between 7 and 17 years; Born, 2003). However, according to Fay (1982), male Pacific walruses usually do not mate until they are physically mature, between 13 and 16 years of age; the same is likely true for Atlantic walruses.

Mating behaviour

Walruses have an extended mating season, the peak of which is January–April, although there might be some regional variation (Fay, 1982; Fay et al., 1984b; Sjare and Stirling, 1996; Born, 2001). In Pacific walruses, oestrus may occur as early as December and as late as August. Most mature bulls are sexually active from November to March, and the younger bulls from December/January to May (Fay, 1982). Female Atlantic walruses are in oestrus from mid-January to late June (Born, 2001), and males can be sexually active from early November until mid-July (Born, 2003).

During the mating season, the sexually-active bulls engage in a ritualised visual and acoustical displays in the water to attract females in oestrus (Fay et al. 1984b; Sjare and Stirling, 1996; Sjare et al., 2003). In addition, male walruses may sometimes engage in intense physical fights in competition for females (Fay et al., 1984b; Fay, 1985). Sjare and Stirling (1996) described the breeding system of Atlantic walruses in the Canadian High Arctic as a 'female-defence polygyny', whereby individual mature males monopolise access to herds with potentially reproductive females in oestrus for extended periods of time. The mating system of Pacific walruses has been described as more of a 'lek' system, in which potent males defend small display territories near a herd of mature females (Fay et al., 1984a). However, it is not clear to what extent this reflects a real difference in organisation during mating in the two subspecies, or simply differences among studies in observation conditions. In the study by Sjare and Stirling (1996), male Atlantic walruses had access to between one and five adult female herd members (although this number of females should not necessarily be interpreted as equivalent to an estimate of 'harem' size). However, there is a significant, direct relationship between harem size and male TBM in phocids and otariids (Lindenfors et al., 2002). According to this relationship, the estimated average 'harem' size of Atlantic walruses in Greenland would be between 12 and 13 females (Born, 2005b).

Reproductive cycle

Implantation of the embryo occurs in late June or early July (Born, 2001). The peak birth period in Atlantic walruses is from late May to early July (Vibe, 1950; Born, 2001; Born et al., 2017). Pregnancy may therefore last around 15 to 16 months (Fay, 1982; Born, 2001). However, the mating season in walruses is prolonged, and therefore extra-seasonal births are not uncommon (Vibe, 1950; Fay, 1982; Born, 2001; Born et al., 2017). According to Mansfield (1958), the reproductive cycle of the female Atlantic walruses in Foxe Basin is basically biennial, but, to an unknown extent, older females may give birth at longer intervals of 3-to-4 years. However, Born (2001) found that adult females in North-West Greenland gave birth every third year. Based on analyses of reproductive organs, rates of fecundity (estimated as the number of calves per year for sexually mature females) range between 0.29 and 0.40 (Mansfield, 1958; Fay, 1982; Garlich-Miller and Stewart, 1999; Born, 2001), suggestive of a reproductive cycle that is basically triennial. In typical pregnancies, only a single calf is born (Born, 2001) and twins are extremely rare (Fay, 1982). Walruses give birth on the sea ice and do not appear to give birth in the water or on land (MacCracken et al., 2017). However, Fay (1982:202) mentions one probable observation of a birth in water reported to him by Alaskan subsistence hunters.

Nursing behaviour

The average Atlantic walrus calf is around 110 cm long at birth (Garlich-Miller and Stewart, 1999; Born, 2001) and weighs 50–60 kg (Mansfield, 1958; Knutsen and Born, 1994). In contrast to other pinnipeds, walruses can nurse when in the water (Miller and Boness, 1983; Fay 1982). This 'aquatic nursing strategy' implies that the calf usually remains with its mother wherever she goes (Fay, 1982), and there is a very strong mother-calf bond (Nansen 1897: 337; Fay, 1982). Walruses nurse their young for between 18 and 24 months, sometimes longer. Despite this extended nursing, some calves begin to also feed on benthic animals at around 5–6 months. A mother walrus feeds during lactation, and the calves accompany their mother (Fay, 1982), although dependent calves can also be attended at the surface by some female members of the group while their mothers forage on the sea floor (Lydersen, 2018) (Fig. 3.4).

FIGUE 3.4

There is a tight bond between mother and calf. Sometimes the calf is riding on the back of its mother when travelling at sea. *Illustration: Elena Kakoshina.*

When weaned at about 2 years of age, the Atlantic walrus calf weighs approximately 250 kg (Knutsen and Born, 1994). This relatively large mass at weaning is likely an advantage in the specialised feeding strategy of walruses; the calf must be able to dive to the seafloor and forage there for several minutes at a time (Fay, 1982). A prolonged lactation period is also an advantage when rearing the calf in an environment where sea ice conditions may vary greatly. For instance, in years when ice conditions are heavy, and the fast ice covering inshore feeding banks does not break up—or breaks up late, thereby limiting the access of walruses to good feeding sites—the effect on the calf may be ameliorated because it can continue to obtain milk from its mother. It is also important for walrus calves to learn where the mollusc banks are located over an entire annual cycle, as the summer and winter feeding grounds can be located in different places (e.g., in West Greenland and East Greenland; Born et al., 1997; Dietz et al., 2014).

Demographic parameters

The sex ratio in Atlantic walrus subpopulations is not well known. In Pacific walruses the sex ratio is even (1F:1M) at birth through to five years of age, and then skewed toward females in older walruses (Fay, 1982, 1985; Taylor and Udevitz, 2015; Taylor et al., 2018). To our knowledge, there are no indications of the sex ratio being skewed toward females in any Atlantic walrus subpopulation, and it has therefore been assumed to be even (Gjertz and Wiig, 1995; Witting and Born, 2005, 2014).

Natural mortality in walruses is also not well established, but is assumed to be low because longevity is relatively high (Fay, 1982; DeMaster, 1984; Fay et al., 1989, 1997). The annual natural mortality rate in adult Pacific walruses has been estimated at 1%–2% (Fay et al., 1994, 1997; Chivers, 1999; Taylor et al., 2018). Presumably, natural mortality in Atlantic walruses is similarly low (Witting and Born, 2014).

The population birth rate (i.e., the fraction of newborn walruses in the total population) has been estimated to be 0.07 (Mansfield, 1966) or 0.11 (Mansfield, 1973) in Atlantic walruses. Counts at haul-outs in the Canadian High Arctic during August suggested a population birth rate of approximately 0.10 (Stewart, 2002). Population birth rate has been estimated at between 0.12 and 0.17 in Pacific walruses (Fedoseev and Gol'tsev, 1969; Fay, 1982).

A finite growth rate (i.e., the rate of increase per individual, per unit time) of approximately 8% per year has been estimated for walrus populations in a phase of growth under favourable environmental conditions with no food limitations (Chivers, 1999). Witting and Born (2014) modelled an annual growth rate of 7.7% for the Baffin Bay subpopulation of Atlantic walruses in 2012. Abundance estimates obtained for the subpopulation of Atlantic walruses at Svalbard indicate that this subpopulation increased at a rate of nearly 8% annually between 2006 and 2012 (Kovacs et al., 2014).

Behaviour

Some of the behavioural traits that make walruses unique among Arctic pinnipeds also make them vulnerable to exploitation by humans. Walruses (1) are highly gregarious and show strong group coherence, haul-out on ice and land, sometimes in large herds and (2) vocalize intensively both in the air and under water, making other group members as well as predators more aware of their presence. Furthermore, (3) walruses have a special foraging niche restricting them to shallow waters close to the coast for most of the year (in particular, Atlantic walruses).

Social behaviour and hauling-out

Walruses are very social animals, and are usually found together in groups. Group size tends to be largest on land, smaller on ice and smallest when they are in water (Fig. 3.5) (Fay, 1982, 1985).

Herds of Pacific walruses numbering in the tens of thousands, occasionally even greater than 100,000, have been known to haul out on land (Fay, 1985; Kavry et al., 2008; Garlich-Miller et al., 2011; Jay et al., 2011; Monson et al., 2013). In comparison, groups of Atlantic walruses at terrestrial haul-outs tend to be much smaller (Gjertz and Wiig, 1995; Born et al., 1997; Lydersen et al., 2008,

FIGURE 3.5

The haul-out Angijaq Island (Cumberland Peninsula, South-East Baffin I., Canada) where 745 Atlantic walruses (males, females, calves) hauled-out on 26 August 2005. *Photo: E.W. Born.*

2012; Stewart et al., 2014b,c). In the autumn of 2020, a mixed herd of approximately 3000 walruses were reportedly found on haul-outs on the Yamal Peninsula (Vasilyeva, 2020) where about 1000 walruses had been recorded in previous years, at the Vaigach and Matveev islands (Gebruk et al., 2020 and references therein). To our knowledge, these are the largest concentrations of Atlantic walruses recorded in recent time.

The number of walruses at a given haul-out site can vary dramatically during the day, between days, and throughout the season depending on weather conditions and intrinsic group dynamics (Salter, 1979; Hills, 1992; Born and Knutsen, 1997; Stewart et al., 2014c; Hamilton et al., 2015; Øren et al., 2018). In general, adverse weather and especially wind-chill negatively affect the tendency to haul out (Born and Knutsen, 1997; Hamilton et al., 2015).

Male Atlantic walruses spend about 30% of their time hauled-out during summer (Born and Knutsen, 1992; Acquarone et al., 2006; Born and Acquarone, 2007; Hamilton et al., 2015), and 10%–15% of their time hauled-out in winter (Hamilton et al., 2015). Haul-out bouts last can easily last 20–40 hours during summer (Born and Knutsen, 1997; Gjertz et al., 2001; Lydersen et al., 2008), and between 5 and 10 hours during winter (Hamilton et al., 2015). Observational studies and the few efforts to satellite-track females suggest a pattern similar to that of males; females typically spending a few days ashore, followed by time at sea for feeding (Kovacs and Lydersen, 2008).

When hauling-out, walruses are usually lying very close together; they also most often travel in groups (Fay, 1985). Dominance order in groups on land and ice is a function of body and tusk size, as well as aggressiveness (Miller, 1975, 1982). Genetic studies indicated that families of closely-related females tend to stay together even when travelling (Andersen and Born, 2000). When young males reach sexual maturity, they join the larger groups of males who tend to stay apart from females (Lydersen, 2018). Atlantic walruses of both sexes and all age classes may haul out together on land during summer (Salter, 1979; Miller, 1982; Miller and Boness, 1983; Gjertz and Wiig, 1994; Born and Acquarone, 2007; Stewart et al., 2014c; Kovacs et al., 2014). This also brings them into contact with polar bears (Øren et al., 2018). When attacked or threatened by predators, such as polar bears or human hunters, walruses usually react by aggressively defending each other (Kiliaan and Stirling, 1978; Stirling, 1984; Born et al., 2017). There are numerous accounts of walruses attacking boats or kayaks during Inuit subsistence hunts, sometimes even killing the hunter who was chasing them (Pedersen, 1962b; Born et al., 2017).

Vocalisation

In contrast to other Arctic pinnipeds, walruses can be very vocal above water at all times of the year. They produce a variety of sounds including grunts, roars, guttural sounds and barks involved in mother-calf recognition (Fay, 1982; Miller, 1985; Charrier et al., 2009, 2011), as well as whistling by males (Verboom and

Kastelein, 1995). Their barks and roars can often be heard kilometres away from where they haul out (Born and Wiig, pers. obs.). Adult males also emit sounds like knocks, taps and bell-like noises underwater (Stirling et al., 1987; Sjare et al., 2003; Verboom and Kastelein, 1995; Mouy et al., 2012). The knock sound emitted by adult males during the mating season can also be produced by females (Schusterman and Reichmuth, 2008) and is likely produced by clicking the molariform teeth in the jaws together (Fay, 1982). Apparently, the barking of female Pacific walruses has a lower frequency and differs in duration than that of female Atlantic walruses (Charrier et al., 2011). Research has indicated that mature female and male walruses are able to discriminate between vocalisations of different herds (Charrier et al., 2011), and that females are able to discriminate the voice of their own calf from that of other calves (Charrier et al., 2009). Walruses in captivity can even be taught to produce novel sounds (Reichmuth and Casey, 2014 and references therein). The hearing range of walruses is 60 Hz to 23 kHz, although this may be broader at high frequencies when underwater (Reichmuth et al., 2020).

Diet and foraging behaviour

Food preferences

Walruses mainly feed on small organisms that are low in the food chain. Therefore, to satisfy their daily energy demands, walruses must eat a large number of individual food items. In one case, more than 6000 prey items were found in the stomach of a Pacific walrus (Fay, 1982).

Male and female walruses have similar diets (Fisher, 1989; Fisher et al., 1992). Although walruses can, and occasionally do, consume a variety of benthic invertebrates (e.g. Maniscalco et al., 2020), they mainly feed on bivalves (Vibe, 1950; Fay, 1982; Sonsthagen et al., 2020). Very few of the bivalves found in walrus stomachs can be identified to species, since the shell itself is usually not eaten. Generally only the bivalve's foot and siphon can be found (Fay, 1982; Sheffield and Grebmeier, 2009; Born et al., 2017). When identifiable, bivalve remains in walrus stomachs are typically the feet of *Serripes groenlandicus* (synonym: *Cardium groenlandicum*), *Ciliatocardium ciliatum* (synonym: *Cardium ciliatum, Clinocardium ciliatum*), *Astarte borealis* and *Macoma calcarea*, and the siphons of *Mya truncata* and *Hiatella arctica* (synonym: *Saxicava arctica*) (Vibe, 1950; Fay, 1982; Born et al., 2017). The bivalves *Mya* spp., *Serripes* spp. and *Hiatella* spp. make up the majority of walrus diets (Vibe, 1950; Fay, 1982; Fisher, 1989; Fisher et al., 1992; Fisher and Stewart, 1997; Sheffield et al., 2001; Born et al., 2003). These three genera of bivalves may constitute up to 87%–98% of the number of consumed invertebrates and 54%–80% of the volume of ingesta (Fay et al., 1984a). In North-West Greenland, *Mya* spp. parts were found to represent 37% and *Hiatella* parts 60% of a single Atlantic walrus' stomach contents, containing a total of

3137 invertebrate food items (Vibe, 1950). When broken down by species and month, *Mya truncata* contributed about 81% and *H. arctica* about 8% of the total gross energy to the diet of Atlantic walruses in Foxe Basin in July (Fisher, 1989; Fisher and Stewart, 1997). In September, *M. truncata* contributed about 60% of the total gross energy to the diet and *S. groenlandicus* contributed about 38%. In Svalbard, Gjertz and Wiig (1992) found siphons of *M. truncata* in 9 out of 14 faeces samples and shell fragments of *Mya* in 4 of the 14 samples. They also found operculae of the common whelk (*Buccinum* sp.) in five of their 14 collected samples. A study of fatty acid composition of walrus blubber from Svalbard (Skoglund et al., 2010) found that they resembled the lipids from *M. truncata* and *Buccinum* spp. Similarly, Scotter et al. (2019) found that nitrogen isotope ($\delta^{15}N$) values in tissue samples of walruses from Svalbard to be consistent with consumption of benthic invertebrates, and carbon isotope ($\delta^{13}C$) values similar to those found in *M. truncata*.

Fay (1982) suggested that walruses select bivalves preferentially and that other food is consumed opportunistically. Genetic studies of prey in gastrointestinal and scat samples of Pacific walruses in Bristol Bay (Alaska) indicated that a major shift in diet occurred between 2014–2015 and 2017–2018 (Maniscalco et al., 2020). According to these researchers, the study indicated that walruses are adaptable to subsist on a wide variety of prey types throughout their range (ibid.).

Walruses may also consume vertebrates including marine mammals, birds and fish. The occasional predation of several species of seals such as ringed seals (*Pusa hispida*) and bearded seals (*Erignathus barbatus*) is well documented, and it has been suggested that predation on seals may play a significant role in the diet (Mansfield, 1958; Lowry and Fay, 1984; Smith et al., 1979; Fay et al., 1990; Gjertz, 1990a; Timoshenko and Popov, 1990; Gjertz and Wiig, 1992; Born et al., 1997, 2017). There is indirect evidence (through trace metal and chlorinated hydrocarbon contaminant levels, $\delta^{13}C$ analysis of tissues) that Atlantic walruses in eastern Hudson Bay regularly prey on seals (Wagemann et al., 1993; Muir et al., 1995). Apparently, the habit of seal-eating occurs in both sexes, but is confined largely to older subadults and adult walruses (Lowry and Fay, 1984). It has been suggested that in seasons and areas where the distribution of seals and walruses overlap considerably, or when ice conditions make the mollusc banks inaccessible to walruses (as in North-East Greenland), seals may well be an important food source for walruses (Born et al., 1997).

Walruses may also feed on dead whales (Born et al., 2017) and on other walruses (Freuchen, 1935). This is supported by the presence of *Trichinella* spp. parasites (Fay, 1982; Born et al., 1982) in some walruses. Additionally, stomach content analyses indicate that walruses sometimes scavenge the carcasses of polar bears or sled dogs, also known to commonly carry the parasite (Born et al., 1982, 2011). Walruses are also known to prey on birds (e.g., black guillemot, *Cepphus grylle*; eider, *Somateria* sp.; thick-billed murre, *Uria lomvia*; pink-footed geese, *Anser brachyrhyncus*; Fay et al., 1990; Gjertz, 1990b; Mallory et al., 2004; Lovvorn et al., 2010; Fox et al., 2010; Giljov et al., 2017). In addition, various fishes (e.g., sand lance, *Ammodytes* sp., and polar cod, *Boreogadus saida*) have been found in walrus stomachs (Fay, 1982; Fay and Stoker, 1982; Born et al., 1994, 1997, 2017).

FIGURE 3.6

Walrus foraging behaviour. *Illustration: Elena Kakoshina.*

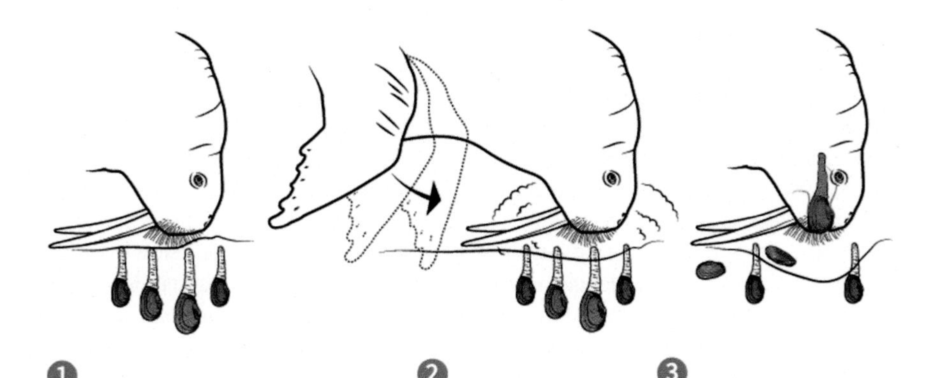

1

To begin feeding walruses angle their body 30-90° from the ocean floor. They explore the area by slowly moving their snouts and detect the presence of prey with erected vibrissae.

2

Walruses expose bivalves by expelling a jet stream of water from their mouths or by waving a flipper.

3

Finally, clams contents are sucked into walrus mouths using a vacuum mechanism. The now empty shells are dropped.

THREE FAVORITE CLAMS CONSUMED

The walrus is unique among marine mammals because it has a very special menu and a special way of obtaining its food. In contrast to the other marine mammals, the walrus is specialized almost completely to live on invertebrate animals that are found on the sea bed. The walruses take many items from the sea bottom: sea cucumber, shrimps, bristle worms, snails, sea urchins, sea anemones, squids, sand eels, polar cod and much more. They prefer first and foremost some certain bivalve species of which three are favorites: Blunt sand gaper *(Mya)*, wrinkled rock borer *(Hiatella)* and Greenland cockle *(Serripes)*.

(mm)
120
100
80
60
40
20
Serripes
Hiatella
Mya
240 220 200 180 160 120 100 80 60 40 20 0
(mm)

FIGURE 3.6

(*Continued*)

Foraging behaviour

Detailed descriptions of the foraging behaviour of walruses have been based on observations of walruses in captivity (Fay, 1982; Kastelein and van Gaalen, 1988; Kastelein and Mosterd, 1989; Kastelein and Wierpkema, 1989; Kastelein et al., 1989, 1990, 1994, 2000) and in the wild (Levermann et al., 2003). Walruses dive directly to the sea floor to feed, and when finished they swim straight to the surface to breathe. During a feeding dive walruses hardly move, or may shift slowly forward as their tusks rest like a sledge on the bottom, with vibrissae maintaining contact with the sea floor. A walrus' hind flippers are used to move forwards and backwards, whereas the front flippers provide stability, or aid directly in finding prey.

Walruses use three different methods of exposing their prey: (1) water jetting, (2) waving with front flippers and (3) digging with the snout (Fig. 3.6). Walruses can remove sediment covering their prey by producing a strong water jet with their mouths (Kastelein and Mosterd, 1989; Kastelein et al., 1994; Kastelein, 2002). They may also wave their front flippers over the substrate—propelling water to remove the sediment, and also to keep their heads clear from stirred up sediment. Walruses have a preference for waving with the right front flipper (Levermann et al., 2003). The muzzle is kept close to the sea floor during the whole feeding process. The snout is used as a digging organ in a pig-like fashion; the prey is located mainly with the mystacial vibrissae, with vibrissae movement and positioning of the vibrissae considered a key source of touch sensations in walruses (Milne et al., 2020). The non-tusk teeth (i.e., the molariform cheek teeth) are worn down to the gums in wild walruses, perhaps by sand moving in the mouth during feeding (Fay, 1982). Eyesight seems to play a role for finding and identifying food items as walruses keep their eyes open during feeding. Retinal anatomy indicates that walruses have colour vision, which may aid in the identification of prey (Kastelein et al., 1993).

Once the prey is located, walruses use their vibrissae to examine and manipulate it. A walrus' large vibrissae on the upper lip are very sensitive and can be moved in unison or individually like fingers (Kastelein and van Gaalen, 1988; Kastelein et al., 1990). By using its vibrissae, a captive walrus was able to distinguish a circular and a triangular object, each of which had an identical surface area of 0.4 cm2, corresponding to approximately 6.3 mm in width.

The mastication muscles in walruses are not very large, as the walrus usually does not chew its prey, instead swallowing soft tissues whole (Fay, 1982). The fleshy parts of bivalve and gastropod molluscs are removed from their shells by suction. In walruses, negative intraoral pressures of at least 90 kPa in air (Fay, 1982: 170) and 119 kPa under water have been recorded (Kastelein et al., 1994). Walruses are very efficient molluscs-eaters, consuming six (Oliver et al., 1983) to nine (Born et al., 2003) bivalves per minute, corresponding to between 40 and 60 bivalves per feeding dive (Kastelein et al., 2000; Born et al., 2003). The empty shells are discarded and can be found on the ocean floor near the feeding site (Oliver et al., 1983, 1985; Born et al., 2003). The foraging activity of walruses can be a significant factor affecting sediment structure as well as biomass and taxonomic composition of bottom-dwelling macrozoobenthos (Nelson and Johnson, 1987; Denisenko et al., 2019).

During foraging, walruses remain submerged approximately 80%–90% of their time (Fay, 1982; Born and Knutsen, 1997). The average duration of a foraging dive is between five and eight minutes (Wiig et al., 1993; Born and Knutsen, 1997; Gjertz et al., 2001; Jay et al., 2001; Acquarone et al., 2006; Born et al., 2003; Born and Acquarone, 2007; Acquarone and Born, 2007; Lowther et al., 2015). The duration of feeding dives is positively correlated with depth (Fay, 1985). An aerobic dive limit of between 10 and 15 minutes has been calculated for walruses (Wiig et al., 1993; Noren et al., 2015), however, maximum dive durations of up to 47 (Lowther et al., 2015) and 49 minutes (Garde et al., 2018), respectively, have been recorded by use of satellite telemetry.

Energy requirements

Rates of food ingestion and metabolism are likely to vary between walruses according to sex, TBM, level of activity, reproductive status and season (Fay, 1982; Kastelein et al., 2000; Acquarone et al., 2006). Young captive Pacific walruses (TBM: 250–750 kg) consumed around 4%–6% of TBM per day according to Fay (1982). In studies of free-ranging adult male Atlantic walruses with a TMB of 1200–1300 kg, daily consumption was estimated to be between 74 and 95 kg of shell-free, wet biomass of bivalve tissues. This amounts to 6%–7% of TBM (Born et al., 2003; Acquarone et al., 2006). Daily energy expenditure in the wild of a walrus with a TBM of 1300 kg has been estimated to be 328–365 MJ/day (Acquarone et al., 2006).

Due to their large body mass, the amount of food eaten by individual walruses is relatively large (Acquarone et al., 2006). However, given that most subpopulations of Atlantic walruses probably remain below their historical levels (Born et al., 1995; Gjertz et al., 1998; Witting and Born, 2014; Laidre et al., 2015), there are no indications at present that population size may be constrained by food availability. Born (2005b) made a rough estimate of the potential amount of benthic food available for the three subpopulations of Atlantic walruses occurring in Greenland (North-West, West and East Greenland) and concluded that there was ample food to support current subpopulation sizes. In a study of the role of walruses in the ecosystem in a North-East Greenland fjord, the annual predation by walruses on the standing bivalve biomass was considered negligible (Born, 2005b; Glud and Rysgaard, 2007). Walrus predation on the standing bivalve biomass in North-West Greenland (within between 5 and 100 m of depth) was estimated to 3.2% per year based on assessments of mean biomass of walrus' preferred prey items (Garde et al., 2018). It was estimated that walruses in Lancaster Sound (High Arctic Canada) consume about 6% of the annual production of bivalves in the area (Welch et al., 1992).

Competitors, predators and pathogens

Competition with bearded seals

Bearded seals have a circumpolar distribution that generally overlaps that of walruses (Cameron et al., 2010 and references therein). Similar to walruses, bearded seals feed primarily on benthic organisms including bivalves. Hence, to a certain extent the two species compete for the same trophic niche (Lowry et al., 1980; Hjelset et al., 1999). However, bearded seals have a much more diverse diet than walruses, and studies of nitrogen and carbon isotopes ($\delta^{15}N$ and $\delta^{13}C$) indicate that bearded seals and walruses do not have a large overlap in prey utilisation (Dehn et al., 2007). Furthermore, unlike walruses which tend to be highly gregarious walruses, bearded seals are largely solitary (Cameron et al., 2010 and references therein). In the competition with bearded seals for a spatially-restricted food resource (i.e., clam beds at shallow waters), walruses probably are at an advantage due to their gregariousness, larger mass (on average TBM of adult walruses is at least three time larger than that of adult bearded seals; Knutsen and Born, 1994; Andersen et al., 1999), tusks and occasional aggressiveness (Born, 2005b).

Natural predators

Apart from humans, the main predators of walruses are polar bears and killer whales. Several reports exist of polar bears attacking walruses, and direct evidence of polar bear predation on walruses has been found in dietary studies (Born et al., 2011; Galicia et al., 2016). Walrus calves and subadults are most vulnerable to predation from bears (e.g., Calvert and Stirling, 1990; Rugh, 1993; Ovsyanikov, 1995, 1996) with observations of polar bears attacking groups of hauled-out walruses and chasing them off the beach may be a hunting strategy aimed at provoking stampedes. The panic may allow bears to snatch calves out of the herd; alternately, they may be able to feed on calves that are trampled to death (Ovsyanikov, 1995). Polar bears sometimes use tools (pieces of ice) for killing walruses (Stirling et al., 2021). Walruses are consumed most often by older male polar bears, and by bears in particular regions, such as Foxe Basin (Thiemann et al., 2007, 2008). While predation rates are unknown, it has been suggested that polar bears are also important predators of walruses in the central Canadian High Arctic (COSEWIC, 2006) and may explain the decreasing population size and poor recovery of Atlantic walruses in West Russia in the 1980s (Popov et al., 1990).

Disturbance by polar bears at terrestrial haul-outs varies between years and seasons (Øren et al., 2018). In wintertime, walruses are most vulnerable to bears when they are frozen out of their breathing holes, when rough ice provides cover for stalking bears or when walruses are forced to rely on a very limited area of open water for breathing and hauling-out (Calvert and Stirling, 1990). In summertime, Gjertz (1990c) observed a large male walrus on the beach at Svalbard that did not pay much attention to a polar bear sneaking around. The bear soon left the area. However, although walruses may remain undisturbed—especially all-male groups—during polar bear visits, the presence of polar bears was found to sometimes significantly disturb walrus herds on Svalbard. It appeared that the effect of polar bears at walrus haul-out sites depended on the age-sex composition of the walrus group. At sites where mother–calf pairs hauled out, bears induced panicked reactions in the herd, causing large and rapid decreases in the number of animals hauled out on shore (Øren et al., 2018). Herd organisation on the beach may reflect a protection strategy against bears. In Svalbard, it was observed that adults lay between the calves and nearby polar bears (Øren et al., 2018).

In contrast to other Arctic pinnipeds, walruses defecate where they haul out on land or ice (Gjertz and Wiig, 1992). Often, a group of walruses may lie in a pool of faeces of which the smell can be picked up from a considerable distance (Born and Wiig, pers. obs.). Polar bears have an acute sense of smell and are able to pick up the scent of a prey from very far away (Pedersen, 1945). Hence, it seems evident that it would be easy for a polar bear to pick up the stench of walrus faeces, as well as the sound. However, these behaviours also indicate that walruses do not regard polar bears as a significant threat to them (Gjertz, 1990c; Øren et al., 2018). Indeed, walruses are aware of the danger represented by polar bears and may sometimes threaten, (Kiliaan and Stirling, 1978) attack and even kill them (Freuchen, 1935; Loughrey, 1959; Pedersen, 1962b). Adult walruses appear able to defend themselves from attacking bears, both as a group and individually, relying on their large body size and actively using their tusks as weapons (Ovsyanikov, 1995). According to Fay (1982: 216), "The bears probably are no match for healthy [walrus] adults in direct combat".

Walruses, primarily young and calves, are also predated upon by killer whales (Zenkovich, 1938; Fay, 1982; Lowry et al., 1987; Melnikov and Zagrebin, 2005; Kryukova et al., 2012). In the Bering Strait region interactions between killer whales and walruses occur in spring, summer and autumn, when the range of the two species can overlap (Fay, 1982). The COSEWIC (2017) review of the status of Atlantic walruses in Canada and adjacent waters concluded that killer whales do not appear to regularly hunt Atlantic walruses, although they might acquire this behaviour and learn to hunt walruses successfully in the future if hunting opportunities increase. However, given the abundance of other prey species that are easier and safer to hunt, it seems unlikely that killer whales will become a major predator of Atlantic walruses in the next few decades. According to Higdon and Ferguson (2009) and Higdon et al. (2014), killer whale occurrence is increasing in the eastern Canadian Arctic. However, information from Inuit indicates that killer whales in this region rarely, if ever, feed on walruses (Ferguson et al., 2012).

Diseases and parasites

A variety of parasites, bacteria and viruses have been observed in walruses. The parasite *Toxoplasma gondii* and antibodies against the bacteria *Brucella* spp. and

Leptospira interrogans serovars (Nielsen et al., 2001; Calle et al., 2002; Scotter et al., 2019), the influenza A virus, calicivirus, phocid herpesvirus (PhHV-1 and PhHV-2), poxvirus and parapoxvirus (Zarnke et al., 1997; Calle et al., 2002; Ganova-Raeva et al., 2004; Melero et al., 2014; Scotter et al., 2019) have all been reported from antibody screening in walrus tissue samples. Apparently, phocine distemper virus (PDV) has been circulating among Atlantic walruses in the western North Atlantic since at least the 1970s (Duignan et al., 2014; COSEWIC, 2017). Serological testing of walruses from Foxe Basin for pathogens found antibodies for canine adenovirus, canine distemper virus, dolphin morbillivirus, PDV and dolphin rhabdovirus (Phillipa et al., 2004). A relatively small percentage of walruses are infested with the helminth nematode *Trichinella* ssp. which causes trichinellosis (trichinosis) in humans (Fay, 1960a,b; Born et al., 1982; Campbell, 1988; Møller et al., 2005, 2007). Infestation of humans and outbreaks of trichinellosis due to eating uncooked or undercooked meat of Atlantic walruses have been reported historically and also recently from Greenland and Canada (Born et al., 1982; Møller et al., 2005, 2007; COSEWIC, 2017). Finally, it should be mentioned that both subspecies of walruses are very often infested with the host-specific anopluran louse *Antarctophthirus trichechi* (Fay, 1982; Leonardi and Palma, 2013). The susceptibility of walruses to mortality from disease is not well understood (COSEWIC, 2017). However, to our knowledge, the negative effects of different pathogens at the population level have not yet been documented in Atlantic walruses.

Conclusion

Due to their restricted habitat requirements (i.e., areas with proper sea-ice conditions and access to benthic food in shallow waters), behaviour (i.e., gregariousness and tendency to haul out in dense herds) and foraging ecology (primarily feeding on benthic molluscs) walruses are unique among Arctic marine mammals. Atlantic walruses occupy very large areas on land and sea ice. They are often absent from their terrestrial haul-outs, and remain out of sight when submerged, making them difficult to study. Gaining information about walruses can be further hampered by the fact that males and females often do not occur in the same areas for much of the year, and that Indigenous selective hunting from which biological samples are commonly obtained, often targets large adults. It is very difficult to obtain estimates of essential vital parameters such as age at first reproduction, fecundity, mortality and survival in Atlantic walruses, and hence, knowledge of the closely-related and more intensively-studied Pacific subspecies is often very helpful.

More detailed information on vital rates, particularly trends over time, is critical to improving our assessment of Atlantic walrus population health and survival. The general lack of such information in most of the subpopulations of Atlantic walruses calls for intensified studies of the biology of this subspecies, particularly considering changing climatic and sea ice conditions.

References

Acquarone, M., Born, E.W., 2007. Estimation of water pool size, turnover rate and body composition of free-ranging Atlantic walruses (*Odobenus rosmarus rosmarus* L.) studied by isotope dilution 2007. Journal of the Marine Biological Association of the United Kingdom 87, 77–84.

Acquarone, M., Born, E.W., Speakman, J., 2006. Walrus (*Odobenus rosmarus*) field metabolic rate measured by the doubly labeled water method. Aquatic Mammals 32, 363–369.

Allen, J.A., 1880. History of North American Pinnipeds. A Monograph of the Walruses, Sea-lions Sea-bears and Seals of North America. U.S. Miscellaneous publications, pp. 1–785.

Allen, B.M., Angliss, R.P., 2015. Alaska marine mammal stock assessments, 2014. U.S. Dep. Commer. NOAA Tech Memo NMFSAFSC-301, 1–304.

Andersen, L.W., Born, E.W., 2000. Indications of two genetically different subpopulations of Atlantic walruses (*Odobenus rosmarus rosmarus*) in west and northwest Greenland. Canadian Journal of Zoology 78, 1999–2009.

Andersen, L.W., Born, E.W., Gjertz, I., Wiig, Ø., Holm, L.E., Bendixen, C., 1998. Population structure and gene flow of the Atlantic walrus (*Odobenus rosmarus rosmarus*) in the eastern Atlantic Arctic based on mitochondrial DNA and microsatellite variation. Molecular Ecology 7, 1323–1326.

Andersen, M., Hjelset, A.M., Gjertz, I., Lydersen, C., Gulliksen, B., 1999. Growth, age at sexual maturity and condition in bearded seals (*Erignathus barbatus*) from Svalbard, Norway. Polar Biology 21, 179–185.

Andersen, A.O., Heide-Jørgensen, M.P., Flora, F., 2018. Is sustainable re-source utilisation a relevant concept in Avanersuaq? The walrus case. Ambio 2018 (47), S265–S280. Available from: https://doi.org/10.1007/s13280-018-1032-0.

Berta, A., Sumich, J., Kovacs, K.M. (Eds.), 2015. Marine Mammals. 3rd Edition Academic Press, pp. 1–738. ISBN: 9780123970022.

Boness, D.J., Bowen, J.D., 1996. The evolution of maternal care in pinnipeds. Bioscience 46, 645–654.

Born, E.W., 1988. Hvalrosstrejfere i Europa (Walrus stragglers in Europe). Flora og Fauna (Århus) 94, 9–14 (In Danish).

Born, E.W., 1992. *Odobenus rosmarus* Linnaeus, 1758. Walross. In: Duguy, R., Robineau, D. (Eds.), Band 6: Meeressäuger. Teil II: Robben—Pinnipedia. Handbuch der Säugetiere Europas. AULA—Verlag, Wiesbaden, pp. 269–299. (In German).

Born, E.W., 2001. Reproduction in female Atlantic walruses (*Odobenus rosmarus rosmarus*) from northwestern Greenland. Journal of Zoology (London) 255, 165–174.

Born, E.W., 2003. Reproduction in male Atlantic walruses (*Odobenus rosmarus rosmarus*) from the North Water (N Baffin Bay). Marine Mammal Science 19, 819–831.

Born, E.W., 2005a. The Walrus in Greenland. Ilinniusiorfik/Education Publishers, Nuuk, Greenland, pp. 1–79, ISBN 87-7975-221-7.

Born, E.W., 2005b. An assessment of the effects of hunting and climate on walruses in Greenland. Greenland Institute of Natural Resources, Nuuk, and University of Oslo, 1–346. ISBN 82-7970-006-4.

Born, E.W., Knutsen, L.Ø., 1992. Satellite-linked radio tracking of Atlantic walruses (*Odobenus rosmarus rosmarus*) in northeastern Greenland, 1989-1991. Zeitschrift für Säugetierkunde 57, 275–287.

Born, E.W., Knutsen, L.Ø., 1997. Haul-out activity of male Atlantic walruses (*Odobenus rosmarus rosmarus*) in northeastern Greenland. Journal of Zoology (London) 243, 381–396.

Born, E.W., Acquarone, M., 2007. An estimation of walrus predation on bivalves in the Young Sound area (NE Greenland). In: Rysgaard, S., Glud, R.N. (Eds.), Carbon Cycling in Arctic Marine Ecosystems: Case Study Young Sound, 58. Monographs on Greenland, pp. 176–191.

Born, E.W., Clausen, B., Henriksen, Sv.Aa, 1982. *Trichinella spiralis* in walruses from the Thule district, North Greenland, and possible routes of transmission. Zeitschrift für Säugetierkunde 47, 246–251.

Born, E.W., Heide-Jørgensen, M.P., Davis, R.A., 1994. The Atlantic walrus (*Odobenus rosmarus rosmarus*) in West Greenland. Monographs on Greenland 40, 1–33.

Born, E.W., Gjertz, I., Reeves, R.R., 1995. Population assessment of Atlantic walrus, Norwegian Polar Institute. 138, 1–100.

Born, E.W., Dietz, R., Heide-Jørgensen, M.P., Knutsen, L.Ø., 1997. Historical and present status of the Atlantic walrus (*Odobenus rosmarus rosmarus*) in eastern Greenland, Monographs on Greenland. 46, 1–73.

Born, E.W., Rysgaard, S., Ehlmé, G., Sejr, M., Acquarone, M., Levermann, N., 2003. Underwater observations of foraging free-living Atlantic walruses (*Odobenus rosmarus rosmarus*) and estimates of their food consumption. Polar Biology 26, 348–357.

Born, E.W., Acquarone, M., Knutsen, L.Ø., Toudal, L., 2005. Homing behaviour in an Atlantic walrus (*Odobenus rosmarus rosmarus*). Aquatic Mammals 31, 23–33.

Born, E.W., Heilmann, A., Kielsen Holm, L., Laidre, K., 2011. Polar bears in Northwest Greenland—an interview survey about the catch and the climate. Monographs on Greenland Man and Society 41. Museum Tusculanum Press, University of Copenhagen, pp. 1–232.

Born, E.W., Stefansson, E., Mikkelsen, B., Laidre, K.L., Andersen, L.W., Rigét, F.F., et al., 2014. A note on a walrus' European odyssey. NAMMCO Scientific Publications 9, 75–91. Available from: http://doi.org/10.7557/3.2921.

Born, E.W., Heilmann, A., Kielsen Holm, L., Laidre, K.L., Iversen, M., 2017. Walruses in West and Northwest Greenland, An interview survey about the catch and the climate. Monographs on Greenland, Vol. 355. Museum Tusculanum Press, Copenhagen, pp. 1–256, Man and Society Vol. 44.

Brodie, P.F., 2000. Field studies of the comparative mechanics of skin and blubber from walrus (*Odobenus rosmarus rosmarus*). In: Mazin, J.-M., de Buffrénil, V. (Eds.), Secondary Adaptations of Tetrapods to Life in Water. Verlag Dr. Friedrich Pfeil, München, pp. 339–344.

Calle, P.P., Seagars, D.J., McClave, D., Senne, C., House, J.A., 2002. Viral and bacterial serology of free-ranging Pacific walrus. Journal of Wildlife Diseases 38, 93–100.

Calvert, W., Stirling, I., 1990. Interactions between polar bears and overwintering walruses in the central Canadian High Arctic. Bears: Their Biology and Management 8, 351–356.

Cameron, M.F., Bengston, J.L., Boveng, P.L., Jansen, J.K., Kelly, B.P., Dahle, S.P., et al., 2010. Status review of the bearded seal (*Erignathus barbatus*). U.S. Department of Commerce. National Oceanic and Atmospheric Administration National Marine Fisheries Service Alaska Fisheries Science Center. NOAA Technical Memorandum NMFS-AFSC-211, 1–246.

Campbell, W.C., 1988. Trichinosis revisited—another look at modes of transmission. Parasitology Today 4, 83–86.

Chapskii, K.K., 1940. Distribution of walrus in the Laptev and East Siberian Seas. Problemy Arktiki 6, 80–94 (Translated by Wokroucheff, D., 1958).

Charrier, I., Aubin, T., Mathevon, N., 2009. Mother–calf vocal communication in Atlantic walrus: a first field experimental study. Animal Cognition 13, 471–482. Available from: https://doi.org/10.1007/s10071-009-0298-9.

Charrier, I., Burlet, A., Aubin, T., 2011. Social vocal communication in captive Pacific walruses *Odobenus rosmarus divergens*. Zeitschrift für Säugetierkunde 76, 622–627. Available from: https://doi.org/10.1016/j.mambio.2010.10.006.

Chivers, S., 1999. Biological indices for monitoring population status of walrus evaluated with an individual-based model. In: Garner, G., Amstrup, S., Laake, J., Manly, B., McDonald, L., Robertson, D. (Eds.), Marine Mammal Survey and Assessment Methods. Balkema, A.A., Rotterdam, Brookfield, The Netherlands, pp. 239–247. Proceedings of the Symposium on Surveys, Status & Trends of Marine Mammal Populations. Seattle, Washington, USA, 25–27 February 1998.

Clarke, P. Bring back the bear and the bison. NewStatesman 20 December 1999. https://www.newstatesman.com/node/193260.

Collett, R., 1912. Norges Pattedyr (Mammals of Norway). H. Aschehoug & Co. Publishers, Kristiania, Norway, pp. 1–744.

COSEWIC, 2006. COSEWIC Assessment and Update Status Report on the Atlantic Walrus *Odobenus rosmarus rosmarus* in Canada. Committee on the Status of Endangered Wildlife in Canada, Ottawa, pp. 1–65.

COSEWIC, 2017. COSEWIC Assessment and Status Report on the Atlantic Walrus *Odobenus rosmarus rosmarus*, High Arctic Population, Central-Low Arctic Population and Nova Scotia-Newfoundland-Gulf of St. Lawrence Population in Canada. Committee on the Status of Endangered Wildlife in Canada, Ottawa, pp. 1–89. Available from: http://www.registrelep-sararegistry.gc.ca/default.asp?lang = en&n = 24F7211B-1.

Costa, D.P., 1993. The relationship between reproductive and foraging energetics and the evolution of Pinnipedia. In: Boyd, I.L. (Ed.), Marine Mammals: Advances in Behavioural and Population Biology. Symposia of the Zoological Society of London, pp. 293–314. No. 66.

Cronin, M.A., Hills, S., Born, E.W., Patton, J.C., 1994. Mitochondrial DNA variation in Atlantic and Pacific walruses. Canadian Journal of Zoology 72, 1035–1043.

Dehn, L.-A., Sheffield, G.G., Follmann, E.H., Duffy, L.K., Thomas, D.L., O'Hara, T.M., 2007. Feeding ecology of phocid seals and some walrus in the Alaskan and Canadian Arctic as determined by stomach contents and stable isotope analysis. Polar Biology 30, 167–181. Available from: https://doi.org/10.1007/s00300-006-0171-0.

DeMaster, D., 1984. An analysis of a hypothetical population of walruses. In: Fay, F., Fedoseev, G. (Eds.), Soviet-American Cooperative Research on Marine Mammals, Vol. 1. National Oceanographic Atmospheric Administration Technical Report NMFS 12, Washington, D.C., USA, pp. 77–80. Pinnipeds.

Denisenko, S.G., Denisenko, N.V., Chaban, E.M., Gagaev, S.Y., Petryashov, V.V., Zhuravleva, N.E., et al., 2019. The current status of the macrozoobenthos around the Atlantic walrus haul-outs in the Pechora Sea (SE Barents Sea). Polar Biology 42, 1703–1717. Available from: https://doi.org/10.1007/s00300-018-02455-3.

Dietz, R., Born, E.W., Stewart, R.E.A., Heide-Jørgensen, M.P., Stern, H., Rigét, F., et al., 2014. Movements of walruses (*Odobenus rosmarus*) between Central West Greenland

and Southeast Baffin Island, 2005-2008. NAMMCO Scientific Publications 9, 53–74. Available from: http://doi.org/10.7557/3.2605.

Duignan, P.J., van Bressem, M.-F., Baker, J.D., Barbieri, M., Colegrove, K.M., de Guise, S., et al., 2014. Phocine Distemper Virus: Current knowledge and future directions. Viruses 6, 5093–5134. Available from: https://doi.org/10.3390/v6125093.

Fay, F.H., 1960a. Structure and function of the pharyngeal pouches of the walrus (*Odobenus rosmarus* L.). Mammalia 2, 361–371.

Fay, F.H., 1960b. Carnivorous walrus and some Arctic zoonoses. Arctic 13, 111–122.

Fay, F.H., 1981. Walrus, *Odobenus rosmarus* (Linnaeus, 1758). In: Ridgway, S.H., Harrison, R.J. (Eds.), Handbook of Marine Mammals, the Walrus, Sea Lions, fur Seals, and Sea Otter. Academic Press, New York, pp. 1–23.

Fay, F.H., 1982. Ecology and biology of the Pacific Walrus, *Odobenus rosmarus* divergens, Illiger. North American Fauna 74, 1–279. United States Department of the Interior, Fish and Wildlife Service 1982.

Fay, F.H., 1985. *Odobenus rosmarus*. Mammalian Species 238, 1–7.

Fay, F.H., Stoker, S.W., 1982. Reproductive success and feeding habits of walruses taken in the 1982 spring harvest, with comparisons from previous years. Final Report to Eskimo Walrus Commission, September 1982. University of Alaska, Institute of Marine Science, 1–87.

Fay, F.H., Bukhtiarov, Y.A., Stoker, S.W., Shults, L.M., 1984a. Foods of the Pacific walrus in winter and spring in the Bering Sea. In: Fay, F.H., Fedoseev, G.A. (Eds.), Soviet-American Cooperative Research on Marine Mammals, Vol. 1. NOAA Technical Reports NMFS 12, Washington, D.C., pp. 81–88. Pinnipeds.

Fay, F.H., Ray, G.C., Kibal'chich, A.A., 1984b. Time and location of mating and associated behavior of the Pacific walrus, *Odobenus rosmarus divergens* Illiger. In: Fay, F.H., Fedoseev, G.A. (Eds.), Soviet-American Cooperative Research on Marine Mammals, Vol. 1. NOAA Technical Reports NMFS 12, Washington, D.C., pp. 89–99. Pinnipeds.

Fay, F.H., Kelly, B., Sease, J.L., 1989. Managing the exploitation of Pacific walruses: a tragedy of delayed response and poor communication. Marine Mammal Science 5, 1–16. Available from: http://doi.org/10.1111/j.1748-7692.1989.tb00210.x.

Fay, F.H., Sease, J.L., Merrick, R.L., 1990. Predation on a ringed seal, *Phoca hispida*, and a black guillemot, *Cepphus grylle*, by a Pacific walrus, *Odobenus rosmarus divergens*. Marine Mammal Science 6, 348–350.

Fay, F., Burns, J., Stoker, S., Grundy, J., 1994. The struck-and-lost factor in Alaskan walrus harvests, 1952-1972. Arctic 47, 368–373.

Fay, F.H., Eberhardt, L.L., Kelly, B.P., Burns, J.J., Quakenbush, L.T., 1997. Status of the Pacific walrus population, 1950–1989. Marine Mammal Science 13, 537–565. Available from: http://doi.org/10.1111/j.1748-7692.1997.tb00083.x.

Fedoseev, G.A., Gol'tsev, V.N., 1969. Vozrastno-polovaya struktura i vosproizvoditel'-naya sposobnost' populyatsii tikhookeanskogo morzha (Age-sex structure and reproductive capacity of the population of the Pacific walrus). Zoologicheskii Zhurnal 48, 407–413. Translated by F.H. Fay, 1-9. Also translated as Can Transl Fish Aquat Sci 5503, 1990.

Ferguson, S.H., Lariviere, S., 2004. Are long penis bones an adaption to high latitude snowy environments? Oikos 105, 255–267.

Ferguson, S.H., Higdon, J.W., Westal, K.H., 2012. Prey items and predation behavior of killer whales (*Orcinus orca*) in Nunavut, Canada based on Inuit hunter interviews. Aquatic Biosystems 8, 3. Available from: https://doi.org/10.1186/2046-9063-8-3.

Fisher, K.I., 1989. Food habits and digestive efficiency in walrus, *Odobenus rosmarus* (Master of Science Thesis), University of Manitoba, Winnipeg, Canada, 1–88.

Fisher, K.I., Stewart, R.E.A., 1997. Summer foods of Atlantic walrus, *Odobenus rosmarus rosmarus*, in northern Foxe Basin, Northwest Territories. Canadian Journal of Zoology 75, 1166–1175.

Fisher, K.I., Stewart, R.E.A., Kastelein, R.A., Campbell, L.D., 1992. Apparent digestive efficiency in walruses (*Odobenus rosmarus*) fed herring (*Clupea harengus*) and clams (*Spisula* sp.). Canadian Journal of Zoology 70, 30–36.

Fox, A.D., Fox, G.F., Liaklev, A., Gerhardsson, N., 2010. Predation of flightless pink-footed geese (*Anser brachyrhynchus*) by Atlantic walruses (*Odobenus rosmarus rosmarus*) in southern Edgeøya, Svalbard. Polar Research 29, 455–457.

Frei, K.M., Coutu, A.N., Smiarowski, K., Harrison, R., 2015. Was it for walrus? Viking Age settlement and medieval walrus ivory trade in Iceland and Greenland. World Archaeology 47, 439–466.

Freitas, C., Kovacs, K.M., Ims, R.A., Fedak, M.A., Lydersen, C., 2009. Deep into the ice: over-wintering and habitat selection in male Atlantic walruses. Marine Ecology Progress Series 375, 247–261.

Freuchen, P., 1935. Mammals, Part II. Field notes and personal observations. Report of the Fifth Thule Expedition, 1921–1924, 2, 68–278.

Freuchen, P., Salomonsen, F., 1959. *The Arctic Year*. Jonathan Cape, Thirty Bedford Square, London, 1–440.

Galicia, M.P., Thiemann, G.W., Dyck, M.G., Ferguson, S.H., Higdon, J.W., 2016. Dietary habits of polar bears in Foxe Basin, Canada: possible evidence of a trophic regime shift mediated by a new top predator. Ecology and Evolution 6, 6005–6018.

Ganova-Raeva, L., Smith, A.W., Fields, H., Khudyakov, Y., 2004. New Calicivirus isolated from walrus. Virus Research 102, 207–213.

Garde, E., Hansen, R.G., Zinglersen, K., Ditlevsen, S., Heide-Jørgensen, M.P., 2018. Diving and foraging characteristics of walrus in Smith Sound. Journal of Experimental Biology 500, 89–99. Available from: https://doi.org/10.1016/j.jembe.2017.12.009.

Garlich-Miller, J.L., Stewart, R.E.A., 1998. Growth and sexual dimorphism of Atlantic walruses (*Odobenus rosmarus rosmarus*) in Foxe Basin, Northwest Territories, Canada. Marine Mammal Science 14, 803–818.

Garlich-Miller, J.L., Stewart, R.E.A., 1999. Female reproductive patterns and fetal growth of Atlantic walruses (*Odobenus rosmarus rosmarus*) in Foxe Basin, Northwest Territories, Canada. Marine Mammal Science 15, 179–191.

Garlich-Miller, J.L., Quakenbush, L.T., Bromaghin, J.F., 2006. Trends in age structure and productivity of Pacific walruses harvested in the Bering Strait region of Alaska, 1952-2002. Marine Mammal Science 22, 880–896.

Garlich-Miller, J.L., MacCracken, J.G., Snyder, J., Meehan, R., Myers, M., Wilder, J.M., et al., 2011. Status Review of the Pacific Walrus (*Odobenus rosmarus divergens*). U.S. Fish and Wildlife Service, Anchorage, AK, pp. 1–155.

Gebruk, A., Mikhaylyukova, P., Mardashova, M., Semenova, V., Henry, L.-A., Shabalin, N., et al., 2020. Integrated study of benthic foraging resources for Atlantic walrus (*Odobenus rosmarus rosmarus*) in the Pechora Sea, south-eastern Barents Sea, Aquatic Conservation: Marine and Freshwater Ecosystems, 1–14.

Giljov, A., Karenina, K., Kochnev, A., 2017. Prey or play: interactions between walruses and seabirds. Acta Ethologica 20, 47–57. Available from: https://doi.org/10.1007/s10211-016-0248-x.

Gjertz, I., 1990a. Ringed seal *Phoca hispida* fright behaviour caused by walrus *Odobenus rosmarus*. Polar Research 8, 317–319.

Gjertz, I., 1990b. Walrus predation of seabirds. Polar Record 26, 317.

Gjertz, I., 1990c. Do polar bears prey on walruses in Svalbard? Fauna 43, 54–56.

Gjertz, I., Wiig, Ø., 1992. Feeding of walrus *Odobenus rosmarus* in Svalbard. Polar Record 28, 57–59.

Gjertz, I., Wiig, Ø., 1994. Past and present distribution of walruses in Svalbard. Arctic 47, 34–42.

Gjertz, I., Wiig, Ø., 1995. The number of walruses (*Odobenus rosmarus*) in Svalbard in summer. Polar Biology 15, 527–530.

Gjertz, I., Hansson, R., Wiig, Ø., 1992. The historical distribution and catch of walrus in Franz Josef Land. Norsk Polarinstitutt Meddelelser 120, 67–81.

Gjertz, I., Henriksen, G., Øritsland, T., Wiig, Ø., 1993. Observations of walruses along the Norwegian coast 1969-1992. Polar Research 12, 27–31. Available from: http://doi.org/10.1111/j.1751-8369.1993.tb00419.x.

Gjertz, I., Wiig, Ø., Øritsland, N.A., 1998. Backcalculation of original population size for walruses *Odobenus rosmarus* in Franz Josef Land. Wildlife Biology 4, 223–230.

Gjertz, I., Griffiths, D., Krafft, B.A., Lydersen, C., Wiig, Ø., 2001. Diving and haul-out patterns of walruses *Odobenus rosmarus* on Svalbard. Polar Biology 24, 314–319.

Glud, R.N., Rysgaard, S., 2007. The annual organic carbon budget in Young Sound, NE Greenland. In: Rysgaard, S., Glud, R.N. (Eds.), Carbon Cycling in Arctic Marine Ecosystems: Case Study Young Sound, 58. Monographs on Greenland, pp. 194–203.

Hamilton, C.D., Kovacs, K.M., Lydersen, C., 2015. Year-round haul-out behaviour of walruses (*Odobenus rosmarus*) in the Northern Barents Sea. Marine Ecology Progress Series 519, 251–263. Available from: https://doi.org/10.3354/meps11089.

Harington, C.R., 1966. Extralimital occurrences of walruses in the Canadian Arctic. Journal of Mammalogy 47, 506–513.

Heide-Jørgensen, M.P., Burt, L.M., Guldborg Hansen, R., Hjort Nielsen, N., Rasmussen, M., Fossette, S., et al., 2013. Significance of the North Water Polynya to Arctic top predators. AMBIO 42, 596–610.

Higdon, J.W., Ferguson, S.H., 2009. Loss of Arctic sea ice causing punctuated change in sightings of killer whales (*Orcinus orca*) over the past century. Ecological Applications 19, 1365–1375.

Higdon, J.W., Stewart, D.B., 2018. State of circumpolar walrus (*Odobenus rosmarus*) populations. Prepared by Higdon Wildlife Consulting and Arctic Biological Consultants, Winnipeg, MB for WWF Arctic Programme, Ottawa, ON, 1–100.

Higdon, J.W., Westdal, K.H., Ferguson, S.H., 2014. Distribution and abundance of killer whales (*Orcinus orca*) in Nunavut, Canada—an Inuit knowledge survey. Journal of the Marine Biological Association of the United Kingdom 94, 1293–1304.

Hills, S., 1992. The effect of spatial and temporal variability on population assessment of Pacific walruses (PhD thesis), University of Maine, Orono, December 1992, 1–120, 1992.

Hjelset, A.M., Andersen, A., Gjertz, I., Lydersen, C., Gulliksen, B., 1999. Feeding habits of bearded seals (*Erignathus barbatus*) from the Svalbard area, Norway. Polar Biology 21, 186–193.

Jay, C.V., Hills, S., 2005. Movements of walruses radio-tagged in Bristol Bay, Alaska. Arctic 58, 192–202.

Jay, C.V., Farley, S.D., Garner, G.W., 2001. Summer diving behavior of male walruses in Bristol Bay, Alaska. Marine Mammal Science 17, 617–631.

Jay, C.V., Marcot, B.G., Douglas, D.C., 2011. Projected status of the Pacific walrus (*Odobenus rosmarus divergens*) in the twenty-first century. Polar Biology 34, 1065–1084.

Kasser, J.J.W.,Wiedmer, J.E., 2012. The circumpolar walrus population in a changing world (B.Sc thesis). Thesis number 594000. University of Applied Sciences Van Hall Larenstein, Leeuwarden, The Netherlands, 1–68.

Kastelein, R.A., 2002. Walrus (*Odobenus rosmarus*). In: Perrin, W.F., Würsig, B., Thewissen, J.M.G. (Eds.), Encyclopedia of Marine Mammals, 2nd Edition Academic Press, San Diego, California, U.S.A, pp. 1212–1217.

Kastelein, R.A., van Gaalen, M.A., 1988. The sensitivity of the vibrissae of a Pacific walrus (*Odobenus rosmarus divergens*). Part 1. Aquatic Mammals 14, 123–133.

Kastelein, R.A., Mosterd, P., 1989. The excavation technique for molluscs of Pacific walruses (*Odobenus rosmarus divergens*) under controlled conditions. Aquatic Mammals 15, 3–17.

Kastelein, R.A., Wierpkema, P.R., 1989. A digging trough as occupational therapy for Pacific walruses (*Odobenus rosmarus divergens*) in human care. Aquatic Mammals 15, 9–17.

Kastelein, R.A., Gerrits, N.M., 1990. The anatomy of the walrus head (*Odobenus rosmarus*). Part 1: the skull. Aquatic Mammals 16, 101–119.

Kastelein, R.A., Wierpkema, P.R., Slegtenhorst, C., 1989. The use of molluscs to occupy Pacific walruses (*Odobenus rosmarus divergens*) in human care. Aquatic Mammals 15, 6–8.

Kastelein, R.A., Stevens, S., Mosterd, P., 1990. The tactile sensitivity of the mystacial vibrissae of a Pacific walrus (*Odobenus rosmarus divergens*). Part 2: masking. Aquatic Mammals 16, 78–87.

Kastelein, R.A., Zweypfenning, R.C.V.J., Spekreijse, H., Dubbeldam, J.L., Born, E.W., 1993. The anatomy of the walrus head (*Odobenus rosmarus*). Part 3: the eyes and their function in walrus ecology. Aquatic Mammals 19, 61–92.

Kastelein, R.A., Muller, M., Terlouw, A., 1994. Oral suction of a Pacific walrus (*Odobenus rosmarus divergens*) in air and under water. Zeitschrift für Säugetierkunde 59, 105–115.

Kastelein, R.A., Schooneman, N.M., Wiepkema, P.R., 2000. Food consumption and body weight of captive Pacific walruses (*Odobenus rosmarus divergens*). Aquatic Mammals 26, 175–190.

Kavry, V.I., Boltunov, A.N., Nikiforov, V.V., 2008. New coastal haulouts of walruses (*Odobenus rosmarus*)—response to the climate changes. International Conference of Marine Mammals of the Holarctic V, Odessa, Ukraine, 249–251.

Keighley, X., Palsson, S., Einarsson, B.F., Petersen, A., Fernández-Coll, M., Jordan, P., et al., 2019. Disappearance of Icelandic walruses coincided with Norse settlement. Molecular Biology and Evolution 36, 2656–2667. Available from: https://doi.org/10.1093/molbev/msz196.

Kiliaan, H.P.L., Stirling, I., 1978. Observations on overwintering walruses in the eastern Canadian High Arctic. Journal of Mammalogy 59, 197–200.

King, J.E., 1964. Seals of the World. British Museum (Natural History), London, pp. 1–154.

Knutsen, L.Ø., Born, E.W., 1994. Body growth in Atlantic walruses (*Odobenus rosmarus rosmarus*) from Greenland. Journal of Zoology (London) 234, 371–385.

Kovacs, K.M., 2016. *Odobenus rosmarus* ssp. *rosmarus*. The IUCN Red List of Threatened Species 2016: e.T15108A66992323. https://doi.org/10.2305/IUCN.UK.2016-1.RLTS.T15108A66992323.en. Downloaded on 01 June 2019.

Kovacs, K.M., Lydersen, C., 2008. Climate change impacts on seals and whales in the North Atlantic Arctic and adjacent shelf seas. Science Progress 92, 117–150. Available from: https://doi.org/10.3184/003685008X324010.

Kovacs, K.M., Aars, J., Lydersen, C., 2014. Walruses recovering after 60 + years of protection in Svalbard, Norway. Research Note. Polar Research 33, 26034. Available from: http://doi.org/10.3402/polar.v33.26034.

Kryukova, N.V., Kruchenkova, E.P., Ivanov, D.I., 2012. Killer Whales (*Orcinus orca*) hunting for walruses (*Odobenus rosmarus divergens*) near Retkyn Spit, Chukotka. Zoologicheskii Zhurnal 91, 734–745. Biology Bulletin 39, 768–778.

Laidre, K.L., Stirling, I., Lowry, L.F., Wiig, Ø., Heide-Jørgensen, M.P., Ferguson, S.H., 2008. Quantifying the sensitivity of Arctic marine mammals to climate-induced habitat change. Ecological Applications 18(sp2), S97–S125.

Laidre, K.L., Regehr, E.V., 2017. Arctic marine mammals and sea ice. In: Thomas, D.N. (Ed.), Sea Ice, Third edition Wiley-Blackwell, pp. 516–522. ISBN: 978-1-118-77838-8.

Laidre, K.L., Stern, H., Kovacs, K.M., Lowry, L.F., Moore, S.E., Regehr, E.V., et al., 2015. Arctic marine mammal population status, sea ice habitat loss, and conservation recommendations for the 21st century. Conservation Biology 29, 724–737.

Leonardi, M.S., Palma, R.L., 2013. Review of the systematics, biology and ecology of lice from pinnipeds and river otters (Insecta: Phthiraptera: Anoplura: Echinophthiriidae). Zootaxa 3630, 445–466.

Levermann, N., Galatius, A., Ehlmé, G., Rysgaard, S., Born, E.W., 2003. Feeding behaviour of wild walruses (*Odobenus rosmarus*): do they exhibit a tendency towards dextrality? BMC Ecology 3, 9 (23 Oct. 2003). Available from: www.biomedcentral.com/1472-6785/3/9.

Lind, S., Ingvaldsen, R.B., Furevik, T., 2018. Arctic warming hotspot in the northern Barents Sea linked to declining sea-ice import. Nature Climate Change 8, 634–639.

Lindenfors, P., Tullberg, B.S., Biuw, M., 2002. Phylogenetic analyses of sexual selection and sexual size dimorphism in pinnipeds. Behavioral Ecology and Sociobiology 52, 188–193.

Lindqvist, C., Bachmann, L., Andersen, L.W., Born, E.W., Arnason, U., Kovacs, K.M., et al., 2008. The Laptev Sea walrus *Odobenus rosmarus laptevi*: an enigma revisited. Zoologica Scripta 38, 113–224. Available from: https://doi.org/10.1111/j.1463-6409.2008.00364.x.

Lindqvist, C., Roy, T., Lydersen, C., Kovacs, K.M., Aars, J., Wiig, Ø., et al., 2016. Genetic diversity of historical Atlantic walruses (*Odobenus rosmarus rosmarus*) from Bjørnøya and Håøya (Tusenøyane), Svalbard, Norway. BMC Research Notes 9, 112. Available from: https://doi.org/10.1186/s13104-016-1907-8.

Loughrey, A.G., 1959. Preliminary investigation of the Atlantic walrus *Odobenus rosmarus rosmarus* (Linnaeus). Canadian Wildlife Service. Wildlife Management Bulletin Series 1. No. 14, 1–123.

Lovvorn, J.R., Wilson, J.J., McKay, D., Bump, J.K., Cooper, L.W., Grebmeier, M., 2010. Walruses attack spectacled eiders wintering in pack ice of the Bering Sea. Arctic 63, 53–56.

Lowry, L.F., Fay, F.H., 1984. Seal eating by walruses in the Bering and Chukchi Seas. Polar Biology 3, 11–18.

Lowry, L.F., Frost, K.J., Burns, J.J., 1980. Feeding of bearded seals in the Bering and Chukchi Seas and trophic interaction with Pacific walruses. Arctic 33, 330–342.

Lowry, L.F., Nelson, R.R., Frost, K.J., 1987. Observations of killer whales, *Orcinus orca*, in western Alaska: sightings, strandings, and predation on other marine mammals. Canadian Field-Naturalist 101, 6–12.

Lowther, A.D., Kovacs, K.M., Griffiths, D., Lydersen, C., 2015. Identification of motivational state in adult male Atlantic walruses inferred from changes in movement and diving behavior. Marine Mammal Science 31, 1291–1313.

Lund, H.M.K., 1954. The walrus (*Odobenus rosmarus* (L.)) off the coast of Norway in the past and after the year 1900 together with some observations on its migration and "cruising speed". Astarte 8, 1–12.

Lydersen, C., 2018. Walrus *Odobenus rosmarus*. In: Würsig, B., Thewissen, J.G.M., Kovacs, K.M. (Eds.), Encyclopedia of Marine Mammals, 3rd Edition Elsevier, London, pp. 1045–1048.

Lydersen, C., Kovacs, K.M., 2014. Walrus (*Odobenus rosmarus*) research in Svalbard, Norway, 2000-2010. NAMMCO Scientific Publications 9, 175–190. Available from: https://doi.org/10.7557/3.2613.

Lydersen, C., Aars, J., Kovacs, K.M., 2008. Estimating the number of walruses in Svalbard from aerial surveys and behavioural data from satellite telemetry. Arctic 61, 119–128.

Lydersen, C., Chernook, V.I., Glazov, D.M., Trukhanova, I.S., Kovacs, K.M., 2012. Aerial survey of Atlantic walruses (*Odobenus rosmarus rosmarus*) in the Pechora Sea, August 2011. Polar Biology 35, 1555–1562.

MacCracken, J.G., Beatty, W.S., Garlich-Miller, J.L., Kissling, M.L., Snyder, J.A., 2017. Final species status assessment for the Pacific walrus (*Odobenus rosmarus divergens*), May 2017 (Version 1.0). U.S. Fish and Wildlife Service, Marine Mammals Management, Anchorage, AK, 1–297.

Magnus, O., 1555. Historia de gentibus septentrionalibus. Rome 1555. Available at Free Google Books.

Mallory, M.L., Woo, K., Gaston, A.J., Davies, W.E., Mineau, P., 2004. Walrus (*Odobenus rosmarus*) predation on adult thick-billed murres (*Uria lomvia*) at Coats Island, Nunavut, Canada. Polar Research 23, 111–114.

Maniscalco, J.M., Springer, A.M., Counihan, K.L., Hollmen, T., Aderman, H.M., Toyukak, M., 2020. Contemporary diets of walruses in Bristol Bay, Alaska suggest temporal variability in benthic community structure. PeerJ 2020 (3), e8735. Available from: https://doi.org/10.7717/peerj.8735.

Mansfield, A.W., 1958. The biology of the Atlantic walrus *Odobenus rosmarus rosmarus* (Linnaeus) in the eastern Canadian Arctic. Fisheries Research Board of Canada Manuscript Report Series (Biology) Reports 653, 1–146.

Mansfield, A.W., 1959. The walrus in the Canadian Arctic. Fisheries Research Board of Canada Arctic Unit Circular 2, 1–13.

Mansfield, A.W., 1966. The walrus in Canada's Arctic. Canadian Geographical Journal 72, 88–95.

Mansfield, A.W., 1973. The Atlantic walrus *Odobenus rosmarus rosmarus* in Canada and Greenland. In: Seals. Proceedings of a working meeting of seal specialists on threatened and depleted seals of the world, held under the auspices of the Survival Service Commission of IUCN, 18–19 August 1972. University of Guelph, Ontario, Canada. IUCN Publ New Ser Suppl Paper 39, Gland, Switzerland, pp. 69–79.

McLeod, B.A., Frasier, T.R., Lucas, Z., 2014. Assessment of the extirpated Maritimes walrus using morphological and ancient DNA analysis. PLoS ONE 9 (6), e99569. Available from: https://doi.org/10.1371/journal.pone.0099569.

Melero, M., García-Parraga, D., Corpa, J.M., Ortega, J., Rubio-Guerri, C., Crespo, J.L., et al., 2014. First molecular detection and characterization of herpesvirus and poxvirus in a Pacific walrus (*Odobenus rosmarus divergens*). BMC Veterinary Research 10, 1–7.

Melnikov, V.V., Zagrebin, I.A., 2005. Killer whale predation in coastal waters of the Chukotka Peninsula. Marine Mammal Science 21, 550–556.

Miller, E.H., 1975. Walrus ethology. I. The social role of tusks and applications of multidimentional scaling. Canadian Journal of Zoology 53, 590–613.

Miller, E.H., 1982. Herd organization and female threat behaviour in Atlantic walrus *Odobenus rosmarus rosmarus* (L.). Mammalia 46, 29–34.

Miller, E.H., 1985. Airborne acoustic communication in the walrus *Odobenus rosmarus*. National Geographic Resolution 1, 124–145.

Miller, E.H., Boness, D.J., 1983. Summer behavior of Atlantic walruses, *Odobenus rosmarus rosmarus* (L) at Coats Island, N.W.T. (Canada). Zeitschrift für Säugetierkunde 48, 298–313.

Milne, A.O., Smith, C., Orton, L.D., Sullivan, M.S., Grant, R.A., 2020. Pinnipeds orient and control their whiskers: a study on Pacific walrus, California sea lion and harbor seal. Journal of Comparative Physiology A 206, 441–451. Available from: https://doi.org/10.1007/s00359-020-01408-8.

Møller, L.N., Petersen, E., Kapel, C.M.O., Melbye, M., Koch, A., 2005. Outbreak of trichinellosis associated with consumption of game meat in West Greenland. Veterinary Parasitology 132, 131–136.

Møller, L.N., Grove Krause, T., Koch, A., Melbye, M., Kapel, C.M.O., Petersen, E., 2007. Human antibody recognition of Anisakidae and *Trichinella* spp. in Greenland. Clinical Microbiology and Infection 13, 702–708. Available from: https://doi.org/10.1111/j.1469-0691.2007.01730.x.

Monson, D.H., Udevitz, M.S., Jay, C.V., 2013. Estimating age ratios and size of Pacific walrus herds on coastal haulouts using video imaging. PLOS One 8 (7), e69806.

MOSJ, Environmental Monitoring of Svalbard and Jan Mayen, 2020. http://www.mosj.no/en/climate/ocean/sea-ice-extent-barents-sea-fram-strait.html (accessed 26 July 2020).

Mouy, X., Hannay, D., Zykov, M., Martin, B., 2012. Tracking of Pacific walruses in the Chukchi Sea using a single hydrophone. Journal of the Acoustical Society of America 131, 1349–1358. Available from: https://doi.org/10.1121/1.3675008.

Muir, D.C.G., Segstro, M.D., Hobson, K.A., Ford, C.A., Stewart, R.E.A., Olpinski, S., 1995. Can seal eating explain elevated levels of PCBs and organochlorine pesticides in walrus blubber from eastern Hudson Bay (Canada)? Environmental Pollution 90, 335–348.

NAMMCO (North Atlantic Marine Mammal Commission) Annual Report 2005. North Atlantic Marine Mammal Commission, Tromsø, Norway, 1–381, 2006. Available at https://nammco.no/topics/sc-working-group-reports/.

NAMMCO, 2018. Report of the NAMMCO (North Atlantic Marine Mammal Commission) Scientific Working Group on Walrus, October 2018. Available at https://nammco.no/topics/sc-working-group-reports/.

NAMMCO (North Atlantic Marine Mammal Commission) 2019. Report of the Scientific Committee 26th Meeting, October 29 – November 1, 2019. Tórshavn, Faroe Islands, 1–78. Available at https://nammco.no/topics/scientific-committee-reports/.

NAMMCO, 2020. North Atlantic Marine Mammal Commission. Available at https://nammco.no/topics/atlantic-walrus/ (accessed 26 July 2020).

Nansen, F., 1897. Fram over Polhavet. Den norske Polarfærd 1893-1896. Vol. 2. H. Aschehoug and Co.s Forlag, Kristiania, Norway, 1-553. (In Norwegian).

Nelson, C.H., Johnson, K.R., 1987. Whales and walruses as tillers of the sea floor. Scientific American 256, 112–118.

Nielsen, O., Stewart, R.E.A., Nielsen, K., Measures, L., Duignan, P., 2001. Serologic survey of *Brucella* spp. antibodies in some marine mammals of North America. Journal of Wildlife Diseases 37, 89–100.

Noren, S.R., Jay, C.V.V., Burns, J.M., Fischbach, A.S., 2015. Rapid maturation of the muscle biochemistry that supports diving in Pacific walruses (*Odobenus rosmarus divergens*). Journal of Experimental Biology 218, 3319–3329.

Oliver, J.S., Slattery, P.N., Oconnor, E.F., Lowry, L.F., 1983. Walrus, *Odobenus rosmarus*, feeding in the Bering Sea—a benthic perspective. Fishery Bulletin 81, 501–512.

Oliver, J.S., Kvitek, R.G., Slattery, P.N., 1985. Walrus feeding disturbance: scavenging habits and recolonization of the Bering Sea benthos. Journal of Experimental Marine Biology and Ecology 91, 233–246.

Orbicon, Results of aerial surveys of marine mammals at Moriussaq, NW Greenland. Report from Orbicon Consultants, Denmark, 1–26, 2020. Available from https://www.orbicon.dk/expertises/milj%C3%B8.

Øren, K., Kovacs, K.M., Yoccoz, N.G., Lydersen, C., 2018. Assessing site-use and sources of disturbance at walrus haul-outs using monitoring cameras. Polar Biology 41, 1737–1750 (2018). Available from: https://doi.org/10.1007/s00300-018-2313-6.

Ovsyanikov, N.G., 1995. Polar bear predation of walruses on Wrangell Island. Bulletin of the Moscow Association of Natural Scientists, Section of Biology 100, 1–13.

Ovsyanikov, N.G., 1996. Interactions of polar bears with other large mammals, including man. Journal of Wildlife Research 1, 254–259.

Pedersen, A., 1945. Der Eisbär (*Thalarctos maritimus* Phipps). Verbreitung und Lebensweise (The Polar Bear: Distribution and Way of Life). E. Bruun & Co.s Trykkerier, København, pp. 1–166 (in German).

Pedersen, A., 1962a. Ein Unterscheidungsmerkmal zwischen dem Pazifischen Walross *Odobaenus obesus* Illiger und dem Grönlandischen Walross *O. rosmarus* L. (A character of difference between the Pacific walrus *Odobaenus obesus* Illiger and the Greenlandic walrus *O. rosmarus* L.). Zeitschrift für Säugetierkunde 27, 237–239 (in German).

Pedersen, A., 1962b. Das Walross (The Walrus). A. Ziemsen, Wittenberg-Lutherstadt. Germany, pp. 1–62 (in German).

Phillipa, J.D.W., Leighton, F.A., Daoust, P.Y., Nielsen, O., Pagliarulo, M., Schwantje, H., et al., 2004. Antibodies to selected pathogens in free-ranging terrestrial carnivores and marine mammals in Canada. Veterinary Record 155, 135–140. Available from: https://doi.org/10.1136/vr.155.5.135.

Popov, L., Timoshenko, Iu., Wiig, Ø., 1990. Review of history and present status of world walrus stocks: Barents, Kara and White Seas. In: Fay, F.H., Kelly, B.P., Fay, B.A. (Eds.), The Ecology and Management of Walrus Populations. Report of An International Workshop. Marine Mammal Commission Report, Seattle, Washington. USA, pp. 6–14.

Ray, C.E., 1960. *Trichecodon huxlei* (Mammalia: Odobenidae) in the Pleistocene of southeastern United States. Bulletin of the Museum of Comparative Zoology at Harvard University 122, 129–142.

Reeves, R.R., 1978. Atlantic walrus (*Odobenus rosmarus rosmarus*): a literature survey and status report. United States Department of the Interior, Fish and Wildlife Service Washington, D.C. Wildlife Research Report 10, 1-41.

Reichmuth, C., Casey, C., 2014. Vocal learning in seals, sea lions, and walruses. Current Opinion in Neurobiology 28, 66–71.

Reichmuth, C., Sills, J.M., Brewer, A., Triggs, L., Ferguson, R., Ashe, E., et al., 2020. Behavioral assessment of in-air hearing range for the Pacific walrus (*Odobenus rosmarus divergens*). Polar Biology 43, 767–772. Available from: https://doi.org/10.1007/s00300-020-02667-6.

Richard, P.R., Campbell, R.R., 1988. Status of the Atlantic walrus, *Odobenus rosmarus rosmarus*, in Canada. Canadian Field-Naturalist 102, 337–350.

Ritchie, J., 1921. The walrus in British waters. Scottish Naturalist 5-9, 77–86.

Rugh, D.J., 1993. A polar bear kills a walrus calf. Northwestern Naturalist 74, 23–24.

Salter, R.E., 1979. Site utilization, activity budgets, and disturbance responses of Atlantic walruses during terrestrial haul-out. Canadian Journal of Zoology 57, 1169–1180. Available from: http://doi.org/10.1139/z79-149.

Salter, R.E., 1980. Observations on social behaviour of Atlantic walruses (*Odobenus rosmarus* (L.)) during terrestrial haul-out. Canadian Journal of Zoology 58, 461–463.

Scotter, S.E., Tryland, M., Nymo, I.H., Hansen, L., Harju, M., Lyderman, C., et al. 2019. Contaminants in Atlantic walruses in Svalbard part 1: Relationships between exposure, diet and pathogen prevalence, Environmental Pollution 244, 9–18.

Schusterman, R.J., Reichmuth, C., 2008. Novel sound production through contingency learning in the Pacific walrus (*Odobenus rosmarus divergens*). Animal Cognition 11, 319–327. Available from: http://doi.org/10.1007/s10071-007-0120-5.

Sheffield, G., Grebmeier, J.M., 2009. Pacific walrus (*Odobenus rosmarus divergens*): Differential prey digestion and diet. Marine Mammal Science 25, 761–777. Available from: https://doi.org/10.1111/j.1748-7692.2009.00316.x.

Sheffield, G., Fay, F.H., Feder, H., Kelly, B.P., 2001. Laboratory digestion of prey and interpretation of walrus stomach contents. Marine Mammal Science 17, 310–330.

Sjare, B., 1993. Vocalization and breeding behaviour of Atlantic walruses in the Canadian High Arctic (PhD thesis), University of Alberta, Edmonton. Canada. ISBN:0315820357 (*fide* Stirling and Thomas, 2003).

Sjare, B., Stirling, I., 1996. The breeding behavior of Atlantic walruses, *Odobenus rosmarus rosmarus*, in the Canadian High Arctic. Canadian Journal of Zoology 74, 897–911.

Sjare, B., Stirling, I., Spencer, C., 2003. Structural variation in the songs of Atlantic walruses breeding in the Canadian high Arctic. Aquatic Mammals 29, 297–318.

Skoglund, E.G., Lydersen, C., Grahl-Nielsen, O., Haug, T., Kovacs, K.M., 2010. Fatty acid composition of the blubber and dermis of adult male Atlantic walruses (*Odobenus rosmarus rosmarus*) in Svalbard, and their potential prey. Marine Biology Research 6, 239–250.

Smith, T.G., Hammill, M.H., Doidge, D.W., Carter, T., Sleno, G.A., 1979. Marine mammal studies in southeastern Baffin Island. Canadian Manuscript Report Fisheries and Aquatic Science 1552, 1–70.

Sonsthagen, S.A., Jay, C.V., Cornman, R.S., Fischbach, A.S., Grebmeier, J.M., Talbot, S.L., 2020. DNA metabarcoding of feces to infer summer diet of Pacific walruses. Marine Mammal Science 36, 1–16. Available from: https://doi.org/10.1111/mms.12717.

Spiridonov, V.A., Gavrilo, M.V., Krasnova, E.D., Nikolaeva, N.G. (Eds.), 2014. Atlas of Marine and Coastal Biological Diversity of the Russian Arctic. WWF Russia, Moscow, pp. 1–64. ISBN: 9785-9902786-2-2.

Star, B., Barrett, J.H., Gondek, A.T., Boessenkool, S., 2018. Ancient DNA reveals the chronology of walrus ivory trade from Norse Greenland. Proceedings of the Royal Society B 285, 20180978. Available from: http://doi.org/10.1098/rspb.2018.0978.

Stewart, R.E.A., 2002. Review of Atlantic walrus (*Odobenus rosmarus rosmarus*) in Canada. CSAS—Canadian Science Advisory Secretariat Research Document 2002 (091) 1-20. (*fide* COSEWIC 2017).

Stewart, R.E.A., 2008. Redefining walrus stocks in Canada. Arctic 61, 292–398. Available from: http://www.jstor.org/stable/40513028.

Stewart, R.E.A., Stewart, B.E., 2005. Comparison of between-tooth age estimates of Atlantic walrus (*Odobenus rosmarus rosmarus*). Marine Mammal Science 21, 346–354.

Stewart, R.E.A., Kovacs, K.M., Acquarone, M., 2014a. Introduction: Walrus of the North Atlantic (2014). NAMMCO Scientific Publications 9, 7–12.

Stewart, R.E.A., Born, E.W., Dunn, J.B., Koski, W.R., Ryan, A.K., 2014b. Use of Multiple Methods to Estimate Walrus *(Odobenus rosmarus rosmarus)* Abundance in the Penny Strait-Lancaster Sound and West Jones Sound Stocks, Canada. NAMMCO Scientific Publications 9, 95–116. Available from: http://doi.org/10.7557/3.2608.

Stewart, R.E.A., Born, E.W., Dietz, R., Ryan, A.K., 2014c. Estimates of minimum population size for walrus around Southeast Baffin Island, Nunavut. NAMMCO Scientific Publications 9, 141–157. Available from: http://doi.org/10.7557/3.2615.

Stirling, I., 1984. A group threat display given by walruses to polar bear. Journal of Mammalogy 65, 352–353.

Stirling, I., 1997. The importance of polynyas, ice edges, and leads to marine mammals and birds. Journal of Marine Systems 10, 9–21.

Stirling, I., Calvert, W., Spencer, C., 1987. Evidence of stereotyped underwater vocalizations of male Atlantic walruses (*Odobenus rosmarus rosmarus*). Canadian Journal of Zoology 65, 2311–2321.

Stirling, I., Laidre, K.L., Born, E.W., 2021. Do wild polar bears (*Ursus maritimus*) use tools when hunting walruses (*Odobenus rosmarus*)? Arctic. 74, 175–187.

Taylor, R.L., Udevitz, M.S., 2015. Demography of the Pacific walrus (*Odobenus rosmarus divergens*): 1974–2006. Marine Mammal Science 31, 231–254. Available from: https://doi.org/10.1111/mms.12156.

Taylor, R.L., Udevitz, M.S., Jay, C.V., Citta, J.J., Quakenbush, L.T., Lemons, P.R., et al., 2018. Demography of the Pacific walrus (*Odobenus rosmarus divergens*) in a changing Arctic. Marine Mammal Science 34, 54–86. Available from: https://doi.org/10.1111/mms.12434.

Tempel, J.T.L., Atkinson, S., 2020. Pacific walrus (*Odobenus rosmarus divergens*) reproductive capacity changes in three time frames during 1975–2010. Polar Biology 43, 861–875.

Thiemann, G.W., Budge, S.M., Iverson, S.J., Stirling, I., 2007. Unusual fatty acid biomarkers reveal age- and sex-specific foraging in polar bears (*Ursus maritimus*). Canadian Journal of Zoology 85, 505–517.

Thiemann, G.W., Iverson, S.J., Stirling, I., 2008. Polar bear diets and Arctic marine food webs: insights from fatty acid analysis. Ecological Monographs 78, 591–613.

Timoshenko, Iu, Popov, L.A., 1990. On predatory habits of Atlantic walrus. In: Fay, F.H., Kelly, B.P., Fay, B.A. (Eds.), The Ecology and Management of Walrus Populations. Marine Mammal Commission Report, Seattle, Washington. USA, pp. 177–178.

Udevitz, M.S., Taylor, R.L., Garlich-Miller, J.L., Quakenbush, L.T., Snyder, J.A., 2013. Potential population-level effects of increased haul-out related mortality of Pacific walrus calves. Polar Biology 36, 291–298.

Vasilyeva, M., 2020. Russian scientists discover huge walrus haul-out in Arctic circle, Reuter. Science News. November 6.

Verboom, W.C., Kastelein, R.A., 1995. Rutting whistles of a male Pacific walrus (*Odobenus romarus divergens*). In: Kastelein, R.A., Thomas, J.A., Nachtigall, P.E. (Eds.), Sensory Systems of Aquatic Mammals. De Spil, Woerden, The Netherlands, pp. 287–298.

Vibe, C., 1950. The marine mammals and the marine fauna in the Thule District (Northwest Greenland) with observations on the ice conditions in 1939-1941. Monographs on Greenland 150, 1–115.

Vorontsov, A.V., Goryaev, Yu.I., Yezhov, A.V., 2007. Results of observations for marine mammals along the Northern Sea Route. In: Matishov, G.G. (Ed.), Biology and Oceanography of the Northern Sea Route: The Barents and Kara Seas. Nauka, Moscow, pp. 161–172. (*fide* Spirodonov et al., 2014) (In Russian).

Wagemann, R., Muir, D.C.G., Stewart, R.E.A., 1993. Trace metals and organic contaminants in walruses from the Canadian Arctic and northern Quebec. In: Stewart, R.E.A., Richard, P.R., Stewart, B.E. (Eds.), Report of the 2nd Walrus International Technical and Scientific (WITS) Workshop. Can Fish Aquat Sci Tech Rep, Winnipeg, Manitoba, Canada, p. 67.

Welch, H.E., Bergmann, M.A., Siffred, T.D., Martin, K.A., Curtis, M.F., Crawford, R.E., et al., 1992. Energy flow through the marine ecosystem of the Lancaster Sound Region. Arctic Canada. Arctic 45, 343–357.

Wiig, Ø., Gjertz, I., 1996. Body size of male Atlantic walruses (*Odobenus rosmarus rosmarus*) from Svalbard. Journal of Zoology (London) 240, 495–499.

Wiig, Ø., Gjertz, I., Griffiths, D., Lydersen, C., 1993. Diving patterns of an Atlantic walrus (*Odobenus rosmarus rosmarus*) near Svalbard. Polar Biology 13, 71–72.

Wiig, Ø., Born, E.W., Stewart, R.E.A., 2014. Management of Atlantic walrus (*Odobenus rosmarus rosmarus*) in the arctic Atlantic. NAMMCO Scientific Publications 9, 315–341. Available from: http://doi.org/10.7557/3.2855.

Witting, L., Born, E.W., 2005. An assessment of Greenland walrus populations. ICES Journal of Marine Science 62, 266–284.

Witting, L., Born, E.W., 2014. Population dynamics of walrus in Greenland. NAMMCO Scientific Publications 9, 191–218.

Yurkowski, D., Carlyle, C., Amarualik, U., Lange, B., 2019. Novel observations of Atlantic walruses (*Odobenus rosmarus rosmarus*) in Archer Fjord, northern Ellesmere Island, Nunavut, Canada. Polar Biology 42, 1193–1198. Available from: https://doi.org/10.1007/s00300-019-02499-z.

Zarnke, R.L., Harder, T.C., Vos, H.W., Ver Hoef, J.M., Osterhaus, A.D.M.E., 1997. Serologic survey for phocid herpesvirus-1 and -2 in marine mammals from Alaska and Russia. Journal of Wildlife Diseases 33, 459–465.

Zech, M., 2014. Europe's oldest walrus dead at 40. NL Times, Netherlands. Available from: https://nltimes.nl/2014/05/29/europes-oldest-walrus-dead-40.

Zenkovich, B.A., 1938. About the Killer Whale, *Grampus orca* L. Priroda (Mosc.) 4, 109–112 (*fide* Kryukova et al., 2012).

CHAPTER

Stocks, distribution and abundance

4

Eva Garde and Rikke G. Hansen

Greenland Institute of Natural Resources, Nuuk, Greenland

Chapter Outline

Introduction

The walrus has a discontinuous circumpolar Arctic distribution. It is divided into two subspecies; the Atlantic walrus (*Odobenus rosmarus rosmarus*), which lives in the Atlantic and Arctic Oceans from the eastern Canadian Arctic eastward to Greenland, Norway (Svalbard) and Russia (Franz Josef Land and the Pechora, Kara and Barents seas, Fig. 4.1), and the Pacific walrus (*Odobenus rosmarus divergens*), which lives in the Pacific Ocean and is found from the Laptev Sea, eastward to the Beaufort Sea, and southward to the Bering Sea (Fay, 1985). The two subspecies are separated by a gap in their distribution between the eastern Chukchi Sea and the central Canadian Arctic. Occasionally, solitary Atlantic walruses are found far from their current range, for example in Iceland and Europe (Born et al., 2014). The Pacific walrus is more abundant than the Atlantic, with an estimated population size of more than 200,000 animals (IUCN Redlist, 2019). Atlantic walruses are estimated to total 35,000–40,000 animals.

Walruses prefer shallow coastal water habitats, as they forage primarily on the sea floor where they usually dive to depths from 10 to 100 m to reach bivalve

The Atlantic Walrus. DOI: https://doi.org/10.1016/B978-0-12-817430-2.00011-X

CURRENT STOCKS AND DISTRIBUTION OF THE ATLANTIC WALRUS (SUMMER)

North Polar Stereographic projection
Open water season, Aug-Sept

INTERACTIVE VERSION OF THE MAP

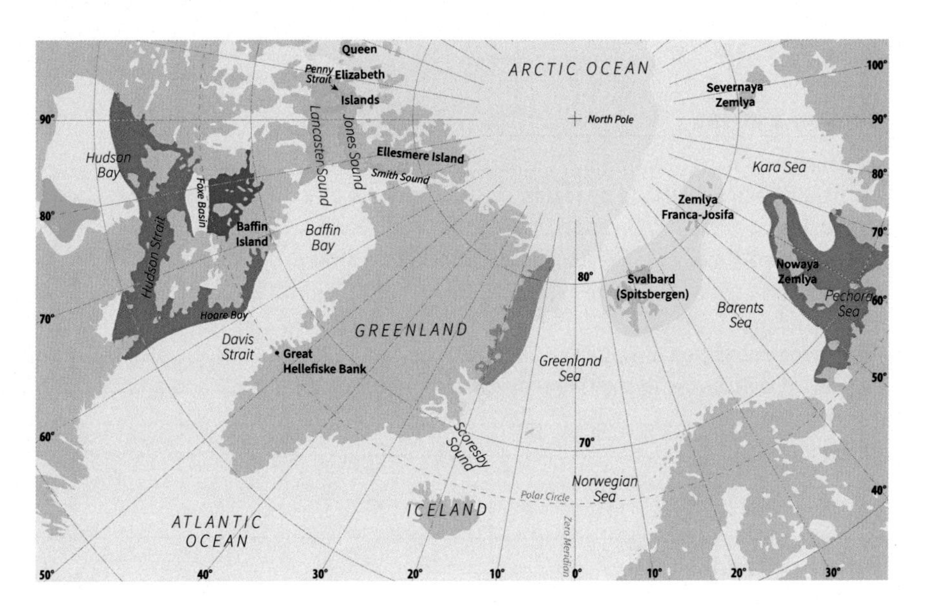

Sea ice concentration
2020 Mar 23
Credits: Copernicus Marine Service

FIGURE 4.1

Current stocks and the summer and winter distribution of the Atlantic walrus.

molluscs, their favoured prey (Born et al., 2003; Garde et al., 2018). Walruses are associated with pack ice, and for much of the year, ice floes are the main platform on which they rest between foraging trips (Photo 4.1). When suitable sea ice is unavailable, walruses haul-out on land, typically situated on low shores with easy access to water for feeding and to escape from predators or other disturbances (Born et al., 1995). Pacific walruses are known to haul-out in large herds of thousands of animals at terrestrial sites, whereas Atlantic walruses are mostly found in smaller groups. During summer, walruses spend the majority of their time in coastal areas where they have access to food and available haul-out sites on land or ice floes. In winter the sea ice covers much of the summer habitats, which restricts walruses to areas with open water off the coast, with access to feeding areas (Born et al., 1995). Walrus populations make seasonal migrations to and from their summer and winter grounds in response to the cyclic disappearance and formation of sea ice. In northern Baffin Bay, walruses undertake yearly migrations from wintering grounds in the eastern part of their range in North-West Greenland to the summering grounds in the western part of their range in the North-East Canadian Archipelago (mostly the east coast of Ellesmere Island). Even though North-West Greenland is largely covered in sea ice during winter and spring, walruses have access to open water and food owing to the occurrence of the North Water Polynya – a highly productive open-water area which acts as key habitat for many species of marine mammals and sea birds (Heide-Jørgensen et al., 2017). When the sea ice retreats in spring and early summer, the walruses

PHOTO 4.1

Walruses hauling out on ice close to Etah, North-West Greenland.

Credit: Carsten Egevang, the Greenland Institute of Natural Resources.

swim across the narrow Smith Sound to summer on the Canadian side. In autumn, when the sea ice starts to form again, most of the walruses return to their Greenlandic wintering grounds. Other populations, for example the population found on Svalbard and Franz Josef Land stay in the same area throughout the year.

The number of total extant Atlantic walruses is estimated to be around 35,000–40,000 (Laidre et al., 2015; Higdon and Stewart, 2018), at least half of the historic population size. Throughout their distribution walruses have been heavily depleted, but the exact extent of this depletion is not completely understood. Since 1952, walruses at Svalbard have been fully protected, and in 1956 walruses were protected in the western Russian Arctic. In Arctic Canada and Greenland, Inuit still hunt walruses for subsistence and in Canada a limited sport hunt involving Inuit outfitters is also allowed (Wiig et al., 2014). No commercial hunt of walruses is allowed in any jurisdiction (Wiig et al., 2014). In Canada, commercial hunting of walruses was banned in 1928 (Stewart et al., 2014b), and in Greenland, legislation involving hunting season and hunting methods has regulated the catch until 2006 where a quota system was introduced (Born et al., 1995; Wiig et al., 2014). 'Pre-exploitation' or 'original' population sizes of Atlantic walruses (the abundance of walruses per population before the onset of commercial harvesting by Europeans during the period from 1500 to 1900 CE) have been coarsely estimated in the different regions of walrus distribution across numerous studies (Witting and Born, 2014; Weslawski et al., 2000; Lydersen and Kovacs, 2014; Stewart et al., 2014b; Garde et al., 2018). Pre-exploitation population estimates are based on available catch statistics of the commercial hunt and use various population models each with a range of assumptions. The catch statistics originate from sources which report walrus catch numbers along with information on, for example, nationality and number of ships and hunters, how the hunt was performed and which weapons were used. As an example of this, Lønø (1972) has gathered catch statistics from Svalbard, Novaya Zemlya and Franz Josef Land. These data were subsequently used by Gjertz et al. (1998) alongside additional original sources on reported catches, to back-calculate the original population size of walruses in Franz Josef Land. It is a difficult task to determine pre-exploitation population sizes from fragmentary catch statistics given that walrus hunting in the European Arctic extended over a long period of time, and was conducted by hunters of many different nationalities (Gjertz et al., 1998). Most pre-exploitation estimates are therefore likely to be underestimates due to gaps in the catch records and under-reporting of catch numbers (Witting and Born, 2014). The total historic estimate of around 78,000 Atlantic walruses should therefore be considered a minimum and the historical population of Atlantic walruses was most likely larger than this.

The current geographic range of the Atlantic walrus is mapped in terms of populations and stocks (Fig. 4.1, Table 4.1). Populations comprise units in a biological and evolutionary sense. The North Atlantic Marine Mammal Commission (NAMMCO) defines **populations** as *'animals with a greater likelihood of breeding with others in the group than with others outside the group'*. In comparison, **stocks**

Table 4.1 Current abundance of Atlantic walrus stocks.

ABUNDANCE ESTIMATES FOR ATLANTIC WALRUS STOCKS

Population	Stock	Abundance estimate	Reference
High Arctic Canada / Baffin Bay	Baffin Bay	1279 (CI: 938-1744; winter)	NAMMCO (2018)
	Penny Strait-Lancaster Sound	727 (CI: 623-831; summer)	Stewart et al. (2014b)
	West Jones Sound	503 (CI: 473-534; summer)	Stewart et al. (2014b)
Central / Low Arctic Canada	Foxe Basin	14,093 (SE: 0.16; summer)	Hammill et al. (2016a)
	Hudson Bay-Davis Strait	7100 (CI: 4100-10,800; summer)	Hammill et al. (2016b)
	Hudson Bay-Davis Strait	1408 (CI: 922-2150; winter)	Heide-Jørgensen et al. (2013)
	South and East Hudson Bay	200 (CI: na-400)	Hammill et al. (2016b)
East Greenland	Northeast Greenland	279 (CI: 226-345; summer)	NAMMCO (2018)
Svalbard and Franz Josef Land	Svalbard and Franz Josef Land	3886 (CI: 3553-4262; summer)	Kovacs et al. (2014)
Southern Barents and Kara Sea	Pechora Sea	3943 (CI: 3605-4325; summer)	Lydersen et al. (2012)
	Pechora Sea	3117 (CI: 1571-5371; winter)	Semyonova et al. (2015)
	Kara Sea	1355 (tot l count; summer)	Semyonova et al. (2015)

are units defined for the purpose of hunting and setting of quotas, which typically (but not always) match the biological units, and are defined by NAMMCO as '*sub-divisions within a population that are largely defined by their interaction with humans, i.e. by hunting areas*' (Stewart et al., 2014a). For management purposes, stocks are treated as segregated units that have fidelity to particular locations; they form part of a precautionary approach aimed at reducing the risk of overexploitation. The name of a stock tends to refer to the areas where the animals are located during summer. The different walrus stocks have been differentiated using a variety of methods, including genetic analyses (e.g. Shafer et al., 2014; Andersen et al., 1998), satellite tracking using tags deployed on walruses (e.g. Dietz et al., 2014; Lydersen and Kovacs, 2014; Heide-Jørgensen et al., 2017) and by the use of isotope data and trace element profiles (Outridge et al., 2003).

Distribution and abundance by region

Canada

The western extent of the Atlantic walrus range is the eastern Canadian Arctic where walruses are found from James Bay in the southern part of Hudson Bay, to Smith Sound in the far north, and from the Canada–Greenland international boundary in Davis Strait, to the longitudinal centre of Canada (Fig. 4.1; Shafer et al., 2014). The current total population size of walruses in the eastern Canadian Arctic is estimated at approximately 23,000 walruses (Table 4.1; Stewart et al., 2014c,d,e; Laidre et al., 2015). Two populations of walruses occur in the eastern Canadian Arctic: the High Arctic population, and the Central/Low Arctic population. These two populations are further subdivided for management purposes in Canada into three and four stocks respectively. The three stocks in the High Arctic population include Baffin Bay, Penny Strait–Lancaster Sound and the West Jones Sound stock. The Central/Low Arctic population includes the northern Foxe Basin, the central Foxe Basin, the Hudson Bay–Davis Strait, and the South and East Hudson Bay stocks. For management purposes the two Foxe Basin stocks are combined. The Hudson Bay–Davis Strait stock is shared with Greenland, as is the Baffin Bay stock (Shafer et al., 2014; NAMMCO—North Atlantic Marine Mammal Commission, 2018). The Baffin Bay stock previously consisted of walruses from three distinct areas; Baffin Bay, West Jones Sound and Penny Strait–Lancaster Sound, but a recent satellite telemetry study discovered that walruses from Baffin Bay migrated to areas of West Jones Sound and Penny Strait–Lancaster Sound. Following on from this novel information, a single Baffin Bay stock has subsequently been agreed upon (Heide-Jørgensen et al., 2017; NAMMCO—North Atlantic Marine Mammal Commission, 2018). The estimated abundance of this stock (summer) is 1499 walruses [95% confidence interval (CI) 1077–2087], with the most recent estimate of 1279 walruses, based on an aerial survey covering only the wintering grounds in North Greenland in April 2018 (CI: 938–1744, Hansen and Heide-Jørgensen, 2018).

The current abundance estimate of 14,093 walruses (SE: 0.16) in Foxe Basin is based on an aerial survey of known haul-out sites conducted in September 2011 (Hammill et al., 2016a). An aerial survey in September 2014 of haul-out sites in the northwestern Hudson Bay and Hudson Strait yielded an abundance estimate of 7,100 walruses (CI: 4100-10,800) for the Hudson Bay–Davis Strait stock (Hammill et al., 2016b). This estimate must be considered a minimum since South-East Baffin Island, where walruses from this stock are also distributed, was not included in the survey. A proportion of the Hudson Bay–Davis Strait stock winters in West Greenland from January through April (Dietz et al., 2014) and walrus in this area were estimated to total 140 from animals (CI: 922-2150) an aerial survey in the winter 2012. A trend analysis indicates a slight increase in the proportion of walruses that winter in Greenland during the period from 1987 to 2017 (Heide-Jørgensen et al., 2013; NAMMCO—North Atlantic Marine Mammal

Commission, 2018). Total abundance of walruses for the small South and East Hudson Bay stock is 200 individuals (Hammill et al., 2016b).

While walruses today are confined to the Arctic waters of eastern Canada, they were previously found as far south as the east coast of Nova Scotia and to approximately 100°W. Commercial hunting from the late 1500s to the late 1700s CE extirpated the Atlantic walrus from southern Quebec and the Atlantic Canada; e.g. walruses from the Gulf of St. Lawrence were hunted to extinction in the 1700s and have not returned (Stewart et al., 2014b). Inuit and earlier peoples of eastern Canada have hunted walruses for centuries for subsistence, but there was no commercial hunt for walruses in the Canadian Arctic until around 1885 CE. As the bowhead hunt declined, whalers increasingly hunted other species, including walrus. In a period of nearly 200 years (from 1820 to 2010), Stewart et al. (2014b) estimated that a minimum catch of over 41,300 walruses was landed in the eastern Canadian Arctic. This estimate was based on walruses landed and on hunting products (e.g. hides, ivory) or descriptors (e.g. Peterhead boatloads), but did not account for underreporting, animals that were killed and lost, gaps in records and did not extrapolate to the whole whaling fleet. Therefore these landed catch estimates indicate the minimum numbers of walruses removed in the eastern Canadian Arctic (Stewart et al., 2014b). For the Baffin Bay stock a pre-exploitation population size of approximately 5000 walruses was estimated based on reported and estimated catch numbers by Greenlandic Inuit from 1900 to 1999 CE, which is approximately four times higher than the current abundance estimates for the stock (Table 4.1; Garde et al., 2018; Hansen and Heide-Jørgensen, 2018). The pre-exploitation abundance of walruses in parts of the eastern Canadian Arctic has been estimated to approximately 34,000 walruses. This estimate does not include Hudson Bay, Foxe Basin and the now extinct southeastern Canadian populations (Garde et al., 2018). Overall, as a consequence of two hundred years of heavy exploitation, walrus abundance for most extant stocks in Canada is still below their historic numbers.

In 1928, the commercial hunt for walruses was banned in Canada and hunting was restricted to Inuit. Today, hunting for walruses by an Inuk or land claims beneficiary in Canadian waters is allowed and nonbeneficiaries can hunt walruses with a licence under the Marine Mammal Regulations or Aboriginal Communal Fishing License Regulation. Sport hunts are also allowed and managed by limiting the number of licences approved annually (Stewart et al., 2014b; Wiig et al., 2014; Chapter 6, this volume).

Greenland

In Greenland, three separate populations or stocks of walruses are recognised based on information on distribution, migration and genetics (Witting and Born, 2005). These are the Baffin Bay stock in North-West Greenland and the Hudson Bay–Davis Strait stock in West Greenland (both shared with Canada) and the East Greenland stock (mainly restricted to North-East Greenland) (Fig. 4.1; Witting and

Born, 2014). The majority of Baffin Bay walruses winter in the eastern part of the North Water polynya along Greenlandic shores where Murchison Sound and Wolstenholme Fjord are important wintering grounds. Here walruses feed intensively on the shallow mussel banks in the area (Heide-Jørgensen et al., 2016). These walruses are an important component of the local hunt for marine mammals and are used as food for humans and dogs, as well as raw material for handicrafts.

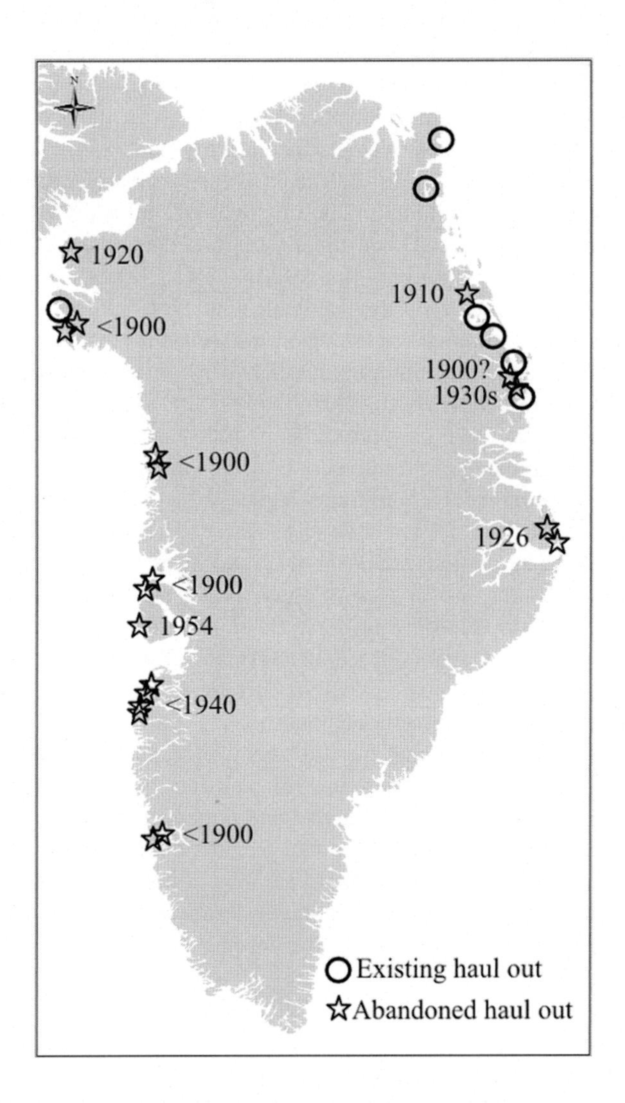

FIGURE 4.2

Map of existing and abandoned haul-out sites in Greenland between 1900 and 2020 CE. Circles represent current (since 2010) terrestrial haul-out sites and stars represent abandoned haul-out site. Numbers are the years when walruses were last seen.

Until the beginning of the 20th century, walruses hauled out on Littleton Island in Smith Sound and at the location of Etah on the Greenland coast opposite Littleton Island (Fig. 4.2; Hayes, 1867; Peary, 1917; Freuchen, 1921). According to Inuit local knowledge walruses also used to haul-out on land at two other yet undefined, places in the Avanersuaq area (North-West Greenland) (Vibe, 1950). None of these terrestrial haul-out sites have been used by walruses for more than a century and they were probably abandoned as a consequence of significant hunting pressure. Instead, walruses haul out on ice floes close to the abandoned terrestrial haul-out sites. Recent sightings (2018 and 2019) of approximately 30 walruses at Edderfugleøerne in Wolstenholme Fjord do however, indicate that walruses have the capacity to shift from hauling out on ice to new terrestrial sites (Photo 4.2). In Greenland, walruses are fully protected from hunting and other anthropogenic disturbances at terrestrial haul-out sites.

A small proportion of the Hudson Bay–Davis Strait stock, also called the West Greenland–South-East Baffin Island "component", is present from autumn to spring off the central West Greenland coast at Great Hellefiske Bank (a highly productive shallow water area with rich mussel banks), and is estimated to be stable at around 1400 walruses (CI: 922–2150; Heide-Jørgensen et al., 2013). In May, the walruses begin their annual migration across Davis Strait to their destination off the coast of southeastern Baffin Island and Hudson Strait, where they spend the summer months

PHOTO 4.2

Walruses hauling out at Edderfugleøerne, Wolstenholme fjord, North-West Greenland, 2018.

Credit: Kasim Wirk, employee at Thule Airbase.

(Born et al., 1994, 1995). In autumn, they return to the same area at Great Hellefiske Bank. The summer range of the West Greenland–South-East Baffin Island "component" has declined during the past 120 years. Before the 1940s, walruses used several haul-out sites on land along the West Greenland coast during summer (NAMMCO—North Atlantic Marine Mammal Commission, 2018), but all of these terrestrial sites were abandoned, possibly due to hunting in the beginning of the 20th century, and to this date, walruses have not returned (Fig. 4.2; Vibe, 1967; Born et al., 1995).

In East Greenland, walruses are found from Scoresby Sound and to the north-eastern part of Greenland (from 73°N to 80°N; Fig. 4.1). There is only limited exchange between walruses from the East Greenland population and neighbouring populations in West and North-West Greenland, and Svalbard–Franz Josef Land (Witting and Born, 2005). A limited number of vagrants, mostly single adult males, can sometimes be encountered south of 73°N.

A local population of walruses was previously found on the northern side of the mouth to Scoresby Sound, but after the town of Ittoqqortoormiit was established in 1925 the local hunt caused the depletion of this population (Born et al., 1997). The current population of walruses in North-East Greenland is small and scattered over a huge area, with walruses found year-round in the North-East Water polynya at Nordostrundingen (NAMMCO—North Atlantic Marine Mammal Commission, 2018). During early summer the solid land-fast ice breaks up, allowing males, and in recent years also females with calves, to migrate from their northern and offshore wintering grounds, to inshore foraging areas close to terrestrial haul-out sites (Born et al., 2005; NAMMCO—North Atlantic Marine Mammal Commission, 2018). The abundance estimate of 279 walruses (CI: 226–345) for the East Greenland stock was based on aerial surveys of walruses summering grounds and the North-East water. The survey was conducted in August 2017 and found 102 walruses on the terrestrial haul-out sites between 73°N and 80°N, and 177 in the North-East Water (NAMMCO—North Atlantic Marine Mammal Commission, 2018). It has been estimated that in the year 1888, which is the year prior to the first historical catches by European sealers, the East Greenland population was 1200 walruses (CI: 810–1300, Witting and Born, 2014).

The three walrus populations in Greenland have been subject to exploitation for centuries by the Norse, European whalers and sealers, and Inuit in Greenland and Canada (Wiig et al., 2014). Since the introduction of fire-arms and motorised vessels from the beginning of the 20th century, Inuit hunted walruses with increasing effort and as a consequence, all three stocks were significantly depleted (Born et al., 1994; Wiig et al., 2014). Quotas for walrus hunting in Greenland (based on recommendations from NAMMCO) were implemented in 2006 and the stocks are now managed sustainably (NAMMCO—North Atlantic Marine Mammal Commission, 2018).

Norway

Walruses are found in the Svalbard archipelago, and studies using satellite tracking and genetics show that they belong to the same population that is found in the

Franz Josef Land archipelago in Russia (Andersen et al., 1998; Wiig et al., 1996). Walruses reside year-round in Svalbard waters, however, most of the males summer around Svalbard and most of the females and calves remain closer to the Franz Josef Land archipelago in the northeastern part of their range. In recent years, more females with calves have been sighted around Svalbard, as animals are expanding further into their former range (Kovacs et al., 2014). Walruses on Svalbard were hunted to near extinction over centuries of unregulated harvest. The hunt for these animals started in the early 1600s CE and lasted until 1952, when walruses in Svalbard were given total protection. In the period of 1954–82, the summering stock was approximately 100 animals and the population has increased since then (Lydersen and Kovacs, 2014). Even though the population is increasing, some areas previously occupied by walruses are still not reestablished. An example is Bjørnøya, the southernmost Island in the Svalbard archipelago, where walruses were exterminated and where sightings are still rare (Kovacs et al., 2014). The most recent abundance estimate (from 2006) for the Svalbard–Franz Josef Land population is 3886 walruses (CI: 3553–4262; Kovacs et al., 2014). Walrus population size prior to exploitation has been estimated to be around 6000–12,500 animals for Franz Josef Land (Gjertz et al., 1998) and 25,000 walruses for the Svalbard archipelago (Weslawski et al., 2000).

Russia

In addition to the shared Svalbard–Franz Josef Land population, another walrus population is found around Novaya Zemlya in the Barents, Kara and Pechora Seas in Russia (the Barents and Kara Seas population; Fig. 4.1; Wiig et al., 2014). A genetic study confirmed that walruses inhabiting the southern Barents Region showed limited but significant genetic distinction from walruses in the Svalbard–Franz Josef Land population (Andersen et al., 2017). Walruses in the Laptev Sea in eastern Russia were once considered a separate subspecies however, but genetic and morphometric analyses documented that the Laptev Sea walruses belong to the Pacific subspecies (Lindqvist et al., 2009). A study conducted in 2011–2014 showed that walruses from the Barents and Kara Seas population spent most of the year, possibly the whole year, in the South-East Barents Sea/Pechora Sea (Semyonova et al., 2015). A recent study confirmed that male walruses tagged with satellite transmitters during summer (average duration of tags was 47 days with a max. of 155 days) on the western coast of Vaygach Island in the Pechora Sea stayed in the eastern part of the Pechora Sea between Pechora Bay, the southern tip of Novaya Zemlya and the western coast of Vaygach Island. Some animals did however, move to areas up to 1500 km away from the tagging location, including the northern part of Novaya Zemlya, the central Kara Sea and the northern part of the Severnaya Zemlya Archipelago (Semenova et al., 2019). In the Pechora Sea, during the time of an aerial survey in the summer of 2011, walruses were found hauled out across three major terrestrial sites (one group on Vaygach Island and two groups on Matveyev Island)

(Lydersen et al., 2012). The abundance estimate from this dedicated survey was 3943 walruses (CI: 3605–4325), and only males were found. All sex and age groups of walruses were however, spotted during another aerial survey in spring 2014 in the Pechora Sea, where the abundance was estimated to 3117 walruses (CI: 1571–5371, Semyonova et al., 2015). In autumn, walruses are found in the Pechora Sea around Amderma, Vaygach, Kolguev, Dolgy and Matveyev islands, but it is not yet fully understood where females and their calves spend summer and autumn (Semyonova et al., 2015). A summer survey in 2013, counting walruses at the terrestrial haul-out sites in the Kara Sea, found both sexes and calves, hauled out on several islands (to the north and northeast of Severny Island), with an estimated total abundance of 1355 walruses (Semyonova et al., 2015).

In Russia, walruses were subject to extensive hunting in the past. The commercial hunt for walruses at Franz Josef Land, and in the Kara and Pechora Seas started during the late 1800s CE (Wiig et al., 2014; Lønø, 1972). From this time up until the late 1920s, Norwegian sealers harvested a considerable number of walruses at Franz Josef Land. In 1956, walruses were protected in the western Russian Arctic, except for a subsistence harvest by locals and expeditions (Wiig et al., 2014). The estimated number of walruses in the Pechora and Kara Sea prior to exploitation is approximately 8000 walruses based upon catch records of the Norwegian hunt in the Pechora and Kara Seas and at Novaya Zemlya from 1880 to 1899 CE (Lønø, 1972; Garde et al., 2018).

Abundance estimation of walrus

Walrus abundance in management areas is often estimated using visual or photographic aerial surveys. Visual surveys use human observers to count and record walruses while flying over haul-out sites or along predestinated transect lines (Heide-Jørgensen et al., 2016). Photographic surveys regularly take photos while flying, with the images subsequently analysed for the number of walruses that were flown over (Lydersen et al., 2012). Recently, satellite imagery and camera traps have been shown to be feasible as tools for monitoring known haul-out sites (Box. 4.1; Semyonova et al., 2015).

When count surveys are repeated at regular intervals, of usually every 5–10 years, trends in abundance can be projected. Depending on the season and area, various small fixed-wing airplanes or helicopters are used for the aerial surveys. Walruses are either counted as an absolute count at the terrestrial haul-out sites in summer, or at sea in any season (counted in the water or when hauled out on ice floes). When counted at sea, a method called distance sampling is used. The survey area is divided into smaller areas (strata), transect lines are placed evenly across the strata and the transects are flown when weather conditions are favourable (Buckland et al., 2001). Each observer concentrates on the area ahead of, and beside, the plane. When a walrus or a group of walruses is spotted, the

Box 4.1 Walrus monitoring.

Present trends in walrus populations are assessed by modelling population trajectories for the different stocks. This is of particular importance in areas where hunting is still undertaken (such as Canada and Greenland) but also in areas with depleted stocks (such as Svalbard, Norway). Monitoring is typically conducted on individual stocks and includes: robust abundance estimates, tagging of walruses with satellite transmitters to establish year long distribution patterns, catch histories of the populations (if available), and collection of tissue samples for life history and genetic studies. Aerial surveys are used in all areas for updating abundance estimates. Tagging of walruses with satellite transmitters is the preferred method to obtain information on stock structure, migration routes and distribution at summering and wintering grounds.

observation is recorded along with additional information (for example group size and number of calves). Aerial surveys for multiple species at a time can also be conducted; requiring observers experienced in species recognition. When an animal or group is spotted, the observer takes a measurement of the perpendicular distance (distance at 90 degrees to the direction of travel) to the animal or group. These distances are later used in statistical analyses to estimate the width of the strip from the track line where animals are certain to be seen (Heide-Jørgensen et al., 2016). Some surveys also use a double observer set-up to account for perception bias; the proportion of walruses that were available but not detected by either the front or rear observer. The density of animals in the entire survey area is calculated by extrapolation of the density of animals in the estimated strip width in each surveyed stratum. Not all walruses are however, available for the observers to spot. For example when surveying terrestrial haul-out sites walruses could be foraging at sea at the time the plane is flying over the site. Also, when flying along transects over water walruses are often submerged below the sea surface, and thus cannot be seen by observers. To account for this availability bias, the abundance estimates are corrected with data estimated using dive data from satellite transmitters deployed on walruses that indicate the proportion of time walruses spend at the terrestrial haul-out sites or at the surface when at sea. A measure of the uncertainty or variance (usually a 95% confidence limit) of the total abundance estimate is predicted including all the mentioned biases.

Tagging of walruses with satellite transmitters

Information on animal movement is essential for understanding stock structure including: migration routes and summer/winter distributions, putative overlap of shared stocks subject to hunting, and area use, for example preferred sites for foraging, breeding and rearing the young. These are central components used in forming management advice of sustainable hunting levels of utilised species, as well as for protection of species-specific key habitats. Satellite telemetry is an important tool to obtain this information on stock structure (Dietz et al., 2014; Heide-Jørgensen et al., 2017; Garde et al., 2018).

Attaching a satellite transmitter on an animal as large as a walrus is not a straightforward process given a walrus' long tusks and the number of equally large fellow herd members that may surround the individual animal, all while floating on an ice pan or swimming in ice-cold water (Stewart et al., 2014a). Different types of satellite tags and deployment methods can be used and the procedure can be done with or without the use of chemical immobilisation. Chemical immobilisation may be necessary in studies that require contact with the animals, for example attaching instruments or physiology experiments. Immobilisation is not without risk for both the animal and the people performing it. Acquarone et al. (2014) described 69 immobilisations with etorphine HCl by remote darting of 41 adult walruses on both land and ice floes, from North-West and North-East Greenland and Franz Josef Land in the period 1989–2001. Full immobilisation was achieved in 58 cases (84%), the animals were insufficiently restrained in 6 cases (9%) and 5 animals died (7%). Tagging of walruses using remote deployments can also be done without previous immobilisation, which has proven successful in several studies. Dietz et al. (2014) used three different types of tags; three were implanted into the skin of unrestrained walruses while the animals

PHOTO 4.3

Satellite tagging of walruses from the Baffin Bay stock at Etah, North-West Greenland where a local hunter use a harpoon to remotely deploy the tag.

Credit: Carsten Egevang, the Greenland Institute of Natural Resources.

were either hauled out on pack ice or swimming, and one type was mounted on the tusk of immobilised animals. The first three were remotely deployed using a modified airgun, arrows and crossbows, traditional hunting harpoons or a CO_2-powered rifle. The tusk tags were deployed on immobilised walruses during the tagging operations. Once deployed, tag duration varied considerably lasting from 7 to 128 days. More recent studies have reported maximum durations of 278 days for remotely deployed tags (Photo 4.3; Heide-Jørgensen et al., 2017) and more than a year for tusk-mounted tags (Lydersen and Kovacs, 2014).

Perspectives

The Atlantic walrus has experienced major demographic shifts owing to human exploitation, resulting in local extinctions, range reduction, and abandonment of terrestrial haul-out sites (Box. 4.2). Walrus research from the last couple of decades has demonstrated that in historic, as well as modern times, humans have heavily influenced walrus populations in terms of changed migration routes and population size which might also have affected population structure and the distribution of sex-segregated groups (Gotfredsen et al., 2018).

The lack of a comprehensive catch history for many areas in the original range of Atlantic walruses makes it difficult to assess the status of current populations, and to place population changes in a historical context. An increasing trend in population size may not, for example, provide evidence of significant recovery if the population is still a small fraction of its former size. One approach to establish appropriate baselines is to collect data on catches over time that can then be used to estimate the pre-exploitation population sizes required to have supported the levels of removal. Historical catch data are needed to model population trajectories, assess population status, set recovery targets, and inform management of sustainable subsistence and sports hunting (Stewart et al., 2014b).

Box 4.2 Local extinction of walruses following human arrival.

Walruses are particularly susceptible to overexploitation given their propensity to return to the same feeding grounds and to haul-out in large groups (Kovacs et al., 2014). The historical distribution of walruses used to cover a larger geographical area than at present, and included the Quebec region of Canada, mainland Norway and Iceland (Stewart et al., 2014b; Keighley et al., 2019; Barrett et al., 2020). No walrus populations inhabit Icelandic waters today, but occasionally individual animals visit the island. The zooarchaeological record is, however, rich in walrus bone and ivory finds from early Norse (Viking Age) contexts concentrated in western and northwestern Iceland, suggesting the past existence of walruses in Iceland (Frei et al., 2015). Radiocarbon dating and genetic analyses of these finds show a continuous presence of a separate Icelandic walrus population up until the arrival of the Norse (870 CE), who likely hunted the population to extinction (Keighley et al., 2019). It has even been proposed that the trade of walrus ivory was at least partly responsible for foundation of the Norse settlements in Greenland in the 980s CE, after Iceland's walrus population was hunted to extinction (Keighley et al., 2019).

Walruses require ice or land for hauling out and their foraging habitat may extend 40–100 km from the haul-out site (Hammill et al., 2016b). This requirement restricts habitat options and walruses therefore show strong site fidelity to established haul-out sites (DFO, 2019). Understanding walrus area utilisation is essential for assessing how animals behave when exposed to environmental stressors, for example the importance of terrestrial haul-out sites following a decline in suitable ice habitat. When hauled out on land, all populations of walruses seem especially sensitive to human activities such as hunting and boat traffic (Øren et al., 2018). The abandonment of terrestrial haul-out sites in Greenland was mainly due to hunting. There are still no clear signs of walruses returning to any of the traditional haul-out sites even though there has been an increased focus on avoiding disturbance of walruses on land. Abandoned haul-out sites provide important lessons about the sensitivity of walruses, and emphasise the need for avoiding human activities in areas of importance to walrus. Avoiding human activities in the proximity of abandoned haul-out sites, might facilitate the return of walruses in the future.

References

Acquarone, M., Born, E.W., Griffiths, D., Knutsen, L.Ø., Wiig, Ø., Gjertz, I., 2014. Evaluation of etorphine reversed by diprenorphine for the immobilisation of free-ranging Atlantic walrus (*Odobenus rosmarus rosmarus* L.). NAMMCO Scientific Publications 9, 345–360.

Andersen, L.W., Born, E.W., Gjertz, I., Wiig, Ø., Holm, L.E., Bendixen, C., 1998. Population structure and gene flow of the Atlantic walrus (*Odobenus rosmarus rosmarus*) in the eastern Atlantic Arctic based on mitochondrial DNA and microsatellite variation. Molecular Ecology 7 (10), 1323–1336.

Andersen, L.W., Jacobsen, M.W., Lydersen, C., Semenova, V.S., Boltunov, A.N., Born, E.W., 2017. Walruses (*Odobenus rosmarus rosmarus*) in the Pechora Sea in the context of contemporary population structure of Northeast Atlantic walruses. Biological Journal of the Linnean Society 122 (4), 897–915.

Barrett, J.H., Boessen, S., Knealea, C.J., O'Connell, T.C., Star, B., 2020. Ecological globalisation, serial depletion and the medieval trade of walrus rostra. Quaternary Science Reviews 229, 106–122. Available from: https://doi.org/10.1016/j.quascirev.2019.106122.

Born, E.W., Heide-Jørgensen, M.P., Davis, R.A., 1994. The Atlantic walrus (*Odobenus rosmarus rosmarus*) in West Greenland. Medd. Grønland. Biosci. 40, 1–33.

Born, E.W., Gjertz, I., Reeves, R.R., 1995. Population assessment of Atlantic walrus. Nor. Polarinst. Medd. 138, 1–100.

Born, E.W., Dietz, R., Heide-Jørgensen, M.P., Knutsen, L.Ø., 1997. Historical and present distribution of Atlantic walruses (*Odobenus rosmarus rosmarus* L.) in Eastern Greenland. Medd. Grønland. Biosci. 46, 73.

Born, E.W., Rysgaard, S., Ehlmé, G., Sejr, M., Acquarone, M., Levermann, N., 2003. Underwater observations of foraging free-living Atlantic walruses (*Odobenus rosmarus rosmarus*) and estimates of their food consumption. Polar Biology 26, 348–357.

Born, E.W., Acquarone, M., Knutsen, L.Ø., Toudal, L., 2005. Homing behaviour in an Atlantic walrus (*Odobenus rosmarus rosmarus*). Aquatic Mammals 31 (1), 23–33.

Born, E.W., Stefansson, E., Mikkelsen, B., Laidre, K.L., Andersen, L.W., Rigét, F., et al., 2014. A note on a walrus' European odyssey. NAMMCO Scientific Publications 9, 75–92.

Buckland, S.T., Anderson, D.R., Burnham, K.P., Laake, J.L., Borchers, D.L., Thomas, L., 2001. Introduction to Distance Sampling: Estimating Abundance of Biological Populations. Oxford University Press, Oxford.

DFO. Mitigation buffer zones for Atlantic walrus (*Odobenus rosmarus rosmarus*) in the Nunavut settlement area. DFO Canadian Science Advisory Secretariat Science Response 2018/055, 2019, Fisheries and Oceans Canada, Ottawa, 27 pp.

Dietz, R., Born, E.W., Stewart, R.E.A., Heide-Jørgensen, M.P., Stern, H., Rigét, F., 2014. Movements of walruses (*Odobenus rosmarus*) between Central West Greenland and Southeast Baffin Island, 2005–2008. NAMMCO Scientific Publications 9, 53–74.

Fay, F.H., 1985. *Odobenus rosmarus*. Mammalian species. American Society of Mammalogists 238, 1–7.

Frei, K.M., Coutu, A.N., Smiarowski, K., Harrison, R., Madsen, C.K., Arneborg, J., 2015. Was it for walrus? Viking Age settlement and medieval walrus ivory trade in Iceland and Greenland. World Archaeology 47, 439–466.

Freuchen, P., 1921. Om hvalrossens forekomst og vandringer ved Grønlands Vestkyst [On the distribution and migration of walruses along the western coast of Greenland]. Videnskabelige Meddelelser Dansk Naturhistorisk Forening 72, 237–249 (in Danish).

Garde, E., Jung-Madsen, S., Ditlevsen, S., Hansen, R.G., Zinglersen, K.B., Heide-Jørgensen, M.P., 2018. Diving behavior of the Atlantic walrus in high arctic Greenland and Canada. Journal of Experimental Marine Biology and Ecology 500, 89–99.

Gjertz, I., Wiig, Ø., Øritsland, N.A., 1998. Back calculation of original population size for walruses *Odobenus rosmarus* in Franz Josef Land. Wildlife Biology 4, 223–230.

Gotfredsen, A.B., Appelt, M., Hastrup, K.B., 2018. Walrus history around the North Water: human-animal relations in a long-term perspective. Ambio 47 (Suppl. 2), 193–212. Available from: https://doi.org/10.1007/s13280-018-1027-x.

Hammill, M.O., Blanchfield, P., Higdon, J.W., Stewart, D.B., 2016a. Estimating abundance and total allowable removals for walrus in Foxe Basin. DFO Canadian Science Advisory Secretariat Research Document 2016/014: v +. Fisheries and Oceans Canada, Ottawa, 37 pp.

Hammill, M.O., Mosnier, A., Gosselin, J.F., Higdon, J.W., Stewart, D.B., Doniol-Valcroze, T., et al., 2016b. Estimating abundance and total allowable removals for walrus in the Hudson Bay-Davis Strait and south and east Hudson Bay stocks during September 2014. DFO Canadian Science Advisory Secretariat Research Document 2016/036: v +. Fisheries and Oceans Canada, Ottawa, 37 pp.

Hansen, R.G., Heide-Jørgensen, M.P., 2018. Abundance of walrus in the North Water 2018. In: NAMMCO SC/25/14-WWG 06.

Hayes, E.E., 1867. The Open Polar Sea. A Narrative of a Voyage of Discovery Towards the North Pole, in the Schooner "United States". Sampson Low, Son and Marson, London.

Heide-Jørgensen, M.P., Laidre, K.L., Fossette, S., Rasmussen, M., Nielsen, N.H., Hansen, R.G., 2013. Abundance of walruses in Eastern Baffin Bay and Davis Strait. NAMMCO Scientific Publications 9, 159–171.

Heide-Jørgensen, M.P., Sinding, M.S., Nielsen, N.H., Rosing-Asvid, A., Hansen, R.G., 2016. Large numbers of marine mammals winter in the North Water polynya. Polar Biology 39 (9), 1605–1614.

Heide-Jørgensen, M.P., Flora, J., Andersen, A.O., Stewart, R.E.A., Nielsen, N.H., Hansen, R.G., 2017. Walrus movements in Smith sound: a Canada–Greenland shared stock. Arctic 70 (3), 308–318.

Higdon, J.W., Stewart, D.B., 2018. State of Circumpolar Walrus (*Odobenus rosmarus*) Populations. Prepared by Higdon Wildlife Consulting and Arctic Biological Consultants, Winnipeg, MB for WWF Arctic Programme, Ottawa, p. 100.

IUCN Redlist – walrus Odobenus rosmarus. https://www.iucnredlist.org/species/15106/45228501 (accessed 06 June 2019).

Keighley, X., Pálsson, S., Einarsson, B.F., Petersen, A., Fernández-Coll, M., Jordan, P., et al., 2019. Disappearance of Icelandic walruses coincided with Norse settlement. Molecular Biology and Evolution 36 (12), 2656–2667. Available from: https://doi.org/10.1093/molbev/msz196.

Kovacs, K.M., Aars, J., Lydersen, C., 2014. Walruses recovering after 60 + years of protection in Svalbard, Norway. Polar Research 33, 1–5.

Laidre, K.L., Stern, H., Kovacs, K.M., Lowry, L., Moore, S.E., Regehr, E.V., 2015. Arctic marine mammal population status, sea ice habitat loss, and conservation recommendations for the 21st century. Conservation Biology 29, 724–737.

Lindqvist, C., Bachmann, L., Andersen, L.W., Born, E.W., Arnason, U., Kovacs, K.M., 2009. The Laptev Sea walrus *Odobenus rosmaruslaptevi*: an enigma revisited. Zoologica Scripta 38 (2), 113–224.

Lønø, O., 1972. The catch of walrus (*Odobenus rosmarus*) in the areas of Svalbard, Novoja Zemlja, and Franz Josef Land. Norsk Polarinstitutt, Årbok 1970, 199–212.

Lydersen, C., Kovacs, K., 2014. Walrus *Odobenus rosmarus* research in Svalbard, Norway, 2000-2010. NAMMCO Scientific Publications 9, 7–12.

Lydersen, C., Chernook, V.I., Glazov, D.M., Trukhanova, I.S., Kovacs, K.M., 2012. Aerial survey of Atlantic walruses (*Odobenus rosmarus rosmarus*) in the Pechora Sea, August 2011. Polar Biology 35 (10), 1555–1562.

NAMMCO—North Atlantic Marine Mammal Commission, 2018. Report of the NAMMCO Scientific Working Group on Walrus, October 2018, pp. 22. https://nammco.no/topics/sc-working-group-reports.

Øren, K., Kovacs, K., Yoccoz, N.G., Lydersen, C., 2018. Assessing site-use and sources of disturbance at walrus haul-outs using monitoring cameras. Polar Biology 41, 1737–1750.

Outridge, P.M., Davis, W.J., Stewart, R.E.A., Born, E.W., 2003. Investigation of the stock structure of Atlantic walrus (*Odobenus rosmarus rosmarus*) in Canada and Greenland using dental Pb isotopes derived from local geochemical environments. Arctic 56 (1), 82–90.

Peary, R.E., 1917. Secrets of Polar Travel. The Century Co, New York.

Semyonova, V.S., Boltunov, A.N., Nikiforov, V.V., 2015. Studying and Preserving the Atlantic Walrus in the South-East Barents Sea and Adjacent Areas of the Kara Sea. 2011–2014 Study Results. Murmansk, World Wildlife Fund (WWF), p. 82.

Semenova, V., Boltunov, A., Nikiforov, V., 2019. Key habitats and movement patterns of Pechora Sea walruses studied using satellite telemetry. Polar Biology 42, 1763–1774.

Shafer, A.B.A., Davis, C.S., Coltman, D.W., Stewart, R.E.A., 2014. Microsatellite assessment of walrus (*Odobenus rosmarus rosmarus*) stocks in Canada. NAMMCO Scientific Publications 9, 15–31.

Stewart, R.E.A., Kovacs, K., Acquarone, M., 2014a. Introduction – Walrus of the North Atlantic. NAMMCO Scientific Publications 9, 7–12.

Stewart, D.B., Higdon, J.W., Reeves, R.R., Stewart, R.E.A., 2014b. A catch history for Atlantic walruses (*Odobenus rosmarus rosmarus*) in the eastern Canadian Arctic. NAMMCO Scientific Publications 9, 219–313.

Stewart, R.E.A., Born, E.W., Dunn, J.B., Koski, W.R., Ryan, A.K., 2014c. Use of Multiple Methods to Estimate Walrus (*Odobenus rosmarus rosmarus*) Abundance in the Penny Strait-Lancaster Sound and West Jones Sound Stocks, Canada. NAMMCO Scientific Publications 9, 95–122.

Stewart, R.E.A., Born, E.W., Dietz, R., Heide-Jørgensen, M.P., Rigét, F.F., Laidre, K., 2014d. Abundance of Atlantic walrus in Western Nares Strait, Baffin Bay Stock, during summer. NAMMCO Scientific Publications 9, 123–140.

Stewart, R.E.A., Born, E.W., Dietz, R., Ryan, A.K., 2014e. Estimates of minimum population size for walrus near Southeast Baffin Island, Nunavut. NAMMCO Scientific Publications 9, 141–157.

Vibe, C., 1950. The marine mammals and the marine fauna in the Thule district (North West Greenland) with observations of the ice conditions in the 1939–41. Meddelelser om Grønland 150.

Vibe, C., 1967. Arctic animals in relation to climatic fluctuations. Meddelelser om Grønland 170 (5), 227.

Weslawski, J.M., Hacquebord, L., Stempniewicz, L., Malinga, M., 2000. Greenland whales and walruses in the Svalbard food web before and after exploitation. Oceanologia 42 (1), 37–56.

Wiig, Ø., Gjertz, I., Griffiths, D., 1996. Migration of walruses (*Odobenus rosmarus*) in the Svalbard and Franz Josef Land area. Journal of Zoology, London 238 (4), 769–784.

Wiig, Ø., Born, E.W., Stewart, R.E.A., 2014. Management of Atlantic walrus (*Odobenus rosmarus rosmarus*) in the arctic Atlantic. NAMMCO Scientific Publications 9, 315–342.

Witting, L., Born, E.W., 2005. An assessment of Greenland walrus populations. International Council for the Exploration of the Sea (ICES) Journal of Marine Science 62, 266–284.

Witting, L.W., Born, E.W., 2014. Population dynamics of walruses in Greenland. NAMMCO Scientific Publications 9, 191–218.

SECTION

Walruses and Indigenous peoples

CHAPTER

Pre-Inuit walrus use in Arctic Canada and Greenland, c.2500 BCE to 1250 CE

5

Christyann M. Darwent[1] and Genevieve M. LeMoine[2]

[1]*Department of Anthropology, University of California, Davis, Calirfornia, United States*

[2]*Peary–MacMillan Arctic Museum, Bowdoin College, Brunswick, Maine, United States*

Chapter Outline

Introduction

This chapter focuses on Pre-Inuit (also known as Paleo-Inuit or formerly, as Paleoeskimo) Indigenous populations of the Eastern Arctic[1] of Canada and Greenland. Although all Inuit and Pre-Inuit populations originated in the Bering Strait region of eastern Siberia and have a distant common ancestor, they are genetically and culturally distinct from one another (Friesen and Mason, 2016; Raghavan et al., 2014). A technological complex known as the Arctic Small Tool tradition (ASTt) was the first specialized maritime-adapted culture to inhabit northern Alaska and the first to settle Arctic Canada and Greenland. The Eastern Arctic, although deglaciated by at least 8500 years ago, was one of the last places

[1] North American Arctic territories east of Alaska (Maxwell 1985).

The Atlantic Walrus. DOI: https://doi.org/10.1016/B978-0-12-817430-2.00007-8

Table 5.1 Pre-Inuit and Inuit chronology for the Eastern Arctic of Canada and Greenland.

	Low Arctic Canada	West, East and South Greenland	High Arctic Canada	North Greenland
1300 CE	Thule	Thule	Thule	Thule
800 CE	Late Dorset		Late Dorset	Late Dorset
1 CE	Middle Dorset			
800 BCE	Early Dorset	Greenlandic Dorset	Early Dorset	Greenlandic Dorset
2000 BCE	Middle/Late Pre-Dorset	Saqqaq	Middle/Late Pre-Dorset	
2500 BCE	Early Pre-Dorset		Early Pre-Dorset/ Independence I	Independence I

Note: *Grey bars indicate an absence of archaeological evidence for human occupation.*
Source: *Adapted from Friesen, T.M., Mason, O.K., 2016. Introduction: archaeology of the North American Arctic. In: Friesen, T.M., Mason, O.K. (Eds.), The Oxford Handbook of the Prehistoric Arctic. Oxford University Press, Oxford, p.12.*

on earth to be inhabited by humans. These initial migrants brought along their technological knowledge of seal hunting and coastal resources, production of ground and polished stone tools, the bow and arrow for hunting terrestrial resources (Tremayne and Rasic, 2016), and dogs as hunting companions (although their use of dogs was sporadic and limited in scope [see Ameen et al., 2019]).

The term Pre-Inuit encompasses two major archaeological traditions: (1) Pre-Dorset, beginning c.2500 BCE; and (2) Dorset, beginning around 800 BCE (Table 5.1). The exact timing of the transition between the two traditions, and the specific technological, housing, hunting and settlement systems used by each, varies somewhat by geographic region (Jensen, 2016; Ryan, 2016). The later Dorset peoples occupied some areas continuously, and other discontinuously until about 750 years ago (c.1250 CE) when the ancestors of modern Inuit moved out of the Bering Strait and into the Eastern Arctic, replacing Pre-Inuit in a poorly understood cultural transition (Appelt, 2016; Friesen, 2020; Park, 2016).

This chapter follows the origin and development of Atlantic walrus (*Odobenus rosmarus rosmarus*) hunting and the use of walrus ivory for tool production among Pre-Inuit populations. We focus on two geographic zones: (1) the *High Arctic*, defined here as the Canadian Arctic Islands north of Lancaster Sound (74°N) and Greenland north of Melville Bay; and (2) the *Low Arctic*, including Baffin Island, the Canadian Arctic, Nunavik (Arctic Québec), Nunatsiavut (coastal Labrador) and south of Melville Bay in Greenland. Walrus hunting by later Inuit inhabitants is the subject of Chapter 6, this volume.

Characteristics of walrus ivory

Walrus tusks are enlarged maxillary canines, the size of which has dramatically altered the shape of their crania, making walruses a highly-distinctive species within the suborder Pinnipedia. Permanent maxillary canines erupt when walruses are about 4 months of age; these canines are initially covered by a cap of enamel, but this is worn away by nearly 2 years of age (King, 1964). Tusks develop in both males and females, although they are longer, more robust and more curved in males (male tusks can weigh up to 5 kg and extend as much as 1 m in length). For Pre-Inuit inhabitants of the Eastern Arctic, these tusks were a valuable osseous raw material source distinct from both bone and antler. (In some animals, when teeth become over-developed, they are referred to as 'tusks' and the material as 'ivory' [King, 1964]).

Tusks are made up of a core of dentine, which is found in all mammalian teeth, and a thin outer layer of cementum (Fig. 5.1). Dentine in walrus ivory differs from other mammalian canines in that it is laid down in two distinct layers. First, or primary, dentine is the outer coating and is homogenous in structure. From the perspective of tool manufacture, this homogenous dentine will exhibit a conchoidal fracture pattern similar to flint or obsidian. Ivory flakes have many of the same features found on stone flakes, including bulbs of percussion and hackle marks (LeMoine and Darwent, 1998:77). Second, or secondary, dentine is irregularly deposited and forms a marbled or mottled structure in the former pulp cavity of the tusk (MacGregor, 1985:18; Penniman, 1952:25). The secondary dentine is translucent and crystalline in appearance and easily visible to the naked eye, which makes it possible to distinguish walrus ivory from that of other species (e.g., elephant and narwhal).

Pre-Dorset walrus use

High Arctic

Initial inhabitants of the High Arctic were highly mobile, with archaeological excavations revealing faunal assemblages reflective of a subsistence economy more or less proportional to locally-available taxa, both marine and terrestrial. These initial inhabitants are often divided into two groups: Early Pre-Dorset and Independence I (c.2500–1900 BCE). The distinction is primarily stylistic and geographic; with Early Pre-Dorset sites found in the Canadian Arctic, and Independence I sites primarily confined to the earliest habitation of northern Greenland, Ellesmere Island and Devon Island. Residents of northern Greenland's interior lakes and long inner fjords focused more on fish and muskoxen (*Ovibos moschatus*), whereas those living in coastal localities focused on ringed seals (*Pusa hispida*). Consistent across all these early archaeological sites is minimal-to-no evidence for active hunting of larger marine prey such as walrus and whale. Ivory flakes and implements, such as needles and carvings, have been identified, but few other walrus skeletal parts are present at these sites.

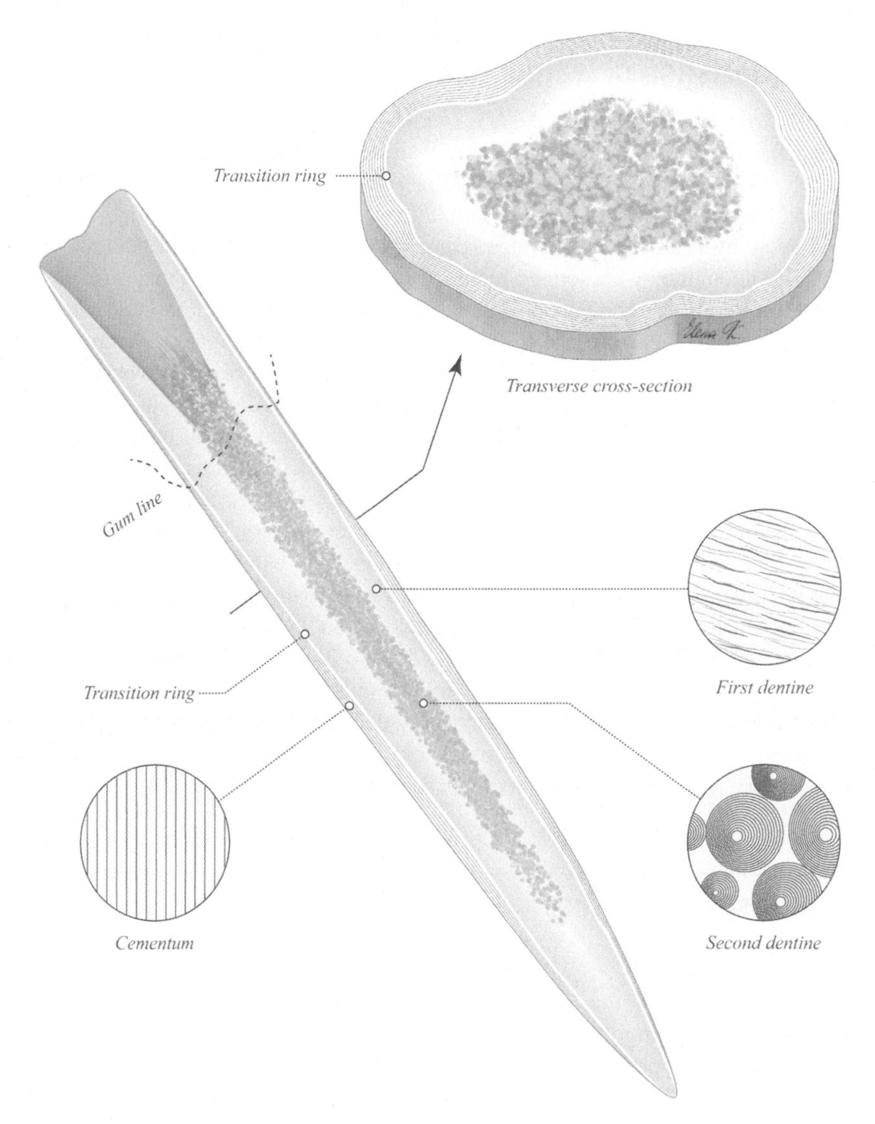

FIGURE 5.1

Walrus tusk cross-section, showing the structure of their modified canines, including cementum, primary (first) and secondary (second) dentine.

Illustration by Elena Kakoshina.

Faunal remains have been identified from over two dozen Early Pre-Dorset and Independence I archaeological sites located north of Lancaster Sound in Nunavut and north of Melville Bay in Greenland (Darwent, 2001, 2003, 2004; McCartney, 1989; McGhee, 1979; Schledermann, 1990). However, most are located well north of the modern range of walrus, and many are found along the

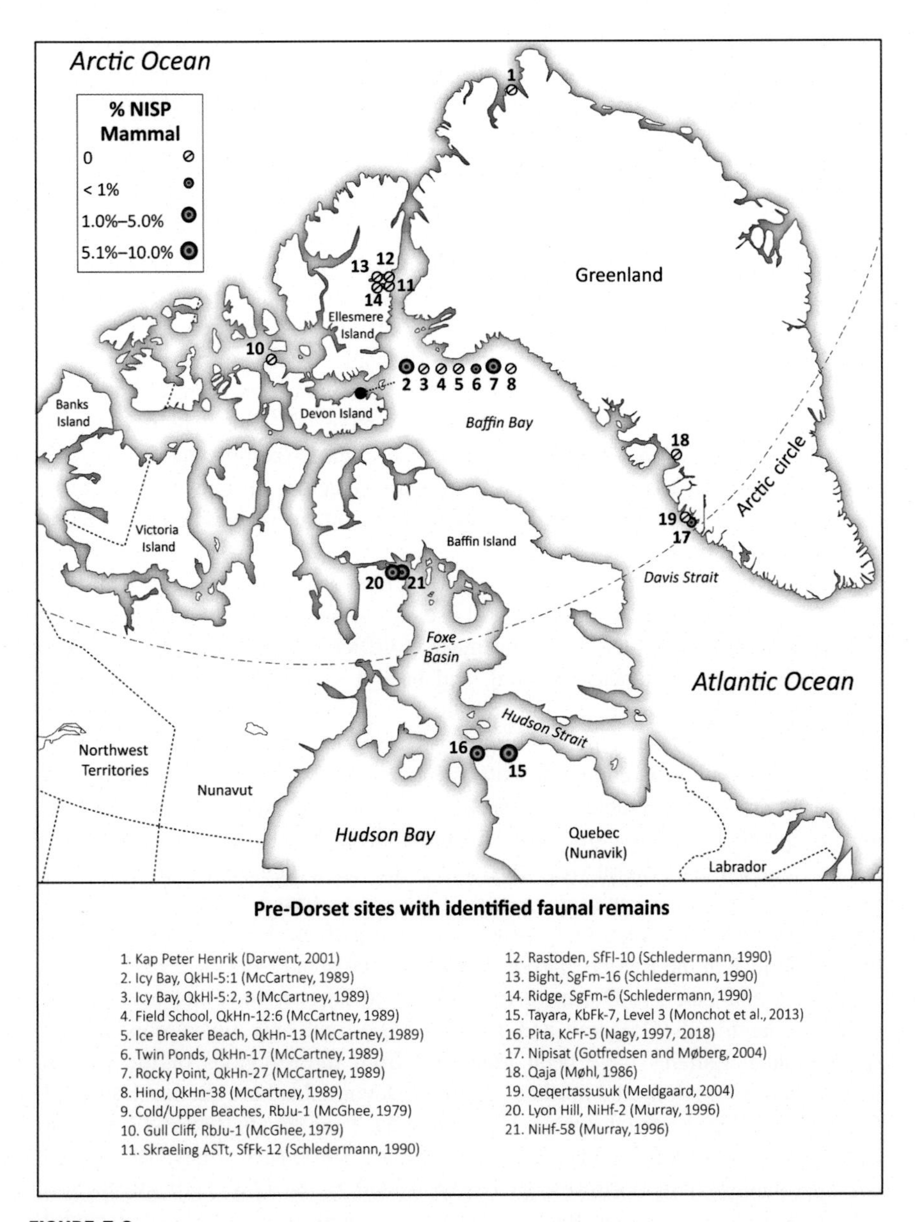

FIGURE 5.2

Map of the Eastern Arctic, showing the location of Pre-Dorset sites and the relative frequency of walrus bone compared to all identified mammal bones in the faunal assemblage. %NISP = relative frequency of the Number of Identified Specimens, or bone fragments.

Figure by John Darwent.

shores of interior lakes and fjords. Fig. 5.2 shows the location of seven Early Pre-Dorset sites located within or near the range of migrating walruses (Kap Peter Henrik; Icy Bay, QkHl-5, Features 2 and 3; Icebreaker Beach, QkHn-13; Hind, QkHn-38; Cold/Upper Beaches, RbJu-1; Skraeling ASTt, SfFk-12; Bight, SgFm-16). Of the 9652 identified mammal bone fragments, only a single walrus bone has been identified in direct association with Early Pre-Dorset cultural material—at the Icebreaker Beach site on northern Devon Island in Lancaster Sound (McCartney, 1989). Although five walrus bones were collected near an Independence I site known as Slikbugten at the entrance to Jørgen Brønlund Fjord in Peary Land (Greenland) (Darwent, 2001:44–45, 2003:343), the association between these bones and the site's tent-ring feature is questionable. There are no direct radiocarbon dates, and the site is located several hundred kilometres north of the modern range of walruses on the northeastern coast of Greenland. Thus, with only a single, clearly-associated walrus bone from Icebreaker Beach (comprising 0.02% of the identified mammal assemblage), and their scarcity across all sites (<0.01%), evidence for active walrus hunting by the earliest inhabitants of these High Arctic latitudes is scant at best.

Despite there being minimal evidence for active walrus hunting, a few Early Pre-Dorset sites have evidence for ivory tool use. Unfortunately, small sample size, poor preservation and incomplete reporting make it difficult to draw reliable conclusions about how the material was acquired, and how it was worked. Icebreaker Beach, with its relatively good organic preservation, including the walrus bone mentioned above, is the richest assemblage. Helmer (1986) describes several ivory pieces, including a possible harpoon foreshaft, a lance head, a carving resembling a bear claw, and, most remarkable, a maskette of a tattooed human face—the earliest known figurative carving in the Eastern Arctic. Other reported finds include an ivory needle fragment and several ivory flakes from the upper component of the Tusk site (SfFk-6) in the Bache Peninsula region of Ellesmere Island (Schledermann, 1990), and eight pieces of worked ivory—including two harpoon heads—from several features at RbJu-1 in the Refuge Harbor area of Devon Island (McGhee, 1979).

Middle and Late Pre-Dorset (c.1900–1100 BCE) sites have not been found in northern Greenland (Grønnow and Jensen, 2003), and only seven sites with preserved faunal remains have been identified from three regions of the Canadian Arctic (Fig. 5.2). Along the northern coast of Devon Island are the Icy Bay (QkHl-5, Feature 1), Field School (QkHn-12, Feature 5), Twin Ponds (QkHn-17, Feature 1 and 4) and Rocky Point (QkHn-27) sites (Helmer, 1991; McCartney, 1989); on the Grinnell Peninsula of northwestern Devon Island is the Gull Cliff (RbJu-1) site (McGhee, 1979); and in the Bache Peninsula of southeastern Ellesmere Island, the Rastoden (SfFl-10) and Ridge (SgFm-6) sites (Schledermann, 1990). Of the 2648 mammal remains identified from these seven sites, only nine walrus bones were identified, and all were from three Middle Pre-Dorset sites in the North Devon Lowlands (0.3%). These sites are adjacent to the Lancaster Sound Polynya, an area that currently hosts migrating walrus herds. No walrus bones have been identified from any Late Pre-Dorset sites. Of the few recorded sites in the Canadian High

Arctic from this period, hunters appear to have narrowed their focus almost exclusively to ringed seals (Darwent, 2001, 2004).

Although there are few assemblages available for analysis, Middle and Late Pre-Dorset peoples in the High Arctic appear to have continued collecting and using walrus ivory, with slightly more ivory tools recovered from sites dating to these later periods. The Ridge site, on eastern Ellesmere Island, stands out in this regard with an ivory carving of a bird, a serrated dart or barb and numerous needles (Schledermann, 1990). Sites in the Devon Lowlands also have occasional ivory specimens, including debitage (manufacturing waste) (Helmer, 1991; McCartney, 1989). Apart from a few reports of ivory 'chips', suggestive of chopping or flaking (McGhee, 1979; Schledermann, 1990), it is not possible to say anything about how these objects were manufactured. Taken together, however, they represent key functional categories including sewing, hunting and 'art', for lack of a better descriptor.

Low Arctic

The Pre-Dorset period in the Low Arctic can be further subdivided into several geographic variants with distinct traditions. Southern Greenland stands out for the Saqqaq culture, a pioneering Pre-Dorset group that occupied the productive fjord systems of coastal Greenland around 2500 BCE and persisted there for over 1000 years. In Canada, Baffin Island and the Foxe Basin region represent a second focus, while Nunavik, Nunatsiavut and Newfoundland make up a third. As Newfoundland and much of southern Nunatsiavut are well outside the current range of Atlantic walruses, these regions are excluded from further discussion.

In West Greenland, Saqqaq sites are clearly part of a unified regional culture with clear adaptations to relatively-rich local marine and terrestrial ecosystems (Grønnow, 2016). Nipisat, a Saqqaq site located near the city of Sisimiut in West Greenland (Fig. 5.2), was periodically occupied in three phases spanning 2000–1000 BCE (Gotfredsen and Møberg, 2004). (This time span is temporally equivalent to Middle/Late Pre-Dorset in the Canadian Arctic.) In recent times, walruses were known to haul out on land along the central coast of West Greenland, with particularly large aggregations in September and October taking advantage of local mollusk beds (Born et al., 1994; Gotfredsen and Møberg, 2004). It appears that inhabitants of Nipisat took advantage of these walrus congregations; by Phase 3 (Late Pre-Dorset), they were hunting juvenile, subadult and adult walruses (Gotfredsen and Møberg, 2004:143). Fragments of the maxillae, or tusk area, of the crania dominate the zooarchaeological finds, particularly in Phase I (59% of identified skeletal parts), when people may have been scavenging beach carcasses rather than actively hunting. By Phase 3, nearly all portions of the walrus are represented, with ribs and phalanges being the most commonly identified part (43.5%), indicative of more extensive use and transport.

Although nearly 15,000 mammal bones were identified from Nipisat, walruses do not appear to have served as a primary economic resource when considered among all other mammals (0.7%) or pinnipeds (1.4%). Ivory, however, appears to have been a locally-worked raw material. Gotfredsen and Møberg (2004) report 174 pieces of worked ivory, including 24 tools and 150 pieces of debitage. Tools include a wedge, harpoon foreshafts, a barbed spear or leister, and possible engraving tools. The quantity of debitage, most of which is from Phase 3, is consistent with a shift from scavenging to active hunting, as suggested by the zooarchaeological data. Capture and transport of walruses back to camp for distribution of meat, skins, bone and ivory would have involved considerable preparation and cooperation of a small crew of hunters. Even more remarkable is the lack of evidence for dogs in Phase 3 to aid in hauling the load from kill and butchery sites back to residential camps.

The Saqqaq site of Qeqertasussuk (c.2400–1400 BCE) is located on an island in the inner part of Sydostbugten, a smaller bay within Disko Bay. Organic tools are abundant, but they are made predominantly from antler, with relatively few made of walrus ivory. Of the 55 recovered harpoon heads, for example, 48 are made of antler and only four of ivory (Grønnow, 2017:75). No walrus bone was identified out of nearly 13,000 mammal bones from this well-preserved and extensively excavated site (Meldgaard, 2004:115). Across the Icefjord at Ilulissat, over 14,000 mammal remains were identified from Saqqaq deposits (2000–900 BCE) at the Qaja site. Although Møhl (1986) notes several ivory harpoon heads, the only walrus bone recovered was a baculum modified into a tool. Given that only six fragments of ivory working debris were recovered from Qeqertasussuk, and none were reported from Qaja (Grønnow, 2017:245), there is no evidence for local hunting of walruses at either of these two sites.

In the Canadian Low Arctic, only four sites have reported data for faunal remains from Pre-Dorset contexts within the limits of Atlantic walrus distribution: two from the Foxe Basin region of Nunavut and two from Nunavik. Murray (1996) analysed archaeofaunal assemblages from the Early Pre-Dorset sites of NiHf-2 (Lyon Hill) and NiHf-58 on Igloolik Island. Despite small faunal assemblages with only 792 identified mammal remains across both sites, 25 walrus bones (3.2%) were identified. Murray (1996:103) suggests that individual hunters as part of small household economic units likely accomplished all subsistence pursuits; however, it is unlikely that lone individuals regularly hunted walrus. That being said, walrus hunting was undertaken more often here than across the entirety of the High Arctic, where walrus remains and ivory objects are rare at any Pre-Dorset site.

In contrast to High Arctic Canada, a number of Early and Late Pre-Dorset sites in the Foxe Basin have produced significant assemblages of ivory tools and debitage. Those from the Igloolik region, excavated originally by Meldgaard (1960a,b), have been described in detail by Houmard (2011, 2015, 2016, 2018). Ivory tools dominate the sample from Early Pre-Dorset features at ParryHill (NiHf-1, 42–51 m terraces) and Lyon Hill (48–42 m terraces), where 40% of organic tools are made from ivory. Harpoon heads are almost all ivory (14 of 16), as are scoops or spoons (five of six). Other ivory tools include shafts and tubes, as well as a single barbed point, a handle and a chisel (Houmard, 2011:99). Ivory debitage is relatively rare in these assemblages, and thus some aspects of the manufacturing sequence remain unknown. Blanks (typically rectangular pieces sized according to the anticipated tool) are most often extracted from tusks by grooving longitudinally and initially shaped either by grooving or percussion; further shaping is by scraping or whittling (Fig. 5.5; Houmard, 2011:108). The open sockets of some harpoon heads were shaped by indirect percussion (Houmard, 2011:109–110, Fig. 31). In a few cases, the distal ends of young tusks seem to have been sawed around the periphery before breaking (Houmard, 2011:110, Fig. 32).

On the Nunavik coast, two Pre-Dorset sites with identified faunal remains have been reported: Level III of the stratified midden deposits at Tayara (KbFk-7) near Qikirtaq (Monchot et al., 2013), and the Pita site (KcFr-5) near Ivujivik (Nagy, 1997, 2018). Of the 1524 mammal bones identified from the Late Pre-Dorset level at Tayara, 138 (9.1%) are of walrus. Ivory use in Pre-Dorset levels at Tayara parallels that recorded from the Igloolik sites, with worked ivory making up about 35% of the organic tool assemblage and dominated by flakes (Houmard, 2011). Harpoons are the most commonly-produced object. Manufacturing techniques are also similar to what was recorded at Igloolik: Initial reduction was undertaken predominantly by grooving, and most pieces were shaped by scraping or whittling rather than by abrasion.

A large assemblage of 13,074 mammal remains with 163 walrus bones (1.2%) were identified at Pita (Nagy, 1997). Ivory tools and debitage make up nearly 60% of the organic artifact assemblage, although there are no further details on tool types or manufacturing methods. Numerous new radiocarbon dates have revealed these bones come from a palimpsest of occupations spanning the entire Pre-Dorset period (Nagy, 2018). Unfortunately, the original data (Nagy, 1997) is presented in such a way that the assemblage cannot be disentangled in light of these new dates, and thus it is unclear if there is any change in the use of walruses and ivory over time (Nagy, 1997).

Dorset walrus use

The beginning of the Dorset period is marked by, among other things, a notable increase in the use of walruses, and the appearance of robust harpoon heads (known as the 'Dorset Parallel' type) to which a thicker, stronger line could be attached (Fig. 5.4; Maxwell, 1976; Taylor, 1968). Several researchers (e.g., Maxwell, 1976; Murray, 1996, 1999) have suggested the Dorset Parallel harpoon head represents a significant economic shift by Early Dorset hunters across the Eastern Arctic, likely originating in the Low Arctic of the Foxe Basin region and spreading north with increased habitation of the High Arctic by the Early Dorset. Others have suggested the increased focus on maritime hunting may have been, in part, a response to climatic changes less favourable to caribou (*Rangifer tarandus*) and muskoxen in the High Arctic (Darwent, 2001; Jensen, 2016). However, the bones of large-bearded seals (*Erignathus barbatus*), unlike those of walrus, remain unchanged relative to the ubiquitous ringed seal.

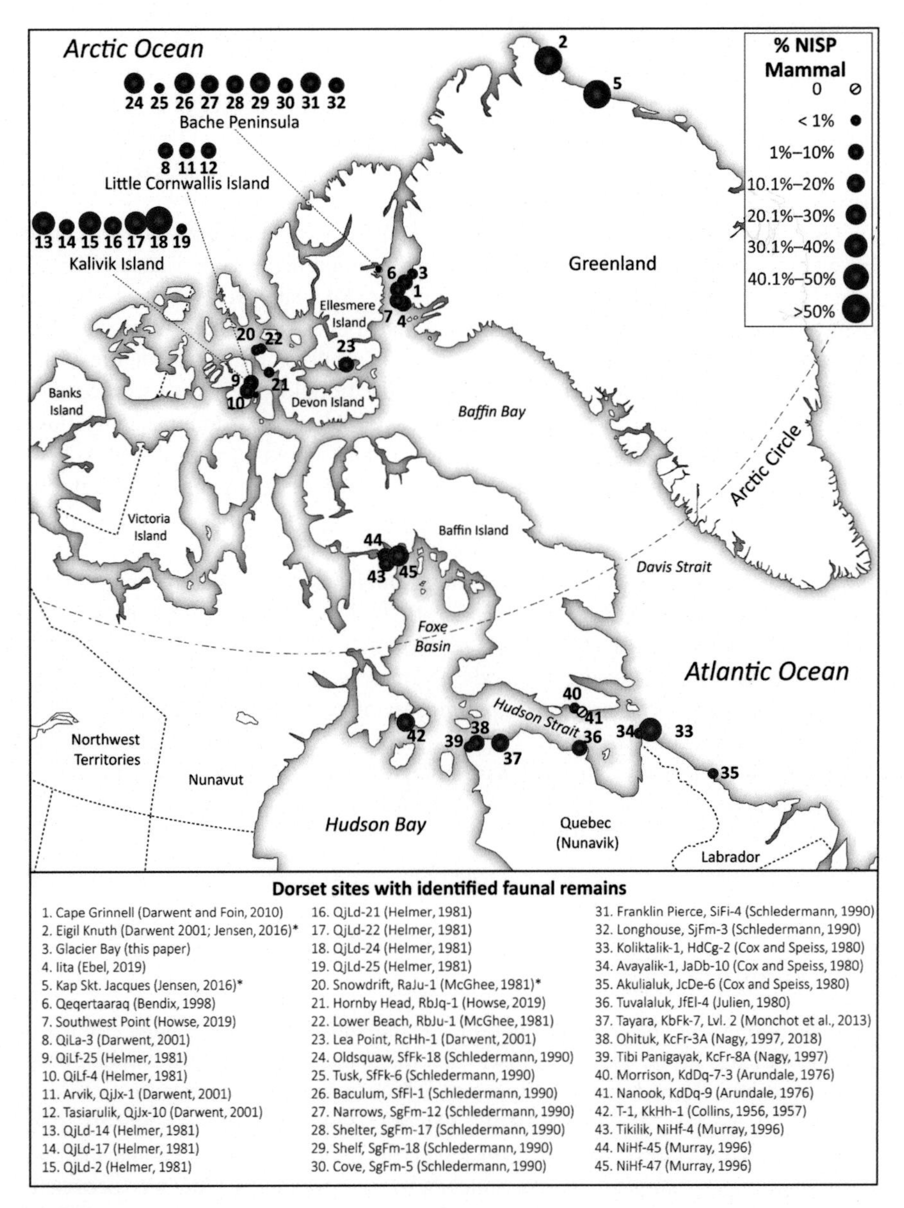

FIGURE 5.3

Map of the Eastern Arctic, showing the location of Dorset sites and the relative frequency of walrus bone compared to all identified mammal bones in the faunal assemblage (%NISP). For three sites, walrus bone was only indicated as being present; the estimated percentage is based on author descriptions (*). %NISP = relative frequency of the Number of Identified Specimens, or bone fragments.

Figure by John Darwent.

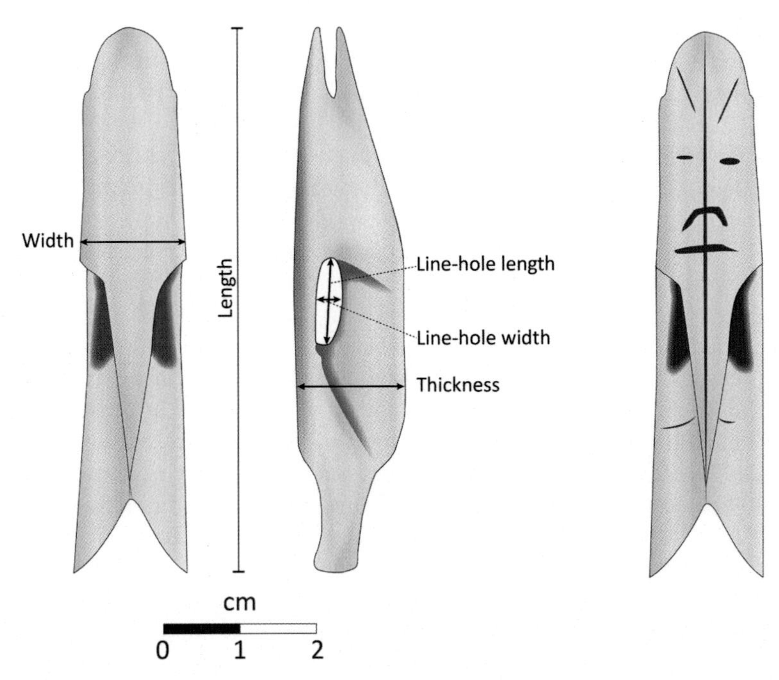

FIGURE 5.4

A Dorset Parallel-type harpoon head collected by the authors in Glacier Bay, Inglefield Land, northwestern Greenland; front (left) and side (centre) view. Line-hole and dimensional measurement locations are taken from Murray (1996). Decorated Dorset Parallel-type harpoon (right) from a hunter's burial at Alarniq (Alarnerk), south of Igloolik Island.

Modified from Lynnerup, N., Meldgaard, J., Jakobsen, J., Appelt, M., Koch, A., Frølich, B. 2003. Human Dorset remains from Igloolik, Canada. Arctic 56(4):349–358. http://pubs.aina.ucalgary.ca/arctic/Arctic56-4-349.pdf: Fig. 6C. Figure by John Darwent.

Neoglacial cooling in the Arctic began sometime after 1500 BCE, but there is significant regional variability in the timing of onset, and it is unclear how this cooling may have affected annual or decadal sea-ice conditions and walrus populations in turn (Finkelstein, 2016). Cooling conditions alone do not explain the shift to active walrus hunting between the Pre-Dorset and Dorset periods. Regardless of the impetus, the distinctive Dorset Parallel harpoon is associated with increased cooperative hunting ventures, enabling successful capture of large, gregarious marine mammals, such as walruses.

High Arctic

Early Dorset in the Canadian High Arctic and Greenlandic Dorset in Greenland (c.800 BCE–1 CE), marked a return of human occupation to these more northerly latitudes. In particular, it marks the end of a long habitation hiatus in northern Greenland (Andreasen, 2000; Grønnow and Jensen, 2003; Jensen, 2016; Knuth, 1967). Excluding sites from areas well outside the range of walruses (e.g., Peary Land, Greenland), 12 Early Dorset sites with faunal remains have been identified across the High Arctic (Darwent, 2001; Helmer, 1981; Jensen, 2016; Schledermann, 1990). In coastal regions of northeastern Greenland, sites dating to the Early Dorset period are situated adjacent to polynyas (Fig. 5.3). As described by Jensen (2016:737), "the open-water area and recurring ledge in the landfast ice extending from Kap Skt. Jacques on Île de France" was so attractive to human settlement that the highest number of known Pre-Dorset or Dorset dwellings ever recorded is located here, and walrus remains are abundant. Seals and walruses (Andreasen, 2000; Jensen, 2016) dominate faunal material recovered from midden excavations at the Eigil Knuth site, which is located on the tip of Holm Land on the outer coast.

Of the 5105 mammal specimens reported from 10 sites across High Arctic Canada (Fig. 5.3), walruses have been recovered from every excavated site, totalling 409 bones (8%). These sites are found in the Crozier Strait region on Kalivik Island (QjLd-2, -14, -21, -22 and -24) and at Markham Point on Bathurst Island (QiLf-4); on Devon Island's Grinnell Peninsula (Lower Beach, RbJu-1); and in the Bache Peninsula of Ellesmere Island (Tusk, SfFk-6; Baculum, SfFl-1; Shelf, SgFm-18). However, the relative frequency of walruses varies widely across the region, from less than 1% at Lower Beach and Tusk, to over 60% at QjLd-24 on tiny Kalivik Island.

These Early Dorset sites have large enough samples of walrus skeletal parts to allow for investigation of carcass transport. At sites from Crozier Strait, frequencies of different skeletal elements indicate that meaty portions of the carcass were preferentially transported from the coast to their dwellings (Darwent, 2001:96–97). Unfortunately, published descriptions of these Crozier Strait sites do not include enough data to determine the extent to which ivory was used to make tools (Helmer, 1981). In contrast, sites on eastern Ellesmere Island, adjacent to the northern extent of the North Water Polynya, have relatively high frequencies of cranial parts; here, it appears that transport of walrus parts was motivated more by the acquisition of ivory than of meat or other walrus products. One ivory artifact was identified from the Tusk site, five from the Baculum site and eight from the Shelf site, the latter of which includes a flaker, a needle and an awl (Schledermann, 1990). Four ivory carvings and three needles were also recovered from several sites in this area known collectively as the Longhouse Ridge complex (Schledermann, 1990) (Fig. 5.4).

The Middle Dorset period saw another abandonment of the High Arctic during a period of colder, climatically unstable conditions. However, humans ventured north again to recolonise the High Arctic around c.800 CE. As stated most clearly by Maxwell (1985:217), "the beginning of the Late Dorset period coincides with, and is probably causally related to, the first warm period in more than 1000 years". The Late Dorset period is marked by changes in social organisation that included communal longhouse structures and cooperative hunting (Friesen, 2007). Stone and gravel caches, although noted in the earliest High Arctic sites (e.g., Knuth, 1967), are much more frequently recorded in Late Dorset contexts (Darwent et al., 2007; LeMoine and Darwent, 2010).

Walrus remains have been identified from every Late Dorset site in the High Arctic except one in Hall Land (Darwent, 2001), northern Greenland (it should be noted that this site is well north of the range of modern walruses). Faunal remains have been recovered from Late Dorset features at Cape Grinnell (Darwent and Foin, 2010), Iita (Darwent et al., 2019; Ebel, 2019), Qeqertaaraq (Appelt and Gulløv, 1999; Bendix, 2000) and Southwest Point (Howse, 2019) in northwestern Greenland. One previously-unpublished dataset that we include here is a Late Dorset semi-subterranean house structure found under a Thule structure in Glacier Bay, Inglefield Land; four walrus bones were identified from an assemblage of 1844 mammal remains (0.2%). Across these five sites, 358 walrus bones have been identified among 7519 mammal remains (4.8%); the amount of walruses varies from 0.2% at Glacier Bay to 10% at Qeqertaaraq (Fig. 5.3). The higher percentages of walruses correlate with local haul outs adjacent to Cape Grinnell, Qeqertaaraq and Southwest Point.

In the Canadian High Arctic, 15 Late Dorset faunal assemblages have been identified from the following sites: QiLf-25, QjLd-17 and QjLd-25 in the Crozier Strait region (Helmer, 1981); Arvik (QjJx-1), Tasiarulik (QjJx-10) and Qila-3 on eastern Little Cornwallis Island (Darwent, 2001); Snowdrift (RaJu-1) and Hornby Head (RbJq-1) on Devon Island's Grinnell Peninsula (Howse, 2019; McGhee, 1981); Lea Point (RcHh-1) near Grise Fjord on southern Ellesmere Island (Darwent, 2001); and Oldsquaw (SfFk-18), Narrows (SgFm-12), Shelter (SgFm-17), Cove (SgFm-5), Franklin Pierce (SiFi-4) and Longhouse (SjFm-3) in Ellesmere Island's Bache Peninsula region (Schledermann, 1990). A total of 638 walrus bones were identified out of 25,656 mammal bones (2.5%) recovered from across all 15 of the above-mentioned sites. In two areas, we can directly compare Early Dorset to Late Dorset walrus hunting. There is no change between Early and Late Dorset at sites in the Bache Peninsula region (13.2%–13.2%) and only a slight decrease at Markham Point (6.2%–4.6%). However, on Kalivik Island, the average walrus contribution drops from 36% to 1.6% as Late Dorset hunters intensified hunting of large quantities of Arctic fox (*Vulpes lagopus*). Similarly, occupants of QiLa-3 on Little Cornwallis Island focused on foxes. A decline in the relative frequency of walruses during the Late Dorset period may relate to an overall broadening of the diet and reduced focus on any one particular resource (Darwent and Foin, 2010; Howse, 2019). This strategy included the capture and storage of large quantities of small-prey species such as foxes, Arctic hares (*Lepus arcticus*) and birds (Darwent, 2004; Ebel, 2019).

We first published on Late Dorset walrus hunting, transport, butchering and ivory-reduction practices with an analysis of faunal and artifact assemblages on Little Cornwallis Island (LeMoine and Darwent, 1998). At Arvik and Tasiarulik, walruses comprise 3.4% and 3.1% of the mammal assemblages respectively, with cranial elements preferentially transported at Arvik and more generalised carcass transport of crania and meat-bearing parts at Tasiarulik

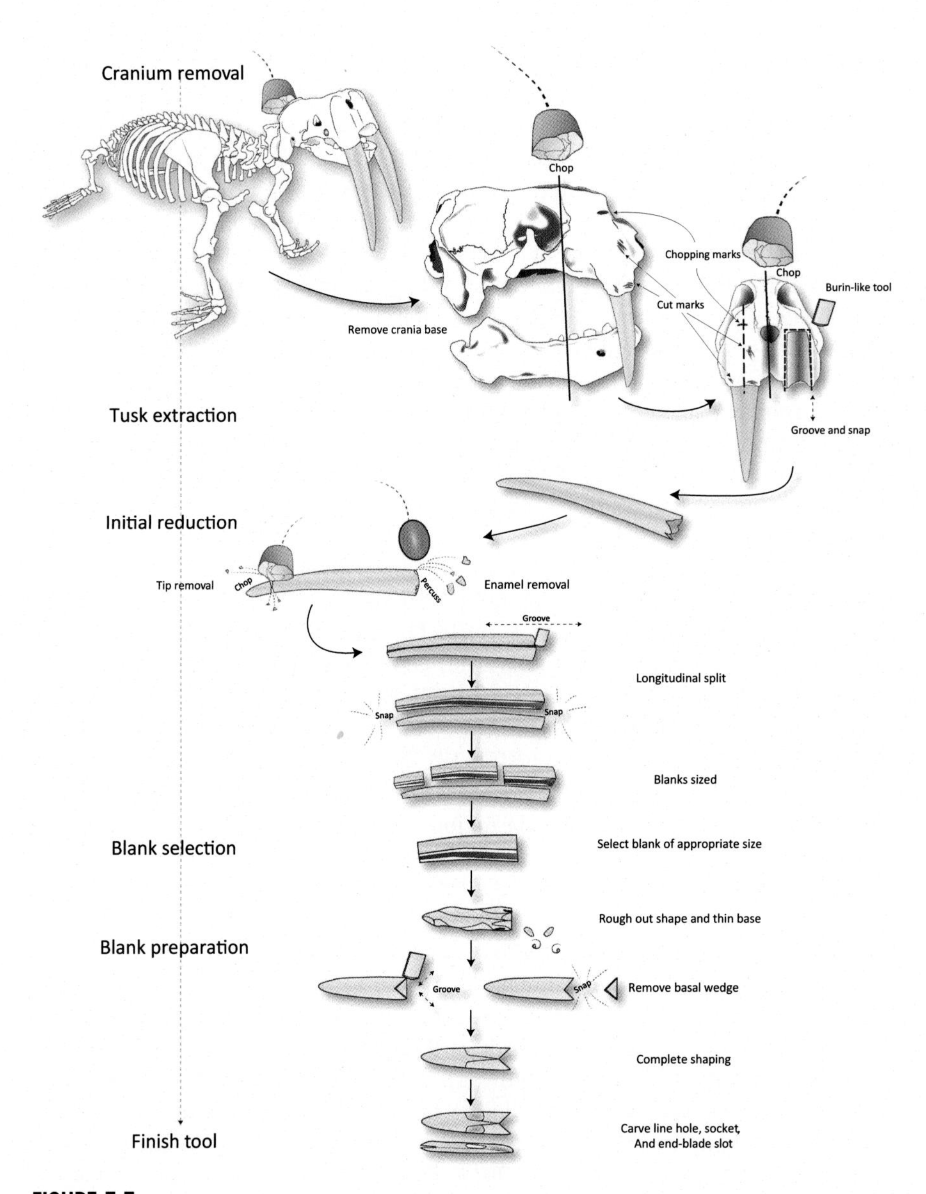

FIGURE 5.5

The reduction sequence, or *chaîne opératoire*, for walrus tusk extraction and ivory tool production.

Figure by John Darwent.

(Darwent, 2001; LeMoine and Darwent, 1998). Also, past occupants of these sites cached walruses in pits. At Arvik, we identified two walrus-crania processing areas. Many crania were broken open with a heavy blow to the top of the skull, or the point behind the orbits where the cranium narrows significantly (Fig. 5.5). "Skulls were typically broken into a minimum of four pieces—right and left maxillae and right and left calva split sagittally [and] often the skull was struck along the side to separate the tusk region from the rest of the crania" (LeMoine and Darwent, 1998:76–77). At these sites, and numerous others across the High Arctic, a combination of grooving and chopping was used to reduce the cranium (particularly the front of the canine alveolus) and to remove the tusk (LeMoine and Darwent, 1998:77–81; Fig. 5.5). Chopping was used to remove the thin root section and the tip, while grooving and snapping were used to reduce the tusk into longitudinal blanks, which were themselves further reduced in size depending on the tool to be created. Blanks were shaped with a combination of adzing, whittling and grooving, and final shaping, which was most often completed by whittling or scraping, or rarely, by abrasion.

When Late Dorset peoples recolonised the High Arctic after c.800 CE, they brought with them a well-established tradition of ivory working. As described previously by LeMoine (2005), Late Dorset people in the High Arctic used ivory for a limited number of objects. Most prominent of these were the fine carvings for which the Late Dorset are so well known. Carvings are the only category of objects in most High Arctic sites where ivory dominates. Ivory was also chosen to make harpoon heads, handles and sled shoes, albeit less often. This summary of walrus ivory use is based on materials from Arvik, with data from other sites in the region (including Snowdrift and Tasiariluk) following the same pattern, which is not surprising given the broad homogeneity typically associated with Late Dorset assemblages (e.g., see Darwent et al., 2019). In all cases, ivory was used frequently, accounting for about 25% of worked organic tools, but not exclusively for any specific type of object.

Low Arctic

Active walrus hunting has been suggested from excavations by Collins (1956, 1957) at the Early Dorset T-1 site (KkHh-1) at Native Point on Southampton Island, where he recovered 438 walrus bones (11.3%; $n = 3873$ mammals), 'abundant' ivory artifacts and larger harpoon heads (Fig. 5.3). Interestingly, Collins (1940) had undertaken previous work on St. Lawrence Island, Alaska. He proposed that the development of complex social organisation in the Bering Strait region began because of the coordination effort and leadership needed to effectively hunt large, migratory congregations of Pacific walrus. This shift to cooperative hunting, accompanied by innovations in large boat construction, arose with the origin of the Northern Maritime tradition (c.200 BCE). The Thule descendants of this tradition are the Inuit who began their move out of the Bering Strait and into the Eastern Arctic after 1000 CE.

On Baffin Island, the only reported Dorset sites within the range of Atlantic walrus stocks are the Early Dorset sites of Morrison (KdDq-7-3) and Nanook (KdDq-9), which are located near the modern Inuit community of Kimmirut (Lake Harbour) on southern Baffin Island. However, of 23,746 mammal remains, only three are identified as walruses (Arundale, 1976), with an additional 56 bones identified as large pinniped (0.1%–0.2%). Arundale (1976) suggests these sites were located too far from walrus haul-out areas for local hunting to have taken place. There are no other reported faunal remains from coastal sites on Baffin Island with which we can compare this assessment, but given increased use of walruses during the Dorset period elsewhere in the Arctic, Arundale's explanation seems likely.

As discussed previously, Dorset Parallel harpoons first appear in Early Dorset sites in both the High and the Low Arctic. As stated by Murray (1996:105) based on her research on Igloolik Island, "there is evidence in Early Dorset, for the development of a new form of harpoon head specifically designed for walrus hunting". Dorset Parallel harpoons are not only made more substantially with a thicker line hole, but are also more frequently decorated. Only 11% of the harpoon heads analysed by Murray (1996:114) at Igloolik across all Pre-Inuit sites had any form of decoration, but of those decorated harpoon heads, 82% were the Dorset Parallel type, and evenly split between Early and Late Dorset contexts. The decorative motifs include diagonal slashes, lines and dashes, skeletal representations and faces, such as this example recovered from a hunter's burial at Alarniq, south of Igloolik Island (Fig. 5.4; Lynnerup et al., 2003; Murray, 1996).

Along with the appearance of this new harpoon type, Murray (1996) recorded 78 walrus bones out of 288 mammal remains (27%) from the Early Dorset site of NiHf-47 on Igloolik Island, which is a significant increase from nearby Pre-Dorset sites. Ivory artifacts were recovered in even greater quantities from Late Pre-Dorset through Middle Dorset sites on Igloolik Island, which Houmard (2011) groups together. These include Qaiqsut (also known as Kaersut; NiHe-1, 20–17 m), ParryHill (NiHf-1, 20–17 m) Freuchen (NiHf-3, 26–20 m) and Jens Munk (NjHa-1, 25–22 m). Ivory made up 23%–48% of the organic tool assemblage, steadily increasing through time. Harpoon heads are commonly, but not exclusively made of ivory (74 of 78), whereas foreshafts are *only* made of ivory (Houmard, 2011:201). Items made predominantly from ivory include barbed points, snow knives and carvings of various sorts (Houmard, 2011:204). Manufacturing techniques are much the same as in Early Pre-Dorset, although more limited. Grooving and scraping or whittling predominate (Houmard, 2011:215), with abrasion occasionally used to smooth surfaces (Fig. 5.5).

The Late Dorset period is best represented by the eight-metre terrace at Qaiqsut, although the sample is small (Houmard, 2011:238). Ivory was used exclusively for sled shoes, animal figurines and shims for handles, as well as for most snow knives, and some harpoon heads, barbed points and chisels. Little debitage was recovered, but manufacturing techniques (primarily grooving and scraping) were much the same as in the previous period. Although the relative frequency of walruses decreases during the Late Dorset period (NiHf-4: 39/501, 7.8%; NiHf-47: 104/2718, 3.8%), they still comprise a higher portion of identified mammals than any Pre-Dorset contexts, and the relative decrease of walruses is accompanied by greater faunal diversity at this time.

Faunal remains were identified from Level II at Tayara on the Nunavik coast, with radiocarbon dates that seem to suggest a composite palimpsest of Early Dorset occupations including butchering activities and a possible communal longhouse structure

(Monchot et al., 2013). Of the 4807 mammal remains identified from Level II, 799 are walruses (16.6%). The most abundant skeletal parts are the skulls and mandibles ($>80\%$), followed by humeri, tibia-fibula and the lumbar vertebrae (Monchot et al., 2013). Metric analysis of the mandibles suggests at least one juvenile, three adult males and two adult females make up the assemblage. The remarkable find of a stone biface fragment embedded in a walrus rib was likely part of a killing lance that dispatched the animal on land after it was initially harpooned. A similar approach to cranial processing to extract the tusks was undertaken here as previously documented by LeMoine and Darwent (1998; Fig. 5.5). Over 50% of the artifacts from Level II at Tayara are made of ivory ($n = 319$, 56%), and they present a similar picture to the Pre-Dorset Level III deposits (Houmard, 2011; Monchot et al., 2013). All harpoon heads are ivory, as are barbed points, foreshafts and figurines of various sorts. Compared to High Arctic assemblages at this time, walruses were used more consistently for both subsistence and as a raw material source at Tayara. This difference may be reflective of site proximity to local walrus haul-outs, or simply higher walrus densities in the Nunavik region.

Three other sites located on the coast of Nunavik have reported Dorset faunal remains, but none with the extensive butchery documented at Tayara: Tuvaaluk (JfEl-4) on Diana Island (Julien, 1980), and Ohituk (KcFr-3A) and Tivi Panigayak (KcFr-8A) near Ivujivik (Nagy, 1997). A small Middle-Late Dorset assemblage of mammal remains ($n = 584$) includes 7% walrus bone at Tuvaaluk (Julien, 1980), which is considerably higher than the mixed Dorset and Middle Dorset deposits at Ohituk (11 of 1449, 0.8%) and Tivi Panigayak (13 of 1076, 1.2%) respectively (Nagy, 1997). Few organic tools are reported from the Dorset deposits near Ivujivik, but ivory represents 30.4% ($n = 7$) of the Ohituk and 62.5% ($n = 5$) of the Tivi Panigayak assemblages.

Few Pre-Inuit sites on the Labrador coast have any bone preservation to speak of, but three Middle Dorset sites from the central and northern coast of Labrador yielded faunal remains: Koliktalik-1 (HdCg-2) on Dog Island in the Nain region; Avayalik-1 (JaDb-10) on Avayalik Island in McLennon Strait; and Akulialuk (JcDe-6) on Killinek Island in Lenz Strait (Cox and Spiess, 1980). Only 31 walrus-bone fragments of 5315 mammals were identified from Koliktalik-1 (0.6%) and only 3 of 618 from Akuliakuk (0.5%). However, the relative frequency of walrus remains from the deep, frozen midden at Avayalik-1 is 35% of the mammal assemblage ($n = 1698$). Although bone, antler and ivory tools were relatively uncommon compared to wood, some 150 tools were recovered. Ivory tools include harpoon heads, amulets, support pieces and handles (Jordan, 1979, 1980). In addition, numerous walrus skulls and the tusks extracted for tool-making were found in caches adjacent to the site, and these represent skeletal parts not recovered from the midden area at Avayalik (Cox and Spiess, 1980:664). Six walrus teeth from the midden deposit were thin-sectioned to assess season of death, and, together with teeth from other mammalian prey, the animals all appear to have been hunted from late winter through summer (Cox and Spiess, 1980). This suggests that the site's occupants hunted walrus seasonally and then stockpiled ivory and meat in caches for use in fall/winter.

Discussion and conclusions

The Eastern Arctic was first populated 4500 years ago, but these Arctic pioneers did not initially hunt walrus. Ivory was widely used for tool manufacture in Pre-Dorset occupations, but for the most part, people appear to have relied on scavenged tusks for this valuable raw material. Dyke et al. (1999:166) have documented the distribution of walrus parts in geological contexts, and they report ivory tusks are, by far, the most common finds, followed by crania with tusks. Walrus remains appear on land under a variety of circumstances, which are described by Dyke et al. (1999:167–168) as "crawlers, sinkers and floaters". That is, animals that have crawled inland, sometimes quite far, and died; those that have died in the water or on the ice and been washed ashore; and those that died at haul-outs onshore. Pre-Dorset people were able to collect tusks when and where they found them, acquiring ivory as a valuable raw material even though they did not actively hunt the animals. There are exceptions to the general pattern we present for Pre-Dorset peoples. Walrus remains are in large enough numbers to suggest the possibility of some small-scale hunting in accessible haul-out areas during Late Pre-Dorset times: Near Sisimiut, West Greenland, near the Jones Sound polynya on Devon Island, in the Igloolik area, and near Qiqirtaq, Nunavik. Still, from the archaeological evidence it is impossible to draw broad conclusions about the general importance of walrus ivory to Pre-Dorset peoples, particularly in the High Arctic.

The abrupt increase in walrus remains from Early Dorset sites across the Eastern Arctic can be attributed to innovations in harpoon technology and social organisation, which was possibly motivated by cooler or more variable climatic conditions (Darwent, 2004; Murray, 1996, 1999). Their subsequent relative decline compared to other pinniped remains in Late Dorset sites seems to be correlated with warming climatic conditions, which may have decreased suitable ice haul-out locations. In the Low Arctic, walruses became a much more significant marine resource during the Dorset period. Cooperative hunting practices must have been in place to be able to acquire these massive animals. With an average walrus male weighing around 900 kg, the participation of several hunters would provide substantially greater chances of a successful and safe hunt. Dorset hunting parties likely coordinated their attacks on walruses at terrestrial haul outs or ice leads. In addition, beginning in Late Pre-Dorset and Early Dorset times, extraction of tusks from walrus crania became a more efficient process.

Processing of ivory involved techniques often associated with flintknapping and ground-stone manufacture, and more objects (in particular, harpoons, needles, foreshafts and artistic carvings) began to be produced predominantly, or exclusively, from ivory. While the Late Dorset period ushered in additional changes in social organisation and complexity (e.g., the appearance of longhouse features), these developments appear not to be driven simply by increased walrus harvesting. Instead, other factors must have been involved. Given the apparent absence of boating technology, cooperation alone did not enable the Dorset to shift their hunting skills to whales. Dorset were at a disadvantage when a new wave of migrants—Inuit hunters—moved into the Eastern Arctic, bringing with them hunting skills that had been honed on the larger Pacific walrus and whales in the Bering Strait.

References

Ameen, C., Feuerborn, T.R., Brown, S.K., Linderholm, A., Hulme-Beaman, A., Lebrasseur, O., et al., 2019. Specialised sledge dogs accompanied Inuit dispersal across the North American Arctic, Proceedings of the Royal Society B, 286. p. 2191929.

Andreasen, C., 2000. Palaeo-Eskimos in Northwest and Northeast Greenland. In: Appelt, M., Bergland, J. (Eds.), Identities and Cultural Contacts in the Arctic. Danish National Museum and Danish Polar Center, Copenhagen, pp. 882–892. Danish Polar Center Publication No. 8.

Appelt, M., 2016. Late Dorset. In: Friesen, T.M., Mason, O.K. (Eds.), The Oxford Handbook of the Prehistoric Arctic. Oxford University Press, Oxford, pp. 783–806.

Appelt, M., Gulløv, H.C., 1999. Late Dorset in High Arctic Greenland: Final Report of the Gateway to Greenland Project. Danish Polar Center, Copenhagen.

Arundale, W.H., 1976. The Archaeology of the Nanook Site: An Explanatory Approach. Unpublished Ph.D. dissertation, Department of Anthropology, Michigan State University.

Bendix, B., 2000. Late Dorset faunal remains from sites at Hatherton Bay, Thule District, Greenland. In: Appelt, M., Bergland, J. (Eds.), Identities and Cultural Contacts in the Arctic. Danish National Museum and Danish Polar Center, Copenhagen, pp. 77–81. Danish Polar Center Publication No. 8.

Born, E.W., Heide-Jørgensen, M.P., Davis, R.A., 1994. The Atlantic Walrus (*Odobenus rosmarus rosmarus*) in West Greenland. Meddelelser Om Grønland, Bioscience 40, 1–33.

Collins, H.B., 1956. The T1 site at native point, Southampton Island, N.W.T. University of Alaska Anthropological Papers 4 (2), 63–89.

Collins, H.B., 1957. Archaeological Investigations on Southampton and Walrus Islands, N.W.T. National Museum of Canada Bulletin 147, 22–61.

Collins, H.B., 1940. Outline of Eskimo prehistory. In: Nichols, F.S.G. (Ed.), Essays in Historical Anthropology of North America. Smithsonian Institution, Washington, DC, pp. 533–592. Smithsonian Miscellaneous Collections Vol. 100.

Cox, S.L., Spiess, A., 1980. Dorset settlement and subsistence in Northern Labrador. Arctic 33 (3), 659–669.

Darwent, C.M., 2001. High Arctic Paleoeskimo Fauna: Temporal Changes and Regional Differences. Unpublished Ph.D. dissertation, Department of Anthropology, University of Missouri-Columbia.

Darwent, C.M., 2003. The zooarchaeology of Peary Land and adjacent areas. In: Grønnow, B., Jensen, J.F. (Eds.), The Northernmost Ruins of the Globe: Eigil Knuth's Archaeological Investigations in Peary Land and Adjacent Areas of High Arctic Greenland, 29. *Meddelelser om Grønland, Man and Society*, pp. 342–395.

Darwent, C.M., 2004. The highs and lows of high Arctic mammals: temporal change and regional variability in Paleoeskimo subsistence. In: Mondini, M., Muñoz, S., Wickler, S.

(Eds.), Colonisation, Migration, and Marginal Areas: A Zooarchaeological Approach. Oxbow Books, Oxford, pp. 62–73.

Darwent, C.M., Foin, J.C., 2010. Zooarchaeological analysis of a late Dorset and an early Thule dwelling at Cape Grinnell, Northwest Greenland. Geografisk Tidsskrift–Danish Journal of Geography 11 (2), 315–336.

Darwent, J., Darwent, C., LeMoine, G., Lange, H., 2007. Archaeological survey of Eastern Inglefield Land, Northwestern Greenland. Arctic Anthropology 44 (2), 51–86.

Darwent, J., LeMoine, G.M., Darwent, C.M., Lange, H., 2019. Late Dorset deposits at Iita: site formation and site destruction in Northwestern Greenland. Arctic Anthropology 56 (1), 96–118.

Dyke, A.S., Hooper, J., Harington, C.R., Savelle, J.M., 1999. The Late Wisconsinan and the Holocene record of Walrus (*Odobenus rosmarus*) from North America: a review with new data from Arctic and Atlantic Canada. Arctic 52, 160–181.

Ebel, E., 2019. Catching birds in the High Arctic. Anthropology News.

Finkelstein, S.A., 2016. Reconstructing Middle and Late Holocene Paleoclimates of the Eastern Arctic and Greenland. In: Friesen, T.M., Mason, O.K. (Eds.), The Oxford Handbook of the Prehistoric Arctic. Oxford University Press, Oxford, pp. 653–671.

Friesen, T.M., 2007. Hearth rows, hierarchies and Arctic hunter-gathers: the construction of equality in the Late Dorset Period. World Archaeology 39, 194–214.

Friesen, T.M., Mason, O.K., 2016. Introduction: archaeology of the North American Arctic. In: Friesen, T.M., Mason, O.K. (Eds.), The Oxford Handbook of the Prehistoric Arctic. Oxford University Press, Oxford, pp. 1–26.

Friesen, T.M., 2020. Radiocarbon evidence for fourteenth-century Dorset occupation in the Eastern North American Arctic. American Antiquity 85 (2), 222–240.

Gotfredsen, A.B., Møberg, T., 2004. Nipisat—A Saqqaq Culture Site in Sisimiut, Central West Greenland. Danish Polar Center, Copenhagen, Meddelelser om Grønland, Man and Society 31.

Grønnow, B., 2016. Independence I and Saqqaq: The first Greenlanders. In: Friesen, T.M., Mason, O.K. (Eds.), The Oxford Handbook of the Prehistoric Arctic. Oxford University Press, Oxford, pp. 711–735.

Grønnow, B., 2017. The Frozen Saqqaq Sites of Disko Bay, West Greenland: Qeqertasussuk and Qajaa (2400-900 BC). Studies of Saqqaq Material Culture in an Eastern Arctic Perspective. Museum Tusculanum Press, Copenhagen, Meddelelser om Grønland, Man and Society 45.

Grønnow, B., Jensen, J.F., 2003. The Northernmost Ruins of the Globe: Eigil Knuth's Archaeological Investigations in Peary Land and Adjacent Areas of High Arctic Greenland. SILA and the Danish Polar Center, Copenhagen, Meddelelser om Grønland No. 329, Man and Society 45.

Helmer, J.W., 1981. Climate Change and Dorset Culture Change in the Crozier Strait Region, N.W.T.: A Test of the Hypothesis. Unpublished Ph.D. dissertation, Department of Archaeology, University of Calgary.

Helmer, J.W., 1986. Face from the past: an early pre-Dorset ivory maskette from Devon Island, N.W.T. Études/Inuit/Studies 10, 179–202.

Helmer, J.W., 1991. Palaeo-Eskimo prehistory of the North Devon Lowlands. Arctic 44 (4), 301–307.

Houmard, C., 2011. Caractérisation chrono-culturelle et évolution du paléoesquimau dans le Golfe de Foxe (Canada): Étude typologique et technologique des industries en matières dures d'origine animale. Unpublished Ph.D. dissertation, Department of History, Université Laval and Université Paris Ouest, Nanterre.

Houmard, C., 2015. L'industrie osseuse de Tayara (KbFk-7, Nunavik) revisitée par la technologie. Études/Inuit/Studies 39 (2), 145–172.

Houmard, C., 2016. L'exploitation technique des ressources animales des premiers peoples de l'Arctique de l'Est canadien (env. 2500 BC-1400 AD). In: Dupont, C., Marchand, G. (Eds.), Archéologie des chasseurs-cueilleurs maritimes de la fonction des habitats à l'organisation de l'espace littoral. Actes de la séance de la Société préhistorique française de Rennes. Société Préhistorique Française, Paris, pp. 261–281.

Houmard, C., 2018. Exploitation des ivoires marins dans les sociétés de l'Arctique de l'Est nord-américain (~2500 BC–1900 AD). L'Anthropologie 122 (3), 546–578.

Howse, L., 2019. Hunting technologies and archaeofaunas: societal differences between Hunter-Gatherers of the Eastern Arctic. Journal of Archaeological Method and Theory 26, 88–111.

Jensen, J.F., 2016. Greenlandic Dorset. In: Friesen, T.M., Mason, O.K. (Eds.), The Oxford Handbook of the Prehistoric Arctic. Oxford University Press, Oxford, pp. 737–760.

Jordan, R.H., 1979. Dorset art from Labrador/80 Folk 21/22, 397–417.

Jordan, R.H., 1980. Preliminary results from archaeological investigations on Avayalik Island, Extreme Northern Labrador. Arctic 33 (3), 607–627.

Julien, M., 1980. Etude préliminare du matériel osseux provenant du site dorsétien DIA.4 (JfEl-4) (arctique oriental). Arctic 33 (3), 553–568.

King, J.E., 1964. Seals of the World. Trustees of the British Museum, London.

Knuth, E., 1967. Archaeology of the Musk-ox Way. Ecole Pratique Des Hautes Études-Sorbonne, Paris.

LeMoine, G.M., 2005. Understanding Dorset from a different perspective: worked antler, bone and ivory. In: Sutherland, P.D. (Ed.), Contributions to the study of the Dorset Palaeo-Eskimos. Canadian Museum of Civilization, Gatineau, pp. 122–145. Archaeological Survey of Canada, Mercury Series Paper 167.

LeMoine, G.M., Darwent, C.M., 1998. The Walrus and the carpenter: late Dorset ivory working in the High Arctic. Journal of Archaeological Science 25, 73–83.

LeMoine, G.M., Darwent, C.M., 2010. The Inglefield Land archaeology project: overview. Geografisk Tidsskrift – Danish Journal of Geography 100 (2), 279–296.

Lynnerup, N., Meldgaard, J., Jakobsen, J., Appelt, M., Koch, A., Frølich, B., 2003. Human Dorset remains from Igloolik, Canada. Arctic 56 (4), 349–358.

MacGregor, A., 1985. Bone, Antler, Ivory and Horn: The Technology of Skeletal Materials since the Roman Period. Croom Helm, London.

Maxwell, M.S., 1976. Pre-Dorset and Dorset artifacts: the view from Lake Harbour. In: Maxwell, M.S. (Ed.), Eastern Arctic Prehistory: Paleoeskimo Problems. Society for American Archaeology, Washington, DC, pp. 58–78. Memoirs of the Society for American Archaeology 31.

Maxwell, M.S., 1985. Prehistory of the Eastern Arctic. Academic Press, Orlando.

McCartney, P.H., 1989. Paleoeskimo Subsistence and Settlement in the High Arctic. Unpublished Ph.D. dissertation, Department of Archaeology, University of Calgary.

McGhee, R., 1979. The Paleoeskimo Occupations at Port Refuge, High Arctic Canada. National Museums of Canada, Ottawa, National Museum of Man, Archaeological Survey of Canada, Mercury Series Paper 92.

McGhee, R., 1981. The Dorset Occupations in the Vicinity of Port Refuge, High Arctic Canada. National Museums of Canada, Ottawa, National Museum of Man, Archaeological Survey of Canada, Mercury Series Paper 105.

Meldgaard, J., 1960a. Origin and evolution of Eskimo cultures in the Eastern Arctic. Canadian Geographical Journal 60, 64–75.

Meldgaard, J., 1960b. Prehistoric culture sequences in the Eastern Arctic as elucidated by stratified sites at Igloolik. In: Wallace, A.F.C. (Ed.), Men and Culture: Selected Papers of the Fifth International Congress of Anthropological and Ethnological Sciences, Philadelphia 1956. University of Pennsylvania Press, Philadelphia, pp. 588–595.

Meldgaard, M., 2004. Ancient Harp Seal Hunters of Disko Bay: Subsistence and Settlement at the Saqqaq Culture site of Qeqertasussuk (2400–1400 BC) West Greenland. Danish Polar Center, Copenhagen, Meddelelser om Grønland, Man and Society 30.

Møhl, J., 1986. Dog remains from a Paleoeskimo settlement in West Greenland. Arctic Anthropology 23 (1–2), 81–89.

Monchot, H., Houmard, C., Dionne, M., Desrosiers, P.M., Gendron, D., 2013. The modus operandi of walrus exploitation during the Palaeooeskimo period at the Tayara Site, Arctic Canada. Anthropozoologica 48 (1), 15–36.

Murray, M.S., 1996. Economic Change in the Paleoeskimo Prehistory of the Foxe Basin, NWT. Unpublished Ph.D. dissertation, Department of Anthropology, McMaster University, Hamilton, ON.

Murray, M.S., 1999. The long-term effects of short-term prosperity—an example from the Canadian Arctic. World Archaeology 30 (3), 466–483.

Nagy, M., 1997. Palaeoeskimo Cultural Transition: A Case Study from Ivujivik (Eastern Arctic). Unpublished Ph.D. dissertation, Department of Anthropology, University of Alberta, Edmonton.

Nagy, M., 2018. Reinterpreting the First Human Occupations of Ivujivik (Nunavik, Canada). Arctic Anthropology 55 (2), 17–43.

Park, R., 2016. The Dorset-Thule transition. In: Friesen, T.M., Mason, O.K. (Eds.), The Oxford Handbook of the Prehistoric Arctic. Oxford University Press, Oxford, pp. 807–826.

Penniman, T.K., 1952. Pictures of Ivory and Other Animal Teeth, Bone and Antler, with a Brief Commentary on their Use in Identification. Pitt Rivers Museum, Oxford, University of Oxford Occasional Papers on Technology No. 5.

Raghavan, M., DeGiorgio, M., Albrechtsen, A., Moltke, I., Skoglund, P., Korneliussen, T.S., et al., 2014. The genetic prehistory of the new world Arctic. Science 345 (6200), 1255832–1255832.

Ryan, K., 2016. The "Dorset Problem" revisited: the transitional and early and middle Dorset periods in the Eastern Arctic. In: Friesen, T.M., Mason, O.K. (Eds.), The Oxford Handbook of the Prehistoric Arctic. Oxford University Press, Oxford, pp. 761–783.

Schledermann, P., 1990. Crossroads to Greenland: 3000 Years of Prehistory in the Eastern High Arctic. Arctic Institute of North America of the University of Calgary, Alberta, Komatik Series No. 2.

Taylor Jr., W.E., 1968. The Arnapik and Tyara Sites: An Archaeological Study of Dorset Culture Origins. Society for American Archaeology, Washington, DC, Memoirs of the Society for American Archaeology No. 22.

Tremayne, A.H., Rasic, J.T., 2016. The Denbigh Flint Complex of Northern Alaska. In: Friesen, T.M., Mason, O.K. (Eds.), The Oxford Handbook of the Prehistoric Arctic. Oxford University Press, Oxford, pp. 349–370.

CHAPTER

Subsistence walrus hunting in Inuit Nunangat (Arctic Canada) and Kalaallit Nunaat (Greenland) from the 13th century CE to present

Sean Desjardins[1,2] and Anne B. Gotfredsen[3]

[1]*Arctic Centre/Groningen Institute of Archaeology, University of Groningen, Groningen, The Netherlands*

[2]*Canadian Museum of Nature, Gatineau, Quebec, Canada*

[3]*Globe Institute, University of Copenhagen, Copenhagen, Denmark*

Chapter Outline

Introduction

Modern inhabitants of both Inuit Nunangat (northern Canada's Inuit territories[1] – land, water and ice) and Kalaallit Nunaat (hereafter, 'Greenland') are descended from culture groups archaeologists refer to as Thule (or Thule Inuit), some of whom migrated relatively rapidly from western and northern Alaska most likely

[1] The designation includes the Inuvialuit Settlement Region in northern Yukon and the western Northwest Territories; Nunavut; Nunavik in northern Quebec and Nunatsiavut in northern and east-central Labrador.

The Atlantic Walrus. DOI: https://doi.org/10.1016/B978-0-12-817430-2.00004-2

in the 13th century CE (Friesen and Arnold, 2008; Friesen, 2016). These pioneers would ultimately settle most of the lands occupied by Tuniit (Dorset Paleo-Inuit), displacing them altogether by the beginning of the 15th century CE. With their (likely) more sophisticated open-water boating technology and specialized marine-mammal hunting toolkit, Thule Inuit arguably had far greater success than Dorset in targeting Atlantic walruses (*Odobenus rosmarus rosmarus*).

The first Thule Inuit migrants out of Alaska continued and remade a rich and long-lasting tradition of hunting bowhead whales (*Balaena mysticetus*) and Pacific walruses (*Odobenus rosmarus divergens*), which gathered in sizable herds on the Bering Strait ice and at terrestrial haul-outs (Hill, 2011a). While much scholarly attention has been paid to the social and economic importance of bowhead whales to Inuit society over time (see Maxwell 1985; McCartney 1979; McCartney and Savelle 1985; Patton and Savelle 2006; Whitridge 2002; among many others), there has been relatively little consideration of the long-term history and importance of Inuit walrus hunting. Today, hunters in dozens of communities across Inuit Nunangat and Greenland continue to hunt walruses in relatively large numbers, and there is great inter-regional diversity in hunting and processing practices; activities that may have once been wide-reaching and ubiquitous have largely given way to distinct regional variations suited to local environmental and social conditions. Even today, the storage and preparation of walrus products are regionally determined. Inuit of Nunavik (Nunavimmiut), though speaking the same language and sharing many of the same cultural traditions as Inuit of Nunavut (Nunavummiut), typically hunt and process walruses in different ways from their northern neighbours. Similar differentiation can be observed among widely-separated walrus-hunting communities in eastern and western Greenland.

The largest and most stable walrus herds favour relatively shallow waters that support their benthic prey, as well as recurring polynyas, which offer both open water for diving and solid ice for hauling out. Important examples of such walrus 'hotspots' that have been heavily exploited by Inuit in the past include (1) northern Foxe Basin, Nunavut (Desjardins, 2013, 2018), (2) the North Water of northern Davis Strait (Gotfredsen et al., 2018) and (3) the Sirius Water near Clavering Island, North-East Greenland (see Gotfredsen, 2010; Grønnow et al., 2011). To more fully explore regional diversity of walrus-hunting practice, we focus our attention on the most prolific hunting traditions that rely upon such environments: those of Amitturmiut (Inuit of northern Foxe Basin, Inuit Nunangat) and communities in Avanersuaq, North-West Greenland. We close with a discussion of the importance of walruses in Pan-Inuit cosmology.

Walrus hunting in Inuit Nunangat

Insights about walrus hunting by Thule and historic Inuit (from approximately 1300 to 1950 CE; hereafter, 'premodern hunting') in Inuit Nunangat can be drawn from zooarchaeological data (animal remains from archaeological sites), ethnographic observations, ethno-historic accounts and *Inuit Qaujimajatuqangit*[2] (*IQ*) (ᐃᓄᐃᑦ ᖃᐅᔨᒪᔭᑐᖃᖕᒋᑦ; 'things long known to Inuit,' or Inuit traditional knowledge based in oral tradition). Though each of these lines of evidence is arrived at differently and rooted in unique understandings of how knowledge is produced, each has a self-evident validity. We believe the long-term cultural continuity of Inuit lifeways and traditions necessitates consideration (if not integration) of all such methods. As such, we include both archaeological data and *IQ* in our discussion of premodern Inuit lifeways.

Walrus meat and organs, both fresh and cached, were an important source of food for not only premodern Inuit but also for their dogs (Ijjangiaq, 1990; Kappianaq, 1992, 1997). Dog teams were essential for travel by sled across dynamic land- and icescapes during winter months. Indeed, perhaps no single factor is more important than the ability to predict and negotiate sea ice in its various forms for successful marine-mammal hunting and associated navigation of the complex terrestrial-marine geography of the Arctic (see Aporta, 2002; Ford et al., 2008). Ambient temperature, tides and winds all contribute to the formation, stability and movement of sea ice and its ability to support important marine prey, such as walruses.

During warmer, ice-free months, *qajait* (ᖃᔭᐃᑦ; 'kayaks', or single-person boats) and *umiat* (ᐅᒥᐊᑦ; sail-powered, multiperson boats) allowed hunters to access walrus herds gathered on moving ice floes. In dispatching the animals, Inuit made use of a specialized hunting toolkit that included throwing harpoons with detachable foreshafts and robust heads (particularly the closed-socketed Thule Type 4), and lances for striking killing blows. Archaeological evidence of such warm-weather hunting in Inuit Nunangat comes partly in the form of shallow pit features (each around 2 m in diameter) likely representing emptied caches. Several hundred such features have been documented on the raised gravel beach ridges of northern Foxe Basin (see Desjardins, 2018). Caching locales were likely chosen according to the presence of suitable beach gravel, and not necessarily according to proximity to residential sites, making it very difficult to explicitly link particular activities at one site (kill/butchery) with another (residential). Still, the proliferation of these pit features along the gravel beaches of walrus-rich northern Foxe Basin, attests to the great success of premodern Inuit hunters, as well as the enhanced food security walruses would have provided residents during the coldest, darkest times of the year.

During cold seasons, fresh walruses supplemented the cached resources acquired during summer. Amitturmiut *IQ* suggests that during cold-weather months, premodern hunters in northern Foxe Basin waited on land for southerly winds to bring walrus-bearing ice floes through the local polynya system to the floe-edge (where the land-fast ice meets open water). *Qamutiit* (ᖃᒧᑏᒃ; sleds) drawn by dog teams facilitated travel across the snow-covered land and ice. Hunters and their dogs would have travelled some distance across both the land-fast ice and the temporarily concentrated ice floes to kill one or more basking

[2] Unless otherwise noted, non-English, italicised terms are in the Inuktut language spoken in Nunavut (specifically, the North Qikiqtaaluk/North Baffin dialect of Inuktitut).

walruses. The hunters and their dogs, now dragging the partially-butchered catch, would return to the safety of the land-fast ice before the ice floes moved back out to sea with the currents (Piugattuk, n.d.; Alaralak, 1990; Kappianaq, 1993; Qamaniq, 2001; Uttak, 2001). This perilous practice would be repeated as the opportunity arose, until the land-fast ice itself began to break up in the spring (Kappianaq, 1992).

Evidence for the lasting nature of these hunting methods comes partly from zooarchaeological data. Desjardins (2018) has compiled previously-recorded faunal data from 32 Thule-to-historic Inuit sites and site-complexes across Inuit Nunangat. Desjardins, 2018's results show that while 19 locales (59%) had walrus remains (see Fig. 6.1), the Atlantic walrus was the top-ranked resource at only two sites according to the number of identified specimens (NISP). Both sites were

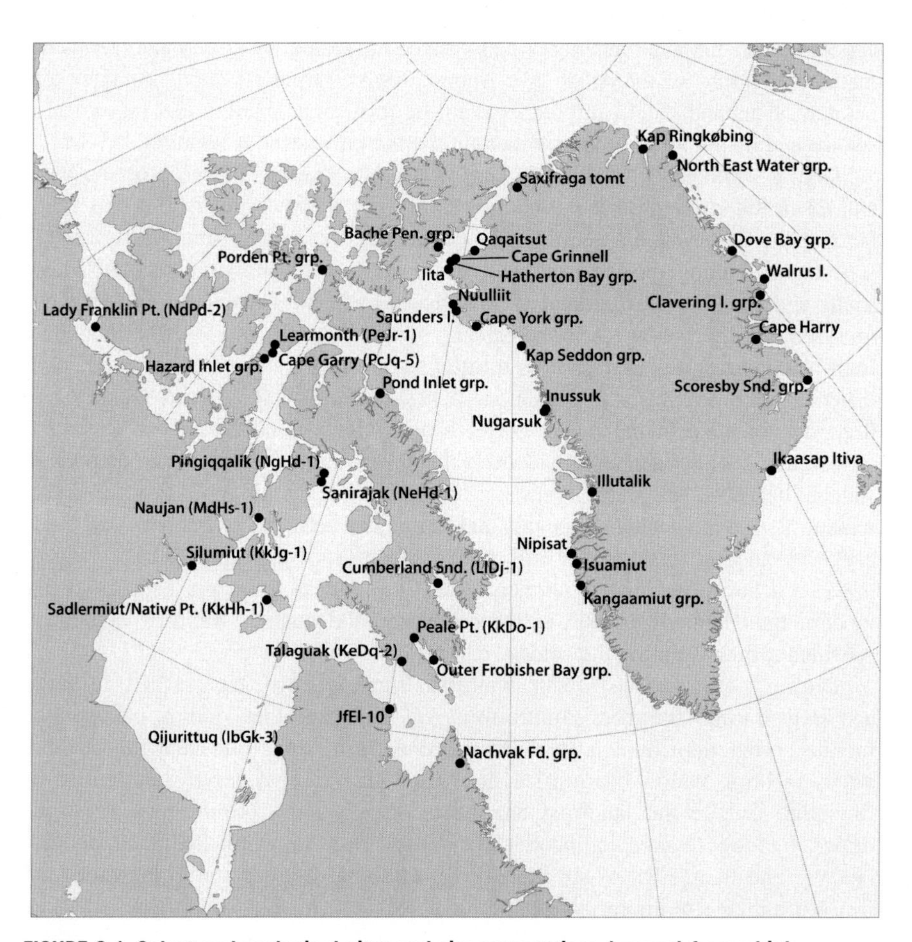

FIGURE 6.1 Select archaeological sites and site aggregations (groups) from which premodern walrus remains have been identified through zooarchaeological analysis. In Inuit Nunangat: Natchvak Fd. Group[a] (Swinarton, 2008), JfEl-10 (Lofthouse, 2003), Qijurittuq

in North-West Foxe Basin and adjacent to the recurring Foxe Basin polynya system: Sanirajak (NeHd-1) (comprising 48.1% of all identified bones) and Pingiqqalik (NgHd-1) (38.2%).[3] It is likely that the bone-laden middens at these and other large premodern winter villages in northern Foxe Basin are the result of similar daring winter forays onto the pack ice. Less time would have been spent

(Desrosiers et al., 2010), Sadlermiut/Native Pt. (Collins, 1956, 1981), Silumiut (Staab, 1979), Naujan (Mathiassen, 1927), Sanirajak (Desjardins, 2013), Pingiqqalik (Desjardins, 2016, 2018), Hazard Inlet grp.[b] (Whitridge, 1992), Cape Garry (Rick, 1980), Learmonth (Taylor and McGhee, 1979; Rick, 1980), Lady Franklin Pt. (Taylor, 1972), Porden Pt. grp.[c] (Park, 1989), Bache Pen. grp.[d] (McCullough, 1989), Pond Inlet grp.[e] (Mathiassen, 1927), Cumberland Snd. (Schledermann, 1975), Peale Pt. (Stenton, 1987), Outer Frobisher Bay grp.[f] (Henshaw, 1995) and Talaguak (Sabo, 1981). In Greenland: Kangaamiut grp.[g] (Degerbøl et al., 1931), Isuamiut, Nipisat (Gotfredsen unpublished), Illutalik (Mathiassen, 1934), Nugarsuk (Møhl, 1979), Inussuk (Mathiassen, 1930), Kap Seddon grp.[h], Nuulliit (Gotfredsen et al., 2018), Cape York grp.[i] (Grønnow et al., 2015), Saunders I. (Grønnow et al., 2017), Iita (Johansen, 2013), Cape Grinnell (Darwent and Foin, 2010), Qaqaitsut (LeMoine and Darwent, 2010), Hatherton Bay grp.[j] (Christensen, 2000), Saxifraga tomt, Kap Ringkøbing (on file, National History Museum of Denmark [NHMD]), North-East Water grp.[k] (on file, NHMD; Gotfredsen unpublished), Dove Bay grp.[l] (on file, NHMD; Thostrup, 1911), Walrus I. (Gotfredsen, 2010; Grønnow et al., 2011), Clavering I. grp.[m] (Degerbøl, 1934; Gotfredsen, 2010), Cape Harry (Degerbøl, 1935), Scoresby Snd. grp.[n] (on file, NHMD; Sandell and Sandell, 1991), Ikaasap Itiva (Møbjerg and Robert-Lamblin, 1990).

[a] Nachvak Village (IgCx-3) and Kongu (IgCv-7); [b] Ditchburn Pt. N, S (PaJs-3), PaJs-4 and PaJs-13; [c] Porden Pt. Brook Village (RbJr-1), Porden Pt. Pond Village (RbJr-4) and RbJr-5; [d] Skraeling I. (SfFk-4), Sverdrup (SfFk-5) and Eskimobyen (SgFm-4); [e] Mittimatalik (PeFr-1) and Qilalukan (PeFs-1); [f] Kamaijuk (KfDe-5), Kuyait (KfDf-2) and Kussejeerarkjuan (KeDe-7); [g] Z.M.K. 13/1930 Utorqaat; Z.M.K. 14/1930 Illutalik; Z.M.K. 15/1930 Uummannat; Z.M.K. 16/1930 Qeqertarmiut; [h] Z.M.K. 80/1979 Illuminerssuit; Z.M.K. 82/1979 Tupersuai; [i] Z.M.K. 44/2015, KNK 3900 Ivsuissoq (Parker Snow Bugt); Z.M.K. 52/2015, KNK 3908 Inersussat, Salve Ø; [j] Z.M.K. 69/1996 KNK 3074 Qeqertaaraq, Hus y142/294; [k] Z.M.K. 48/1993 KNK 2071 Eskimonæsset; Z.M.K. 49/1993 KNK 2073, Henrik Krøyer Holme; Z.M.K. 50/1993 KNK 2072 Sophus Müllers Næs, Amdrup Land; Z.M.K. 54/1993 KNK 2076, Eigil Knuth Site; [l] Z.M.K. 13j/1909 Maroussia; Z.M.K. 13n/1009 Renskæret; Z.M.K. 13p/1909 Rypefjeldet; Z.M.K. 13r/1909 Snenæs; Z.M.K. 13w/1909 Stormbugt, Østkyst; Z.M.K. 13z/1909 Syttenkilometernæsset; [m] Z.M.K. 28/1932 Dødemandsbugten; Z.M.K. 101/2007, KNK 3101 Cla-06 Fladstrand; Z.M.K. 119/2007, KNK 3117 Cla-33 Kap Breusing; Z.M.K. 121/2007, KNK 3119 Cla-36 Dahls Skær; Z.M.K. 124/2007 KNK 3122 Blåklokkenæs; Z.M.K. 61/2008 KNK 3131 Cla-02; Z.M.K. 63/2008 KNK 3130 Cla-63 Tangen; [n] Z.M.K. 76/1984, KNK 3011 Sandells' vinterhus; Z.M.K. 8a/1924 Kap Tobin; Z.M.K. 101/1966 Kap Tobin, 'Varde Pynt'; Z.M.K. 129/1964 Hurry Inlet.

[3] Importantly, most of the assemblages examined came from sites occupied primarily during cold seasons (autumn and winter); due in part to their imposing presence on the Northern landscape, such sites have commanded a disproportionate amount of archaeological attention over the past several decades.

on specialized field butchery for caching meat during the cold-weather months than during summer hunts because the impetus for hunters would have been to return to the safety of the land-fast ice as quickly as possible. The shortened processing time at kill sites meant that greater amounts of skeletal material, as well as a wider variety of both high- and low-utility elements, would have been returned to residential sites during cold-weather months.

This reliance on walruses by the ancestors of Amitturmiut likely provided a degree of food security denied to Inuit groups elsewhere in Inuit Nunangat as the Little Ice Age (approximately 13th to 19th centuries CE) intensified. Increasing sea ice caused the 'Classic' Thule Inuit bowhead whaling industry to largely collapse. In select regions, walruses (second in mass only to bowhead whales) may have effectively replaced the latter species as a source of 'wealth' in food and raw materials.

Recent-historic hunting of walruses followed much the same pattern as in premodern times, especially in northern Foxe Basin, where the cumulative effects of Euro-Canadian colonialism on traditional subsistence regimes took hold relatively late (see Rasing, 2017). During the breakup of land-fast ice in the spring, Amitturmiut used *umiat* to travel sometimes long distances eastward to central Foxe Basin, where walruses regularly hauled out on fields of moving ice (see Paniaq, 1998). Prior to the introduction of rifles, a harpoon (with a toggling head) would be used in tandem with a lance. A hole would be carved in the ice in which to anchor and secure the harpoon line (Inukshuk, 1990), which could be made from durable bearded seal (*Erignathus barbatus*) skin, and was coiled carefully to avoid becoming tangled during the hunt. A basking walrus would be approached slowly and quietly, harpooned, and then set upon with lance blows to either the lower abdomen (which was believed to exhaust the animal quickly) or the heart for a relatively quick kill (Piugattuk, 1990).

Today, walruses complement the modern 'country-food' diet of hunted and gathered resources across the parts of Inuit Nunangat where the natural distribution of walruses and traditional Inuit lands overlap (Fig. 6.2). Rifles have replaced lances as the primary killing instrument, though harpoons with metal heads are still regularly used to secure an animal after it is shot. In addition, skin and canvas sails have given way to square-stern canoes with outboard motors and recreational fishing boats generally less than 10 m in length. These modern, motor-powered boats have dramatically reduced hunting times (from days to several hours), as well as increased the haul that can be safely returned home.

Among Amitturmiut, as soon as the land-fast ice has broken up sufficiently (typically by mid-July) hunters travel east by boat to meet the moving pack ice. Great care is taken to keep noise to a minimum for fear of frightening the walruses, which can be found basking on ice floes in small groups of between two and ten individuals. Most often, a single animal is targeted and shot at the closest reasonable distance (depending upon opportunity), with the neck as the preferred location for a disabling shot. The lead boat then races to the kill site, where the animal is harpooned to prevent it being sunk and lost.

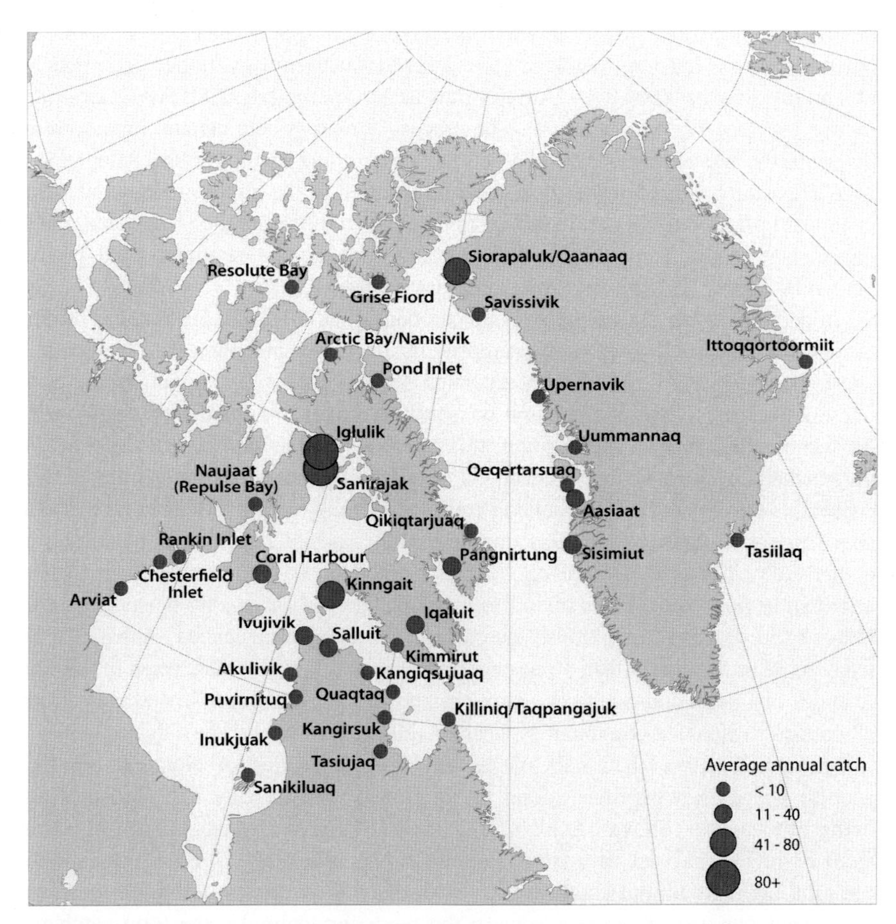

FIGURE 6.2 The extent of the contemporary subsistence walrus hunt in Inuit Nunangat, Nunavik and Greenland by average annual catch; data for Inuit Nunagat: Priest and Usher, 2004 (Nunavut*, 1996/97 to 2000/01); Brooke, 1992 (Nunavik, 1974 to 1991); data for Greenland: Siorapaluk/Qaanaaq, Savissivik, Upernavik, Uummannaq, Qeqertarsuaq, Aassiaat, Sisimiut (Born et al., 2017; CITES, 2018), Ittoqqortormiit and Tasiilaq (Wiig et al., 2014; CITES, 2018).

** In Clyde River, the 5-year average was less than 1 but greater than 0, as a single walrus was recorded as having been harvested in 1996/97 (Priest and Usher, 2004: 198); for the purposes of this analysis, the community's 5-year harvest was 0 animals.*

In the summer of 2012, Desjardins (2016) documented a representative Amitturmiut hunt during which a small hunting party in two small boats acquired a single adult bull walrus in an ice field approximately 75 km east of Iglulik (also known as "Igloolik"). Shot with a small-calibre hunting rifle by the crew's only female member, the animal was then harpooned, rolled into the water and towed to a more stable ice floe, where it was butchered over approximately six hours. The butchery process has long revolved around the production of cylindrical skin pouches of varying lengths known by Amitturmiut as *ungirlaat* (ᐅᖏᕐᖥᑦ) [sing., *ungirlaaq* (ᐅᖏᕐᖥᖅ)][4] (see Figs. 6.3 and 6.4). The process is both systematic and time consuming, with the processing of a single walrus taking up to several hours. After processing, *ungirlaat* are loaded into boats and transported to the shore or floe-edge. Meat is almost immediately buried in loose gravel close to the shoreline (Kappianaq, 1997; Paniaq, 1998; Qamaniq, 2000). After several months, the resulting aged product known as *igunaq* (ᐃᒍᓇᖅ) is dug out and feasted upon.

In addition to the six *ungirlaat* produced during the 2012 hunt, a number of other elements were brought back to the campsite for dog food, including the ribs,[5] sternum and front limbs (rear limbs were incorporated directly into two of the *ungirlaat*). Some intestines, kidneys, liver and cranium with tusks were also returned. The heart (widely considered a delicacy by Amitturmiut) was eaten raw by the members of the crew during butchery. Though walrus stomach contents are often examined and collected, time constraints prevented the 2012 hunting party from doing so. The production of modern *ungirlaat* varies slightly depending upon the habits and interests of the hunter and his or her crew. The goal, as it likely was in premodern times, is to produce as many *ungirlaat* as possible; there is often much prebutchery debate among hunters about the best way to proceed to maximise the cacheable haul of each landed animal. Though today more of a delicacy than a seasonal staple, *igunaq* is still widely eaten in the communities of Iglulik and Sanirajak (formerly known as "Hall Beach"), and is occasionally traded to Nunavut communities where walruses are not locally available.

Hunters in more than half of the small communities of Nunavik regularly acquire walruses, with those along the coasts of Hudson Bay and Hudson Strait being most active (Stewart et al., 2014: 283–284).[6] An 18-year (1974–91, inclusive) review of harvest data from hunters in Nunavik's 16 communities[7] showed the total average annual catch in the region was 72 walruses, although there was significant variation from one year to the next; for example, the total catch was only four walruses in 1978, but 174 in 1986 (Brooke, 1992: 14). This unevenness could be attributed to inclement weather, financial considerations (e.g., access to equipment and money for fuel), or even occasional outbreaks of foodborne illness attributed to raw or undercooked walrus products [walruses are considered the most susceptible of the regularly harvested marine-mammal species to trichinellosis infestation (Proulx et al., 2002; Larrat et al., 2012)].

[4] See Desjardins (2018a) for a detailed description of the *ungirlaat* production and caching processes.

[5] In the past, as today, meat from the ribs could be cut away from the bones and placed within *ungirlaat* (Kappianaq, 1997), or they could be returned in sections and dried (Iqallijuq, 1999; Paniaq, 1998).

[6] Stewart et al. (2014) have compiled available quantitative data from multiple sources on both recent-historic and contemporary Inuit walrus harvests in both Nunavik and Nunavut.

[7] The author includes two small communities—Killiniq and its successor community Taqpangajuk—that are currently disused.

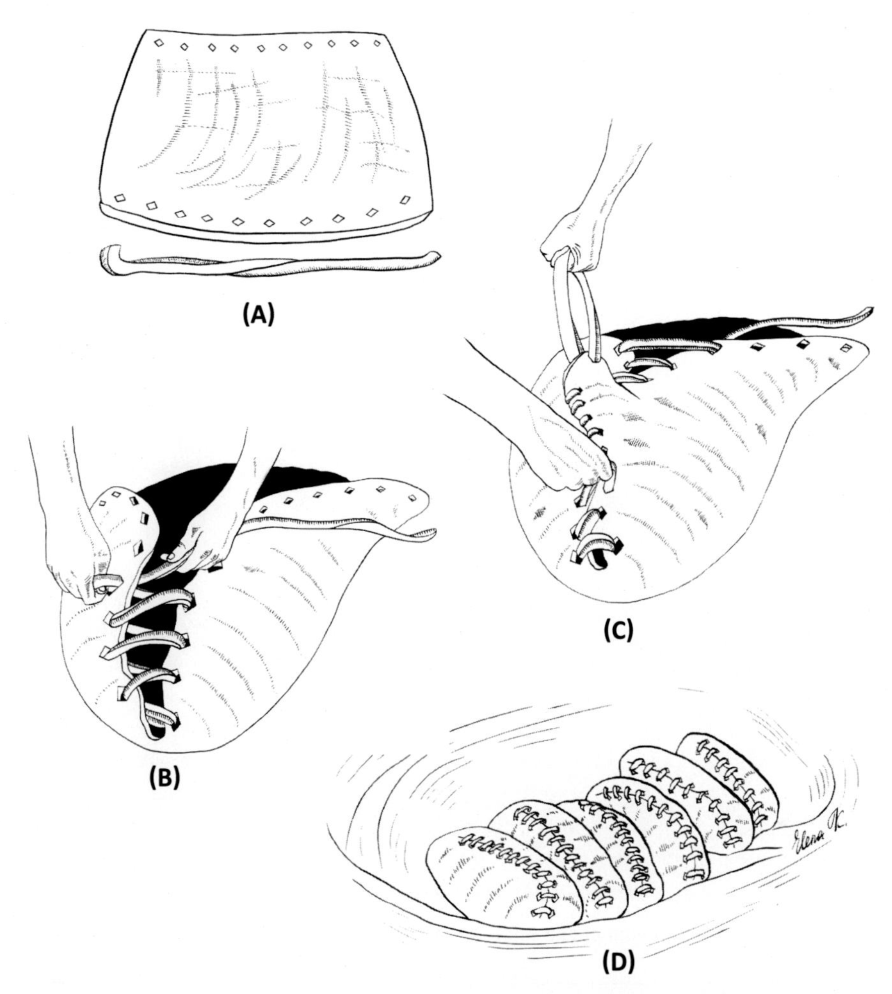

FIGURE 6.3 Typical production of an *ungirlaaq* pouch: (A) a sizable rectangle and a long, thin strip of walrus skin are cut from the animal, and the rectangle is punctured multiple times on two sides; (B) meat and select organs are placed on the interior surface of the rectangle, which is then folded over; the skin strip is woven through the punctures; (C) the weaving strip is periodically tightened to ensure the meat within the *ungirlaaq* is secure; and (D) once the hunting crew has returned to shore, *ungirlaat* are placed in a neat row within the beach gravel (laces-down for the first layer, which sits upon the permafrost) before being covered with gravel and left to age for several months. Illustration by Elena Kakoshina.

FIGURE 6.4 Amitturmiut men cache *ungirlaat* after a successful walrus hunt, northern Foxe Basin, c. 1957.

Photograph by J. Meldgaard; courtesy, National Museum of Denmark, Copenhagen.

Inuit of Nunavik and Nunavut are generally free to hunt walruses without significant restrictions.[8] The Nunavut Wildlife Harvest Study (NWHS) (Priest and Usher, 2004) provides comprehensive quantitative data derived from hunter surveys on subsistence harvesting in all 27 Nunavut communities[9] over a five

[8] Some restrictions apply to individual Inuit; federal community walrus harvest quotas exist only for hunters in the Nunavut hamlets of Sanikiluaq, Coral Harbour (Salliq), Arctic Bay and Clyde River (see COSEWIC Committee on the Status of Endangered Wildlife in Canada, 2017: 43–45).

[9] The authors included two small communities—Umingmaktok and Bathurst Inlet—that are not regularly occupied year-round.

year period (1996/97 to 2000/01).[10] During the NWHS survey period the combined average annual harvest of walruses of all communities was 382 (Priest and Usher, 2004: 804), with the highest numbers by far recorded in the northern Foxe Basin Amitturmiut communities of Iglulik (an average of 152 walruses per year) and Sanirajak (95 per year). These numbers are consistent with more recent catch estimates reported to Fisheries and Oceans Canada by community Hunters and Trappers Organization/Associations (HTO's/HTA's), which quantified the total walrus subsistence harvest across in 2016/17 to be 346 animals. A remarkable 69% of the territory-wide catch was acquired by Amitturmiut, with 110 walruses hunted out of Sanirajak and 129 out of Iglulik [DFO (Fisheries and Oceans Canada), n.d., Appendix I].

Walrus hunting in Greenland

Recurring sizable polynyas in both North-East and North-West Greenland have long formed hotspots of ecological productivity similar to northern Foxe Basin, offering relatively light ice conditions and excellent benthic feeding opportunities for walruses. Such areas did not go unnoticed by the earliest Inuit occupations in Greenland (Fig. 6.1). By the 14th century CE, people of the pioneering 'Ruin Island' phase of Thule Inuit expansion into the North-East had established themselves on both sides of the North Water Polynya, stretching from northern Melville Bay to Hall Land in Greenland along central East Ellesmere Island in Nunavut, Canada (Holtved, 1954; McCullough, 1989).

In the Avanersuaq region of northern Greenland, archaeological evidence from the ten known Ruin Island sites suggests people were organised around communal hunting of large whales and walruses. Each hunting group was presumably headed by a talented *umialik* (hunting crew chief) (see Savelle, 2002). Walrus bones by NISP accounted for around 20% of the identifiable faunal assemblage at the recently-excavated Ruin Island site Nuulliit. (When artefact-production waste is included in the zooarchaeological tally, ivory comprised approximately 20% of the total number of fragments by NISP, while walrus-bone debris totalled 7%.) These results highlight the importance of walruses as a source of ivory and bone for tool production, blubber for fuel, and edible material for both Inuit and their dogs (Gotfredsen et al., 2018, see also Darwent & LeMoine, this volume).

A notable premodern Inuit site in the understudied North-East of Greenland is the aptly-named Walrus Island, situated nearby the large Sirius Water Polynya. Archaeological work at the site (2007–2008) revealed approximately 1700 stone structures (mostly caches, but also tent rings and other light shelters). Radiocarbon dates indicate the site was occupied continuously from the earliest Thule Inuit occupations beginning around 1400 CE in the region until the 19th

[10] In Iqaluit, data were available for only four years (1997/98 to 2000/01).

century CE. The number and distribution of features at the site differs significantly from those at other known archaeological sites in North-East Greenland. Zooarchaeological evidence indicates that walrus remains comprised more than 50% of all bone material by NISP and were primarily associated with the numerous caches (Gotfredsen, 2010; Grønnow et al., 2011). Walrus skeletal element distribution at Walrus Island showed an abundance of cranial bones (such as mandibles, premaxillae and tusk-bearing maxillae), a pattern largely mirrored at Pingiqqalik (see Desjardins, 2018). Also present in high numbers at Walrus Island were meat- and blubber-bearing front and hind limb elements (minus flippers).

The subsistence-settlement system evolving around Walrus Island was likely characterised by the hunting of (1) ringed seals on the land-fast ice during winter; (2) intensive, communal hunting of larger marine mammals, including walruses, in and around the polynya during spring; (3) ice-edge hunting of seals and narwhals during late spring and summer and (4) hunting of caribou in the interior during summer and autumn (Gotfredsen, 2010). Walrus bone elements at the nearby winter settlement Fladstrand (Cla-06) were less common (comprising only 3% of mammal NISP) and more frequently found as the raw material for tools and decorative purposes (tusk fragments, post-canines, and bacula), thus reflecting influx of raw materials and occasional winter walrus hunting (Gotfredsen, 2010; Grønnow et al., 2011). As at Pingiqqalik, the longevity of the occupation can reasonably be attributed to the increased food security of the large-scale walrus caching regime (see Grønnow et al., 2011). Unlike in northern Foxe Basin, however, the productive social–ecological system at Walrus Island and nearby sites appears to have collapsed when the local walrus population declined during extremely cold episodes of the 18th and 19th centuries (associated with the Little Ice Age). By the mid-19 century, North-East Greenland had been abandoned altogether by Inuit (Grønnow et al., 2011).

A recent-historic Inughuit[11] site of particular importance is Inersussat on Saunders Island, North-West Greenland. The site was inhabited in the 19th and early 20th centuries and is known today from the 1903 visit by the Literary Greenland Expedition. Historical sources reveal that many snow houses were present, indicating winter occupation. Still, most of the archaeological features that have left clear traces are from spring, summer and autumn settlements (primarily tent rings and rock shelters) (Grønnow et al., 2017). The two walrus-hunting camps (Walrus Island and Inersussat) show great similarity in their walrus-bone element distribution despite the great geographic distance between the two (approximately 1350 km) and the differences in site age (Fig. 6.5).[12] Walruses were likely hunted at both sites during

[11] The first Europeans named people in the Avanersuaq 'Polar Eskimos'; locals, however, referred to themselves as Inughuit.

[12] MNE (Minimum Number of Elements) denotes the number of skeletal elements used for calculations of the relative distribution of walrus body parts. The %MAU (Minimum Animal Unit) (Lyman, 1996) calculations take into consideration that different skeletal elements in a walrus skeleton are represented in different numbers, e.g., by two scapulae (one left and one right) but 30 ribs (15 left and 15 right).

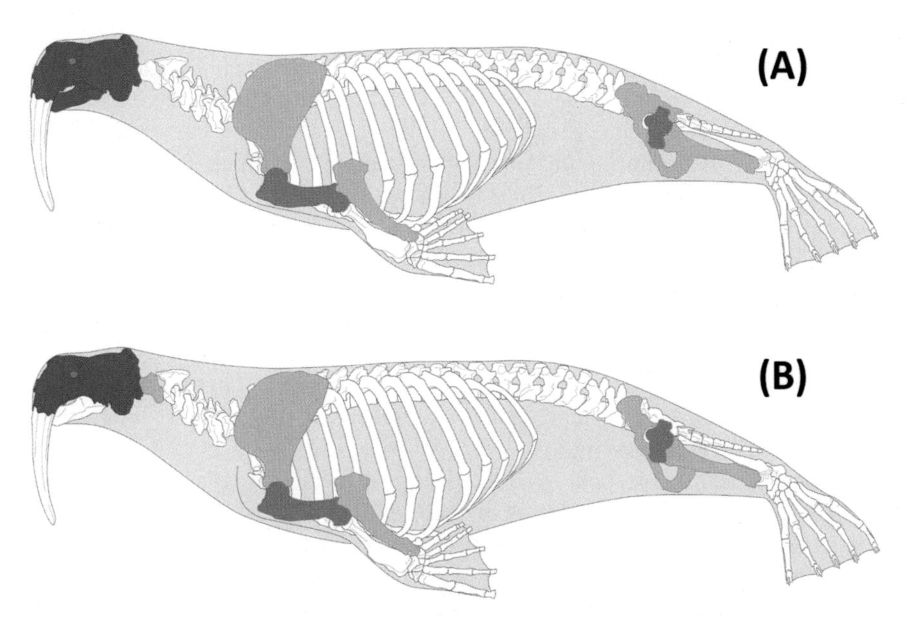

FIGURE 6.5 Walrus element frequency (as %MAU) from (A) Walrus Island (MNE = 294) and (B) Saunders Island (MNE = 302), showing a clear dominance of cranial and high-utility elements. Dark grey>75%; medium grey>50%; light grey>25%, white<25%.

(A) Data from Grønnow et al. (2011) and (B) Gotfredsen unpublished. *Image by M. Coutureau, archaozoo.org.*

winter and early-spring when the land- and seascape were still covered in snow and ice. This would have allowed dog teams to transport large chunks of partially-butchered walruses back to residential camps. Numerous accounts of walrus hunting in leads and cracks in ice formed near Saunders Island attest to this practice, although walruses were also known to gather offshore in the region during the open-water period from October to November (Holtved, 1935, 1936). In winter and early spring most hunting took place on the thin-ice (Grønnow, 2016; Born et al., 2017). When the weather was stable and the land-fast ice had light snow-cover, hunters walked on the thin-ice to acquire walruses by harpooning them when the walruses broke through the ice to create rough breathing holes.[13]

The Danish ethnographer Erik Holtved, who excavated Inuit settlements and lived in North-West Greenland during the 1930s, offered a vivid description of a thin-ice walrus-hunting trip in the mid-20th century. Hunters left their dog teams and sleds at the boundary of thin-ice, and then quietly proceeded in single-file on

[13] Walruses are able to break through ice exceeding 20 cm in thickness (Fay, 1982).

the newly-formed ice (no more than 7 or 8 cm thick) (Holtved, 1967). Upon sighting a walrus, the hunters approached quietly and thrust their harpoons into the animal, securing their long lines to the ice by the means of an ice chisel and a lance. The wounded animal, anchored to the ice, was unable to escape. Hunters then attacked the animal with lances (Holtved, 1967) which today have been almost entirely replaced by rifles (Born et al., 2017: 147ff). Such thin-ice hunting requires significant patience, timing, skill and cooperation. The hunter whose line was the strongest was considered the harpooner, and it was he who received the choicest share of the animal (typically the head and heart) (Holtved, 1967: 100–103). The extent to which thin-ice hunting was performed varied significantly between regions and years, depending upon local weather conditions. The method was considered a 'signature' form of Greenlandic Inuit walrus hunting and was practised from the 19th century CE until around a generation ago.

Faunal remains from both Walrus Island and Inersussat indicate that premodern Inuit regularly hunted walruses of both sexes and across all age groups. This was also broadly true for both residents of the Ruin Island site Nuulliit, as well as Paleo-Inuit assemblages from central West Greenland (Gotfredsen and Møbjerg, 2004; Gotfredsen et al., 2018). Despite this, selective hunting may have been carried out on a smaller scale due to a number of important factors: (1) seasonal changes in walrus behaviour (the sexes are segregated for long periods of each year); (2) accessibility and (3) selection by Inuit for traits unique to sex and/or age. (For example, females were of greater economic value because they yield relatively more blubber and have tusks considered better suited for toolmaking [Born et al. 1994: 22ff]).

Information on historic and premodern Inuit walrus hunting in Greenland can be obtained from descriptions by both explorers and colonial administrators. Although often anecdotal, such accounts provide a generally reliable sense of the scale of hunting at particular points in time, as well as the locations of prime hunting locales, and preferred methods.[14] Quantitative information is available in the Hunters List of Game (HLG), a Danish compendium of harvest data beginning in the 1860's CE in West Greenland. However, catches from North-West Greenland are only reported occasionally to HLG until the 1950's (Andersen et al., 2018). Zooarchaeological assemblages dating to the 19th and early 20th centuries provide additional insights into walrus hunting practices in Greenland during recent-historic times.

Reliable modern catch statistics were not available until the implementation of the *Piniarneq*, a wildlife harvest recording system begun in 1993. Under this system, it is mandatory that each harvested walrus be documented on a 'Special Reporting Form' (Andersen et al., 2018; Born et al., 2017: 35). Over the past two

[14] Overviews of historic walrus hunting for various regions of Greenland are found in Grønnow (2016) (for the northwest), Born et al. (1994) (for the western coast) and Born et al. (1997) and Sandell and Sandell (1991) (for the eastern coast).

decades, good qualitative information has been collected through one-on-one interviews with, and surveys of, Greenlandic Inuit walrus hunters. In a 2010 study by Born et al. (2017), surveys recording Local Ecological Knowledge (LEK) of 76 experienced hunters in walrus-hunting regions between Maniitsoq (central West Greenland) and Siorapaluk (North-West Greenland) demonstrated conclusively that walruses remain highly important in both West and North-West Greenland. Walrus products are regularly and widely consumed by Inuit and their dog teams, dogs being especially important to the subsistence hunting economies of North-West Greenland. Walrus meat, hides and tusks are sold on local open markets known as *Kalaaliaraq* (Danish: *Brættet*), as well as shared via traditional exchange networks (Born et al., 2017: 41ff), as is common in Inuit Nunangat.

Generally, present-day walrus-hunting patterns in Greenland are distinct from those practised by historic and premodern Greenlandic Inuit. Since the introduction of hunting quotas for walruses in 2006, harvests tend to be skewed towards large adult males due to the greater amounts of meat available to feed dog teams. Unlike in Inuit Nunangat, dog teams remain in wide use across Greenland.[15] A decline in walrus hunting has been observed since the early 1990's, particularly following the introduction of harvest quotas Born et al. (2017) has attributed the long-term decline to a number of factors, including:

1. Walruses becoming a 'secondary' prey, or 'by-catch' as walrus hunts are combined with hunts of more economically 'valuable' resources, such as beluga whales (*Delphinapterus leucas*), narwhals (*Monodon monoceros*) and polar bears (*Ursus maritimus*) (also subject to quotas);
2. General decrease in market demand for walrus products;
3. Transition over time from full-time hunting to wage-earning employment (in industries such as fishing) and
4. Effects of modern anthropogenic climate change (Born et al., 2017).

Climate change is resulting in unstable ice and unpredictable weather conditions, thereby changing the amount of time walruses spend at traditional hunting locales and hence access by Inuit to walruses (see Desjardins and Jordan, 2019). In central West Greenland, a clear decrease in walrus hunting in recent years can be attributed to a reduction in the number of dog teams, although the hunt still offers cash income from the sale of trophy skulls, tusks and bacula (Born et al., 2017: 41ff).

The decrease in walruses caught across Greenland is recorded in the *Piniarneq* (from 1993 to 2012), as are clear differences in the significance of walrus hunting among communities in North-West and West Greenland. In Qaanaaq (North-West Greenland) 46 full-time hunters (from a population of around 596) hunted 120 walruses in 1993. Between 1995 and 2006, the annual harvest varied

[15] In addition to their overall robustness, adult males provide superior cranial trophies with longer and fuller tusks; however, some Greenlandic Inuit hunters still prefer younger males and/or adult females due to their more tender meat (Born et al., 2017: 47, 81).

between 30 and 70 walruses, while from 2006 to 2012, the total annual catch ranged between 5 and 20. Similarly, in Sisimiut (West Greenland) 101 full-time hunters (from approximately 5000 total residents) hunted 75 walruses in 1993; in the following years, the average annual catch dropped to well below 100 (Born et al., 2017: 44f). Throughout the 1980's and 1990's, hunters from two of South-East Greenland's communities (Ittoqqortoormiit and Tasiilaq) harvested between 20 and 30 walruses annually (Born et al., 1997).

In North-West Greenland, the primary walrus hunting season runs from January to June, and again from October to November (Born et al., 2017: 40). As in Inuit Nunangat, hunting in Greenland is carried out during warm-weather months on the moving pack ice and in open water during early autumn. Historically walruses were also taken at terrestrial haul-out sites; a practice now forbidden by law throughout Greenland.[16] In cold-weather months, most hunting used to take place on the thin-ice. Thin-ice hunting is rarely practised today, and what remains is under pressure from climate change. Warmer temperatures and increased storminess has led to shorter periods of time while the ice is sufficiently stable for hunting, and increased snowfall can lead to 'squeaking' while stalking walruses, warning them of the hunters' arrival (Born et al., 2017: 76).

Andersen et al. (2018) documented a 2015 summer hunt in Inglefield Land, North-West Greenland. A herd of walruses on the pack ice was approached quietly by boat. At the right moment and from the right distance, the younger of the two hunters jumped suddenly onto the ice, harpooned and then shot a large male walrus. Having landed the animal, this hunter (also the owner of the boat) received the head, tusks and baculum. During the butchery process, both hunters scrutinised and sampled the stomach contents, which consisted mainly of mussels. The stomach contents, along with the heart, were tasted and collected for later consumption. The remainder of the walrus was cut into easily-transportable slabs of flippers, ribs, back, etc., and stacked on the beach where the butchering took place. To lighten the amount of meat and blubber to be transported back to the hunters' home community of Qaanaaq, two *ungirlaat* – created in much the same way as those produced by Amitturmiut – were created. One with meat and blubber that was immediately buried under a heap of rocks, and another made at the hunters' cabin with liver and kidneys, which was also left under rocks to age. (Unlike in northern Foxe Basin, the consumption of *igunaq* is reserved mainly for times of celebration [Andersen et al., 2018]).[17]

[16] Several terrestrial haul-outs were known to be in use in premodern times up to the 1950's in the West and North-West Greenland—as well as in North-East Greenland, where only two remain today; however, recently, walruses have started using other haul-out places in the North-East (Born et al., 1994, 1997; Born, 2012).

[17] According to contemporary Greenland hunting regulations, all edible walrus products must be brought to the home community, or alternatively cached in meat caches (Anonymous, 2006). This makes nonsubsistence trophy-hunting for tusks or complete skulls illegal. This is broadly in keeping with traditional Inuit provisions for using all or most of the walrus.

Hunting strategies in East Greenland differed significantly from those in the West and North-West. In recent-historic times, walruses along the East Greenland coast were harvested in large numbers by European sealers, hunters and trappers (see Gjertz, this volume). Due in part to the commercially-driven historic decimation of the stock, contemporary hunting is of less importance for local Inuit communities than for those elsewhere in Greenland (hunting being somewhat more commonly practised in Ittoqqortoormiit than in Tasiilaq; walruses in these communities were primarily hunted from the floe-edge from February to June [Sandell and Sandell, 1991: 108]). Today, the hunt is carried out from May to September, with (typically) adult male walruses shot in the head, neck or back from small dinghies with outboard motors. Harpoons are rarely used, but a large hook is applied to retrieve sunken walruses from shallow waters (see Born et al., 1997: 40; Born, 2012).

Walruses and Inuit cosmology

In Inuit cosmology, as in that of many Indigenous societies across the circumpolar Arctic, personhood is attributed to many nonhuman animals; especially those figuring in the hunter–prey relationship (see Desjardins, 2017; Hill, 2011b; Laugrand and Oosten, 2014; Nuttall, 2000). This worldview was expressed clearly and succinctly by Knud Rasmussen: "*In former times animals in human form were common [. . .] In olden times, too, everybody could easily turn into animals, and until quite recently shamans have had the same powers*" (1932: 35). Beyond the obvious dietary significance of walruses and the utilitarian value of their ivory,[18] walruses also had, and continue to have, tremendous symbolic and cosmologic importance to Inuit and Tuniit before them (see Hill, 2017 for a thorough discussion of the symbolic significance of walruses for Bering Strait Indigenous peoples).

One way in which the bond between human and nonhuman was solidified and actuated was through the wearing of clothing either made from, or made to look like, specific prey species including walruses (Desjardins, 2017). Exemplary archaeological finds from Nuulliit that may partly reflect this worldview include fragments of three separate gut–skin *annuraat* (ᐊᓐᓄᕌᑦ; waterproof coats) dating to the 14th century CE (Fig. 6.6), the oldest-known such examples recovered in the Arctic. Elongated wedges are sewn into both the front and the back of the

[18] It is increasingly clear premodern Inuit peoples in regions where walruses were locally available made abundant use of ivory. Walrus skeletal element frequencies at Walrus Island and Pingiqqalik showed strong evidence of a focus on tusk extraction and ivory harvesting; both sites featured an abundance of cranial bones, such as mandibles, premaxillae and tusk-bearing maxillae. Desjardins (2013) has emphasised the importance of considering the desirability of ivory in any application of a food/meat utility index for walruses.

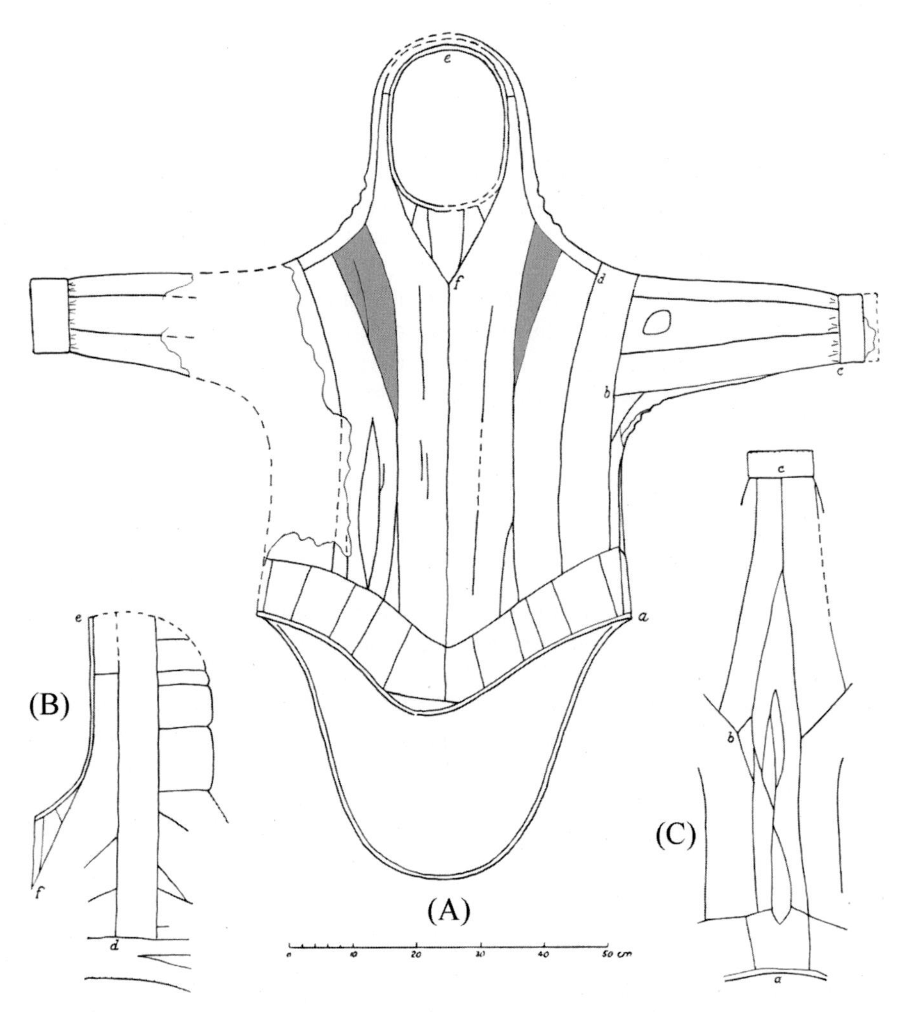

FIGURE 6.6 Multiple views (A–C) of one of three well-preserved gut–skin *annuraat* recovered from House 28 of the Ruin Island site *Nuulliit*. The two curved, elongated wedges on (shaded here for illustration) are interpreted as walrus tusks.

Modified from Holtved (1954). ©Museum Tusculanum Press. Illustration: Erik Holtved.

coat (Holtved, 1954: Figs. 49–54), likely representing walrus tusks extending lengthwise down each side of the garment (Schmidt et al., 2013). In wearing such clothing, a hunter was protected not only from the frigid water but also assumed the strength of the walrus, as well as its ability to hold its breath underwater for extended periods.

A total of 116 recorded traditional Greenlandic Inuit accounts[19] from the 18th to early 20th centuries feature walruses in diverse and meaningful ways: as real beings-in-the-world, as symbols of particular emotions and social situations, and as spiritual collaborators or opponents due to their fierce and sometimes unpredictable behaviour during the hunt (see Gotfredsen et al., 2018). Indeed, walrus hunting has long been considered by many Inuit as a particularly dangerous activity requiring great skill, endurance, physical strength and courage. According to the esteemed late Amitturmiut hunter and elder Piugattuk, "[w]alruses are the most ferocious marine animals" (1990).

The premodern association between walruses and the Inuit afterlife/underworld[20] is manifest in several ways. Like all marine mammals, walruses were traditionally believed to have been created from the severed fingers of the shadowy sea-woman *Sedna/Senna* (ᓴᓐᓇ) (see Laugrand and Oosten, 2010: 108–109). They could serve as important and powerful *tuurngait* (ᑑᕐᖓᐃᑦ), helping spirits whose perspectives, physical forms or characteristics, could be assumed by Inuit shamans for a wide variety of purposes (see Laugrand and Oosten, 2010: 71–72; Sonne, 2004a: IDs 126, 152, 160, 279, 340, 382, 1850).[21] The notion that the human and nonhuman soul is returning to a new body after death (Nuttall, 1994) may still manifest itself in some present-day communities across Inuit Nunangat and Greenland. In this sense, game animals, such as walruses, can be perceived by hunters as a kind of renewable resource (Fienup-Riordan, 2000). One widely held belief is the premodern concept that if afforded respect, prey animals give themselves willingly to hunters. Conversely, when taboos are broken or respect is withheld, game may take revenge by making themselves scarce in the future (Rosing, 1998; Laugrand and Oosten, 2014). In premodern times, this concept may have resulted in self-regulatory hunting practices guided by strict and sometimes idiosyncratic taboos and rituals. Andersen et al. (2018) suggest that this perception of reciprocity, respect and the concept of hunting sustainability still exists among today's subsistence walrus-hunting communities.

---

[19] A comprehensive archive of 2280 online records compiled in 'Sagn & Myter', as well as an extensive introduction to this database, is available in Sonne (2004a,b).

[20] Amitturmiut still maintains a tradition—known widely in Nunavut—that the northern lights are the manifestation of spirits playing a ball game with walrus heads (Arnatsiaq, 1990; Niviattian Aqatsiaq, 1990).

[21] Hunters also engaged in symbolically meaningful activities with walruses during and after the hunt. As with small seals, water would sometimes be poured into a recently-hunted walrus' mouth, a gesture meant to give thanks to the animal (Kappianaq, 1990), presumably thirsty not only because of the rigors of the hunt but also because it lived its life in salt water (Laugrand and Oosten, 2010: 83).

Conclusion

For more than half a millennium, Inuit in Inuit Nunangat and Greenland have looked to walruses to stave off food insecurity, build their toolkits and enrich their symbolic lives. In examining Inuit–walrus relationships over time, we rely not only on a rich archaeological record, but also on a still-vibrant hunting tradition. There is invaluable traditional knowledge about walrus use by Inuit in the recent-historic and more distant past. We believe long-term Inuit–walrus relationships have the capacity to inform on hunter–gatherer resilience in response to changing climates, with sites such as Walrus Island and Pingiqqalik exemplifying food storage strategies likely developed in direct response to the Little Ice Age climate change episode.

For many modern Inuit communities, walrus products remain both economically and culturally important, despite the overall decline of country-food consumption seen in Indigenous societies across the circumpolar Arctic. As the only surviving species in their biological family, walruses have a unique morphology that sets them apart from other species regularly encountered by Inuit. Their great bulk, bold tusks and general gregariousness have likely contributed over time to a coequal symbolic importance attributed by Inuit to the species. Walruses often feature prominently as distinct beings with agency in modern Inuit art, a testament to their premodern status as nonhuman persons imbued with culture, emotions and individuality.

Acknowledgements

This chapter was made possible in part by a Veni research grant held by author S.P.A. Desjardins from the Dutch Research Council [Dutch: *Nederlandse Organisatie voor Wetenschappelijk Onderzoek* (*NWO*)]. Both the authors extend sincere thanks to communities in both Inuit Nunangat and Greenland – in particular, Iglulik, Nunavut. F. Steenhuisen (Arctic Centre, University of Groningen) provided assistance with figures, while M. Appelt (National Museum of Denmark) graciously provided permission to publish Fig. 6.4. We are grateful to Ms. J. Kadlutsiak (Iglulik) for assistance with the Inuktut translations. For invaluable help on modern catch statistics of Greenland, we thank E.W. Born. Archaeofaunal material analysed by author A.B. Gotfredsen was provided by staff at the Natural History Museum of Denmark (NHMD), University of Copenhagen; in particular, we thank K.M. Gregersen (NHMD) for assistance with the database of premodern sites in Greenland.

References

Alaralak, F., 1990. Interview IE 114 for the Igloolik Oral History Project. Inullariit Elders Society and the Igloolik Research Institute, Igloolik.

Andersen, A.O., Heide-Jørgensen, M.P., Flora, J., 2018. Is sustainable resource utilisation a relevant concept in Avanersuaq? The walrus caseIn Hastrup, K., Mosbech, A., Grønnow, B. (Eds.), The North Water Polynya: A High Arctic Oasis under TransformationAmbio 47 (2).

Available from: https://doi.org/10.1007/s13280-018-1027-x special issue. Available from: http://link.springer.com/article/10.1007/s13280-018-1027-x.

Anonymous, 2006. Hjemmestyrets bekendtgørelse nr. 20 af 27. oktober 2006 om beskyttelse og fangst af hvalros [Greenland Home Rule Executive Order No. 20 of 27 October 2006 on the protection and catch of walrus]. Greenland Home Rule Government: 5 pp. (In Danish and Greenlandic; available from Naalakkersuisut/The Government of Greenland, www.nanoq.gl). http://lovgivning.gl/lov?rid = {37A7FCF4-46A0-4B47-B42C-B8385F315D65.

Aporta, C., 2002. Life on the ice: understanding the codes of a changing environment. Polar Record 38 (207), 341–354.

Arnatsiaq, C., 1990. Interview IE 77 for the Igloolik Oral History Project. Inullariit Elders Society and the Igloolik Research Institute, Igloolik.

Born, E.W., 2012. Walrus *Odobenus rosmarus*. In: Boertmann, D., Mosbech, A. (Eds.), The Western Greenland Sea: A Strategic Environmental Impact Assessment of Hydrocarbon Activities. Aarhus University, pp. 113–125. Available from: https://www.dmu.dk/Pub/SR22.pdf.

Born, E.W., Heide-Jørgensen, M.P., Davis, R.A., 1994. The Atlantic walrus (*Odobenus rosmarus rosmarus*) in West Greenland. Meddelelser om Grønland, Bioscience 40, 33 pp.

Born, E.W., Dietz, R., Heide-Jørgensen, M.P., Knutsen, Ø., 1997. Historical and present distribution, abundance and exploitation of Atlantic walruses (*Odobenus rosmarus rosmarus* L.) in eastern Greenland. Meddelelser om Grønland. Bioscience 46, 1–73.

Born, E.W., Heilmann, A., Holm, L.K., Laidre, K., Iversen, M., 2017. *Walruses and the Walrus Hunt in West and Northwest Greenland. Monographs on Greenland, Man and Society*, 44. Museum Tusculanum Press, Copenhagen.

Brooke, L.F., 1992. Report on the 1991 Beluga Whale and Walrus Subsistence Harvest Levels by the Inuit of Nunavik. Prepared by the Department of Fisheries and Oceans, Canada.

Christensen, B.B., 2000. *Analysis of faunal remains from Late Dorset Paleoeskimo sites at Hatherton Bay, Inglefield Land, North Greenland.* MA Thesis, Zoological Museum, University of Copenhagen.

CITES, 2018. CITES (Convention on International Trade in Endangered Species of Wild Fauna and Flora) Non Detrimental Findings – For Havpattedyr i Grønland. Greenland Institute of Natural Resources, CITES Scientific Authority in Greenland.

Collins, H.B., 1956. The T1 site at Native Point, Southampton Island, N.W.T. Anthropological Papers of the University of Alaska 4, 63–89.

Collins, H.B., 1981. Record of Animal Bones Recovered at Sadlermiut, 1954-1955. Canadian Museum of History Ms. 1922.

COSEWIC (Committee on the Status of Endangered Wildlife in Canada), 2017. COSEWIC Assessment and Status Report on the Atlantic Walrus *Odobenus rosmarus rosmarus*, High Arctic Population, Central-Low Arctic Population and Nova Scotia-Newfoundland-Gulf of St. Lawrence Population in Canada. Committee on the Status of Endangered Wildlife in Canada. Ottawa. xxi + 89 pp. http://www.registrelep-sararegistry.gc.ca/default.asp?lang = en&n = 24F7211B-1.

Darwent, C.M., Foin, J.C., 2010. Zooarchaeological Analysis of a Late Dorset and an Early Thule Dwelling at Cape Grinnell, Northwest Greenland. Journal of Geography 110, 115–336.

Degerbøl, M., 1934. Zoological Appendix. Animal bones from the Eskimo settlement in Dødemandsbugten. Clavering Island. A contribution to the immigration history of the Musk Ox and Reindeer in East Greenland. Meddr Grønland 102 (1), 173–180.

Degerbøl, M., 1935. Animal Bones from King Oscar Fjord Region in East GreenlandIn Glob, P.V: Eskimo Settlements in Kempe Fjord and King Oscar Fiord Meddelser om Grønland 102 (2), 93–97.
Degerbøl, M., Hørring, R., Pfaff, I.R., 1931. Animal bones from the Sukkertoppen District, 1930In: Mathiassen, T. (Ed.), Ancient Eskimo Settlements in the Kangaamiut Area Meddelelser om Grønland 91 (1), 134–139.
Desjardins, S.P.A., 2013. Evidence for intensive walrus hunting by Thule Inuit, northwest Foxe Basin, Nunavut, Canada. Anthropozoologica 48, 37–51.
Desjardins, S.P.A., 2016. Food Security, Climate Change and the Zooarchaeology of Neo-Inuit Sea-Mammal Hunting, Northwest Foxe Basin, Nunavut, Canada (Unpublished Ph. D. dissertation). Department of Anthropology, McGill University, Montreal.
Desjardins, S.P.A., 2017. A change of subject: perspectivism and multinaturalism in Inuit depictions of interspecies transformation. Études/Inuit/Studies 41 (1-2), 101–124.
Desjardins, S.P.A., 2018. Neo-Inuit strategies for ensuring food security during the Little Ice Age climate change episode, Foxe Basin, Arctic Canada. Quaternary International. Available from: https://doi.org/10.1016/j.quaint.2017.12.026.
Desjardins, S.P.A., Jordan, P.D., 2019. Arctic archaeology and climate change. Annual Review of Anthropology 48, 279–296.
Desrosiers, P.M., Lofthouse, S., Bhiry, N., Lemieux, A.-M., Monchot, H., Gendron, D., et al., 2010. The Qijurittuq site (IbGk-3), eastern Hudson Bay: an IPY interdisciplinary study. Geografisk Tidsskrift/Danish Journal of Geography 110, 227–243.
DFO (Fisheries and Oceans Canada), n.d. Integrated fisheries management plan (IFMP): Atlantic walrus in the Nunavut settlement area. http://www.dfo-mpo.gc.ca/fisheries-peches/ifmpgmp/walrus-atl-morse/walrus-nunavut-morse-eng.html (Accessed 21 April 2020).
Fay, F.H., 1982. Ecology and Biology of the Pacific walrus, *Odobenus rosmarus divergens* IlligerWashington, DC: U.S. Fish and Wildlife Service North American Fauna 74, 279.
Fienup-Riordan, A., 2000. Hunting Tradition in a Changing World: Yup'ik Lives in Alaska Today. Rutgers University Press, New Brunswick, NJ.
Ford, J.D., Pearce, T., Gilligan, J., Smit, B., Oakes, J., 2008. Climate change and hazards associated with ice use in Northern Canada. Arctic, Antarctic and Alpine Research 40 (4), 647–659.
Friesen, T.M., Arnold, C.D., 2008. The timing of the Thule migration: new dates from the western Canadian Arctic. American Antiquity 73 (3), 527–538.
Friesen, T.M., 2016. Pan-Arctic population movements: the early Paleo-Inuit and Thule Inuit migrations. In: Friesen, T.M., Mason, O. (Eds.), The Oxford Handbook of the Prehistoric Arctic. Oxford University Press, New York, pp. 673–691.
Gotfredsen, A.B., 2010. Faunal remains from the Wollaston Forland – Clavering Ø Region, Northeast Greenland – Thule culture subsistence in a High Arctic Polynya and ice-edge habitat. Geografisk Tidsskrift-Danish Journal of Geography 110 (2), 175–200.
Gotfredsen, A.B., Møbjerg, T., 2004. Nipisat – a Saqqaq Culture Site in Sisimiut, Central West Greenland. Meddelelser om Grønland, Man & Society 31, 243.
Gotfredsen, A.B., Appelt, M., Hastrup, K., 2018. Walrus history around the North Water: Human-animal relations in a long-term perspectiveIn: Hastrup, K., Mosbech, A. Grønnow, B. (Eds.). The North Water Polynya: A High Arctic Oasis under TransformationAmbio 47 (2), 193–212. Available from: https://doi.org/10.1007/s13280-018-1027-x. Available from: http://link.springer.com/article/10.1007/s13280-018-1027-x.

Grønnow, B., 2016. Living at a High Arctic Polynya: Inughuit settlement and subsistence around the North Water during the Thule Station Period, 1910–53. Arctic 69 (Suppl. 1), 1–15.

Grønnow, B., Gulløv, H.C., Jakobsen, B.H., Gotfredsen, A.B., Kauffmann, L.H., Kroon, A., et al., 2011. At the edge: high Arctic walrus hunters during the Little Ice Age. Antiquity 85 (329), 960–977. Available from: https://doi.org/10.1017/S0003598X00068423.

Grønnow, B., Sørensen, M., Gotfredsen, A.B., 2015. Arkæologiske og arkæo-zoologiske registreringer i Kap York-omra°det, NOW projektet, 2014 [Archaeological and Archaeo-Zoological Recordings in the Cape York Area, the NOW-project, 2014]. Feltrapport 35. Sila—Arktisk center ved Etnografisk Samling, Nationalmuseet (in Danish).

Grønnow, B., Appelt, M., Gotfredsen, A.B., Myrup, M., 2017. Arkæologiske registreringer, opmålinger og arkæo-zoologiske undersøgelser på Appat (Saunders Ø) og andre bopladser i Wolstenholme Fjord (Avanersuaq). NOW Projektet, 2016 [Archaeological Recordings, Surveys, and Archaeo-Zoological Investigations at Appat (Saunders Ø) and Other Sites in Wolstenholme Fjord (Avanersuaq). The NOW Project, 2016]. Feltrapport 38. Sila—Arktisk center ved Etnografisk Samling, Nationalmuseet (in Danish).

Henshaw, A.S., 1995. Central Inuit Household Economies: Zooarchaeological, Environmental, and Historical Evidence From Outer Frobisher Bay, Baffin Island. Department of Anthropology, Harvard University, Cambridge, MA.

Hill, E., 2011a. The historical ecology of walrus exploitation in the North Pacific. In: Braje, T., Rick, T.C. (Eds.), Human Impacts on Seals, Sea-Lions and Sea-Otters: Integrating Ecology and Archaeology in the Northeast Pacific. University of California Press, Berkeley, pp. 41–64.

Hill, E., 2011b. Animal as agents: hunting ritual and relational ontologies in Prehistoric Alaska and Chukotka. Cambridge Archaeological Journal 21 (3), 407–426.

Hill, E., 2017. The archaeology and ethnohistory of walrus ritual around Bering Strait. Études/Inuit/Studies 41 (1–2), 73–99.

Holtved, E., 1935. Erik Holtveds dagbog fra Thule Distriktet 1935-1937 [The Diary of Erik Holtved From the Thule District 1935-1937], vol. 1 (in Danish).

Holtved, E., 1936. Erik Holtveds dagbog fra Thule Distriktet 1935-1937 [The Diary of Erik Holtved From the Thule District 1935-1937], vols. 1 and 2 (in Danish).

Holtved, E., 1954. Archaeological investigations in the Thule District, III: Nûgdlît and Comer's Midden. Meddelelser om Grønland 146 (3).

Holtved, E., 1967. Contributions to Polar Eskimo ethnography. Meddelelser om Grønland 182 (2).

Ijjangiaq, M., 1990. Interview IE 138 for the Igloolik Oral History Project. Inullariit Elders Society and the Igloolik Research Institute, Igloolik.

Inukshuk, A., 1990. Interview IE 95 for the Igloolik Oral History Project. Inullariit Elders Society and the Igloolik Research Institute, Igloolik.

Iqallijuq, R., 1999. Interview IE 430 for the Igloolik Oral History Project. Inullariit Elders Society and the Igloolik Research Institute, Igloolik.

Johansen, T.B., 2013. Foraging efficiency and small game: the importance of Dovekie (*Alle alle*) in Inughuit subsistence. Archaeozoologica 48, 75–78. Available from: https://doi.org/10.5252/az2013n1a4.

Kappianaq, G., 1990. Interview IE 155 for the Igloolik Oral History Project. Inullariit Elders Society and the Igloolik Research Institute, Igloolik.

Kappianaq, G., 1992. Interview IE 234 for the Igloolik Oral History Project. Inullariit Elders Society and the Igloolik Research Institute, Igloolik.
Kappianaq, G., 1993. Interview IE 273 for the Igloolik Oral History Project. Inullariit Elders Society and the Igloolik Research Institute, Igloolik.
Kappianaq, G., 1997. Interview IE 427 for the Igloolik Oral History Project. Inullariit Elders Society and the Igloolik Research Institute, Igloolik.
Larrat, S., Simard, M., Lair, S., Bélanger, D., Proulx, J.-F., 2012. From science to action and from action to science: the Nunavik Trichinellosis Prevention Program. International Journal of Circumpolar Health 71. Available from: https://doi.org/10.3402/ijch.v71i0.18595.
Laugrand, F., Oosten, J., 2010. Inuit Shamanism and Christianity. McGill-Queen's University Press, Montreal.
Laugrand, F., Oosten, J., 2014. Hunters, Predators and Prey: Inuit Perceptions of Animals. Berghahn Books, New York.
LeMoine, G.M., Darwent, C.M., 2010. The Inglefield Land Archaeology Project: introduction and overview. Journal of Geography 110, 279–296.
Lofthouse, S., 2003. A taphonomic Treatment of Thule Zooarchaeological Materials From Diana Bay, Nunavik (Arctic Quebec). Department of Anthropology, McGill University, Montreal.
Lyman, R.L., 1996. Vertebrate Taphonomy. Cambridge University Press.
Mathiassen, T., 1927. Archaeology of the Central Eskimos. Gyldendal, Copenhagen.
Mathiassen, T., 1930. Innugsuk – a Mediaeval Eskimo Settelment in Upernavik District, West Greenland. Meddr Grønland 77 (4), 147–342.
Mathiassen, T., 1934. Contributions to the Archaeology of the Disko Bay. Meddelelser om Grønland 93 (2), 1–192.
Maxwell, M.S., 1985. Prehistory of the Eastern Arctic. Academic Press, Orlando, FL.
McCartney, A.P. (Ed.), 1979. Archaeological whale bone: A northern resource. University of Arkansas Anthropological, Fayetteville, Arkansas, Papers no. 1.
McCartney, A.P., Savelle, J.M., 1985. Thule Eskimo whaling in the central Canadian Arctic. Arctic Anthropology 22 (2), 37–58.
McCullough, K.M., 1989. The Ruin Islanders. Early Thule Pioneers in the Eastern High Arctic. In: Mercury Series 141. National Museums of Canada, Ottawa.
Møbjerg, T., Robert-Lamblin, J., 1990. The Settlement at Ikaasap Ittiva, East Greenland. An ethnoarchaeological investigation. Acta Archaeologica 60, 229–262.
Møhl, J., 1979. Description and analysis of the bone material from Nugarsuk: an eskimo settlement representative of the Thule Culture in West Greenland. In: McCartney, A.P. (Ed.), Thule Eskimo Culture: An anthropological retrospective. National Museum of Man. Mercury Series, Archaeological Survey of Canada, Paper No. 88, pp. 380–394.
Niviattian Aqatsiaq, S., 1990. Interview IE 79 for the Igloolik Oral History Project. Inullariit Elders Society and the Igloolik Research Institute, Igloolik.
Nuttall, M., 2000. Becoming a hunter in Greenland. Etude/Inuit/Studies 24 (2), 33–45.
Paniaq, H., 1998. Interview IE 431 for the Igloolik Oral History Project. Inullariit Elders Society and the Igloolik Research Institute, Igloolik.
Park, R.W., 1989. Porden Point: An Intrasite Approach to Settlement System Analysis. Department of Anthropology, University of Alberta, Edmonton.
Patton, K., Savelle, J.M., 2006. The symbolic dimensions of whale bone use in Thule winter dwellings. Études/Inuit/Studies 30 (2), 137–161.

Piugattuk, N., 1990. Interview IE 136 for the Igloolik Oral History Project. Inullariit Elders Society and the Igloolik Research Institute, Igloolik.

Piugattuk, N., n.d. Interview IE 54 for the Igloolik Oral History Project. Inullariit Elders Society and the Igloolik Research Institute, Igloolik.

Priest, H., Usher, P.J., 2004. The Nunavut Wildlife Harvest Study. Report Prepared by the Nunavut Wildlife Management Board, Iqaluit & P.J. Usher Consulting Services, Ottawa.

Proulx, J.-F., MacLean, J.D., Gyorkos, T.W., Leclair, D., Richter, A.K., Serhir, B., et al., 2002. Novel prevention program for trichinellosis in Inuit communities. Clinical Infectious Diseases 34 (11), 1508–1514.

Qamaniq, N., 2000. Interview IE 438 for the Igloolik Oral History Project. Inullariit Elders Society and the Igloolik Research Institute, Igloolik.

Qamaniq, N., 2001. Interview IE 472 for the Igloolik Oral History Project. Inullariit Elders Society and the Igloolik Research Institute, Igloolik.

Rasing, W., 2017. Too Many People: Contact, Disorder, Change in an Inuit Society: 1822-2015. Nunavut Arctic College Media, Iqaluit, p. 568.

Rasmussen, K., 1932. Intellectual Culture of the Copper Eskimos. Nordisk Forlag, Copenhagen, Report of the Fifth Thule Expedition 1921-249.

Rick, A.M., 1980. Non-cetacean vertebrate remains from two Thule winter houses on Somerset Island, N.W.T. Canadian Journal of Archaeology 4, 99–117.

Rosing, J., 1998. Fortællinger om INUA. Tidsskriftet Grønland 5, 155–174 (in Danish).

Sabo III, G., 1981. Thule Culture Adaptations on the South Coast of Baffin Island, N. W. T. (Unpublished Ph.D. dissertation). Michigan State University, East Lansing, MI.

Sandell, H.T., Sandell, B., 1991. Archaeology and environment in the Scoresby Sund fjord. Ethno-archaeological investigations of the last Thule culture of Northeast Greenland. Meddelelser om Grønland, Man & Society 15.

Savelle, J., 2002. The Umialiit-Kariyit whaling complex and prehistoric Thule Eskimo social relations in the Eastern Canadian Arctic. Bulletin of National Museum of Ethnology 27 (1), 159–188.

Schledermann, P., 1975. Thule Eskimo Prehistory of Cumberland Sound, Baffin Island, Canada. National Museum of Man, Ottawa.

Schmidt, A.-L., Gottlieb, B., Gulløv, H.C., Jensen, C., Petersen, A.H., 2013. Pelse fra Nord. [Furs from the North]. Nationalmuseets Arbejdsmark 64–79 (in Danish).

Sonne, B., 2004a. Grønlandske sagn & myter/Sonnes base. [Greenlandic legends & myths/ Sonne's database]. http://arktiskinstitut.dk/vidensdatabaserne/groenlandske-sagn-myter-sonnes-base/

Sonne, B., 2004b. Vejledning og introduktion til Grønlandske sagn & myter. [Guide and introduction to Greenlandic legends & myths]. http://arktiskinstitut.dk/vidensdatabaserne/groenlandske-sagn-myter-sonnes-base/

Staab, M.L., 1979. Analysis of faunal material recovered from a Thule Eskimo site on the island of Silumiut, N.W.T., Canada. In: McCartney, A.P. (Ed.), Thule Eskimo Culture: An Anthropological Retrospective. National Museum of Man, Ottawa, pp. 349–379.

Stenton, D.R., 1987. Recent archaeological investigations in Frobisher Bay, Baffin Island, N.W.T. Canadian Journal of Archaeology 11, 13–48.

Stewart, D.B., Higdon, J.W., Reeves, R.R., Stewart, R.E.A., 2014. A catch history for Atlantic walrus (*Odobenus rosmarus rosmarus*) in the eastern Canadian Arctic. NAAMCO Scientific Publication 9, 219–313.

Swinarton, L.E., 2008. Animals and the Precontact Inuit of Labrador: An Examination Using Faunal Remains, Space and Myth. Department of Anthropology and Archaeology, Memorial University of Newfoundland, St. John's.

Taylor Jr., 1972. An Archaeological Survey Between Cape Parry and Cambridge Bay, N.W.T., Canada. National Museum of Man, Ottawa.

Taylor Jr., W.E., McGhee, R., 1979. Archaeological Material From Creswell Bay, N.W.T., Canada. National Museum of Man, Ottawa.

Thostrup, C.B., 1911. Ethnographic description of the Eskimo Settlements and Stone Remains in North East Greenland. Meddelelser om Grønland 44 (4), 183–351.

Uttak, L., 2001. Interview IE 487 for the Igloolik Oral History Project. Inullariit Elders Society and the Igloolik Research Institute, Igloolik.

Whitridge, P., 2002. Social and ritual determinants of whale bone transport at a classic Thule winter site in the Canadian Arctic. International Journal of Osteoarchaeology 12 (1), 65–75.

Whitridge, P.J., 1992. Thule Subsistence and Optimal Diet: A Zooarchaeological Test of a Linear Programming Model. Department of Anthropology, McGill University, Montreal.

Wiig, Ø., Born, E.W., Stewart, R.E.A., 2014. Management of Atlantic walrus (*Odobenus rosmarus rosmarus*) in the arctic Atlantic. NAMMCO Scientific Publications 9, 315–339. Available from: http://doi.org/10.7557/3.2855.

SECTION

III

European walrus use from the Norse to present

CHAPTER

7

Early European and Greenlandic walrus hunting: Motivations, techniques and practices

Jette Arneborg

National Museum of Denmark, Middle Ages, Renaissance and Numismatics, Copenhagen, Denmark

Chapter Outline

Introduction

In northern and western Europe, the use of walrus ivory in decorative arts can be traced back at least as far as the 9th century CE. (Curnow, 2000, p. 293; Roesdahl, 2005, p. 185). Around this time, walrus ivory was being sourced along the Arctic coasts of Scandinavia and Russia. The ivory was then traded westwards into the markets of northern and western Europe, but also eastwards via the Muslim lands, into the Orient and even into the Far East (Tegengren, 1962, p. 7ff.). This chapter examines the role played by western European societies in the active hunting of walruses across the remote North Atlantic periphery. Chronologically, it spans the 9th to the mid-15th centuries CE; that is, from the earliest written descriptions of the lucrative trade in walrus products, through to

The Atlantic Walrus. DOI: https://doi.org/10.1016/B978-0-12-817430-2.00008-X

the eventual decline and abandonment of the Norse colonies in Greenland. By this time, Greenland had taken over from Arctic Russia as the main supplier of walrus ivory to the luxury markets of North-West Europe.

The hunting of walruses for ivory and hide is first mentioned in the early travel accounts of Ohthere, a North Norwegian chieftain, who made a journey to the Arctic territories of European Russia, and later relayed his experiences to the Anglo-Saxon King Alfred at the end of the 800's. Ohthere's account suggests that large and impressive walrus tusks were already circulating as prestigious objects and were highly sought after in northern European markets and trade networks as a valuable raw material for decorative pieces of artwork. In addition to the ivory, walrus hide was also valued as a material for producing tough and waterproof ropes.

The demand for walrus products—and especially ivory—grew rapidly. The early stages of the 'boom' in ivory coincided with the Norse settlement of Iceland around 870 CE and subsequent arrival of Icelandic settlers in Greenland at the end of the 900's. Both events allowed new access to North Atlantic walruses to meet European commercial demands. For the remote colonies of Iceland and Greenland, the trade in ivory generated valuable opportunities to supply the markets of northern Europe. Much of the demand appears to have been driven by the leaders of emerging kingdoms and the Roman Catholic Church elite. Both were keen to use magnificent and valuable ivory artworks in events and ceremonies that signalled their wealth and reinforced their claims to power and authority.

Othere's northern travels: Early descriptions of the walrus ivory trade

Around 890 CE, after returning from his northern voyages, the North Norwegian chieftain Ohthere reported to the English King Alfred of Wessex that he lived as the northernmost of all Norwegians and that he had ventured even further North-East, following the coastline up into Arctic Norway and beyond. There, he had encountered diverse peoples living along the route, including the *Finnas* and *Beormas* (for the full text in Old English and in Danish, see: https://heimskringla.no/wiki/Ottars_og_Ulfstens_Rejseberetninger). While he describes how this journey had been motivated by a more general desire to become familiar with these unknown lands, he also makes explicit reference to walruses, as "they have very fine bone in their teeth'. Until this time, Scandinavian societies had been traditionally oriented towards farming, and the dynamics of power in local societies was closely linked to ownership of land. It is also clear in Ohthere's account that he owned holdings of land and was a settled farmer (Bately, 2007b, p. 46; Näsman and Roesdahl, 2003, p. 290ff.). However, it is interesting to note that his main income was derived from taxing the *Finnas*, who paid their dues in "animals' skins and of birds' feathers and whale's bone and of those ships' ropes" (Bately, 2007b, p. 40ff. Stylegar, 2015, p. 182f.). The *Finnas* appear to correspond to Scandinavia's Sami population (Valtonen, 2007, p. 106f.). It has been suggested by Tegengren (1962) that the Sami were probably intermediaries in the ivory trade as they were mainly engaged in reindeer herding rather than marine-mammal hunting, and that they traded ivory and other Arctic goods with the *Beormas*—also mentioned by Ohthere—who probably correspond to the coastal communities living along the shores of the White Sea in northern Russia (Hofstra and Samplonius, 1995, p. 239ff.; Makarov, 2007; Stylegar, 2015, p. 183; Tegengren, 1962, p. 33ff.; Urbanczyk, 1992, p. 55). It appears that Arctic products such as ivory, hide and pelts were all highly valued in Europe. Ohthere describes trading journeys from northern Scandinavia that ventured as far South as Hedeby in Jutland (now Germany), from where the goods were exchanged deeper into Europe (see Chapter Eight, this volume).

Norse exploitation of Icelandic walrus stocks

Today there are no native walrus stocks in Icelandic waters. However, various walrus bones, teeth and tusks have been found during archaeological excavations of several early Norse sites in Iceland. The presence of walrus remains suggests that there may have been a local population of walruses around Iceland prior to Norse colonisation in the late 800's. This scenario has now been confirmed by a range of different analyses, including ancient DNA, which also suggest that overhunting may have played a major role in the rapid extinction of this local population (McGovern, 2011, p. 2; Keighley et al., 2019).

Most finds of walrus bones come from the west coast of Iceland, from Breiðafjörd in the north down to Reykjavík in the south, with a large concentration around Akranes, just north of Reykjavík (Frei et al., 2015; Keighley et al., 2019; McGovern, 2011, p. 2). One of the more remarkable finds was made in the city of Reykjavík itself, in Aðalstræti, where archaeologists excavating a house from the earliest period of settlement found intact tusks and bones from at least three fully grown walruses. Three tusks were found inside the house, and the remains of a crushed scapula (shoulder blade) and partially-articulated vertebral column were embedded into the turf wall (McGovern, 2011). The tusks appeared to have been carefully removed from the skull, probably in order to sell them. In contrast, the scapula and vertebral column had been built into the fabric of the wall and would have been readily visible to passers-by, hinting at some kind of ritual or symbolic practice, though there are no other parallels with this kind of behaviour. It may even have been done to 'advertise' the fact that walrus ivory was being processed by carvers living in the house (McGovern, 2011).

The Icelandic finds of walruses include limb bones, skull pieces and valuable tusks (McGovern, 2011), which suggests that carcasses were being processed for meat and to extract ivory. The arrival of Norse to Iceland occurred at approximately at the same time as Ohthere lived in northern Norway, indicating that

Icelanders could have supplied walrus ivory to well-established trading networks. Such networks already linked the Arctic hunting grounds of northern Russia to the markets of Europe (Lebecq, 2007).

It remains difficult to reconstruct the intensity of hunting efforts during the earliest phase of Icelandic settlement or how these activities were conducted. However, it does appear that these initial hunting efforts were probably well-organised and peaked early before going into steady decline (Frei et al., 2015). The settlers managed to establish a social system with status based primarily on ownership of land. Icelandic legislation from the 1100's highlights that walrus exploitation was not central to the economy, but that opportunistic hunting provided additional income to hunters as well as to the farmers who owned the coastline where animals were brought ashore. For example, according to the two collections of laws, *Grágás* from the beginning of the 1100's and *Jónsbók* from the late 1200's, foxes, and the occasional walrus or polar bear, could be hunted by anyone, in any locale when found on Icelandic shores. Half the value of the walrus went to the landowner of the beach where the animal was landed, and the other half to the hunter, while the furs from foxes and polar bears went directly to the hunter (Benediktsson, 1981a, p. 552).

Motivations for the Norse settlement of Greenland

Two regions in South-West Greenland—the Eastern Settlement (around 60–61°N and the Western Settlement (around 64°N)—were settled by the Norse who made the journey over from Iceland in the late 900's. In many ways, the Norse colonies in Greenland attempted to replicate the well-established Icelandic society, where status was based on ownership of land and access to resources. However, the remote location of the Greenlandic Norse settlements, combined with the harsher climate generated many practical challenges. As a result, subsistence relied on hunting of seal and wild reindeer, as well as the more typical Norse practices of animal husbandry. In addition, the settlers needed access to iron and other goods which could only be obtained through importation via external trade networks.

The wealthiest members of the Greenlandic Norse population sought to build churches on their farms, and by 1124 CE, a Bishopric was established at *Garðar*, which was one of the richest farms of the Eastern Settlement (present-day Igaliku). In the early years, this Bishopric was incorporated into the Danish ecclesiastical province, which was overseen by the archbishop at Lund, but this shifted to Trondheim (Nidaros) in 1152 CE after the establishment of the Norwegian ecclesiastical province.

Around the same time that the Norse were becoming established in South-West Greenland, the elite families of Novgorod were consolidating their control over the trade in Arctic products from northern Russia, especially the most valuable commodities of walrus ivory and sable, fox and other furs. The focus of trade also shifted to supplying the demands of lucrative eastern markets, which funnelled Russian goods towards the Near East, as well as along the Silk Road that led into Central and East Asia. This shift in trade probably reduced the supply of Arctic products to North Norwegian traders who were supplying the expanding markets of western Europe (Brisbane et al., 2012, p. 9; Keller, 2010; Makarov, 2012; Star et al., 2018; Wallerström, 1995, p. 256ff.).

The seafaring Norse, with growing access to the rich walrus hunting grounds across the North Atlantic, plus established settlements in Iceland and new outposts in Greenland, were well-placed to exploit this gap in supply. Moreover, they could focus on the large walrus populations in West, North-West and maybe even East Greenland with direct transport of ivory into the well-established Icelandic, Scandinavian and European trade networks. There are long-running debates about whether the primary motivations for Icelanders establishing colonies in Greenland was acquisition of land to establish farms and social status, or whether the farming merely served to support more extensive hunting expeditions that sourced walrus ivory from further North (Frei et al., 2015; Star et al., 2018). Both factors probably played a role, and the establishment of successful farmsteads in the South combined with the search for valuable Arctic trade goods, were both important for the overall socioeconomic viability of these remote European outposts (Fig. 7.1).

In the description of his stay at the Eastern Settlement at the end of the 1300's, the Norwegian cleric Ívar Bárðarson describes how the Norse Greenlanders fished and hunted on the East coast of Greenland, the primary prey being birds and eggs, whales and 'white' bears (Jónsson, 1930, p. 19ff.). While Ívar mentions whales, he does not discuss walruses, and even though the hunters

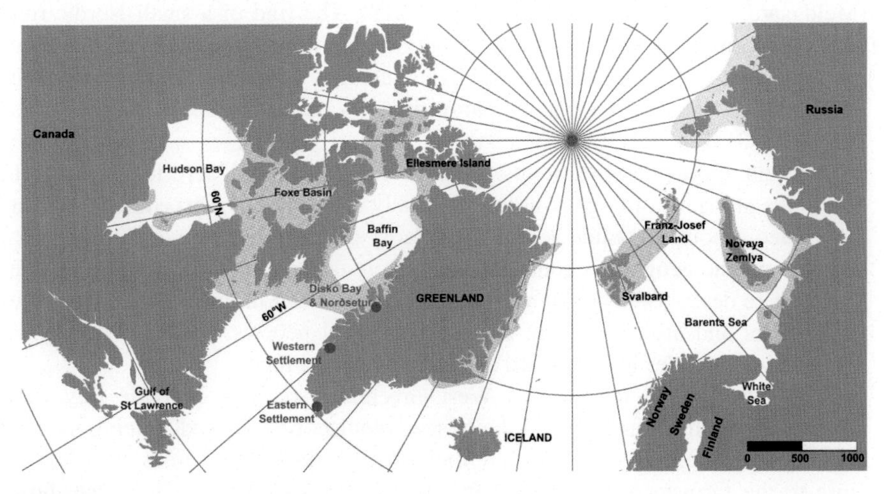

FIGURE 7.1

Norse settlements in Greenland (blue circles). Purple areas indicate the range of the Atlantic walrus; yellow areas indicate the range of the Laptev walrus.

FIGURE 7.2

The small rune stone found in a cairn on the small rocky island of Kingittorsuaq, near Upernavik in North Greenland. According to the text on the stone, the three Norse hunters, Erlingr Sighvartr's son, Bjarni þorðr's son and Eindriði Oddr's son, were on the island on the Saturday before Rogation Day, which is in the early spring. Length of stone: 10 cm.

Courtesy of the National Museum of Denmark, licence CC-BY-SA.

probably did encounter occasional walruses on the East coast, it appears that most walruses were probably captured on the West coast. In this period, there were probably significant concentrations of walruses around Kangaamiut, with even larger populations between Sisimiut and Aasiaat, and in and around Disko Bay (Muus et al., 1981, p. 405ff.). A solidly-built stone house on the north side of the Nuusuuaq peninsula in Disko Bay confirms that Norse Greenlanders were operating in this region, and they probably used the structure as a store house (Meldgaard, 1995, p. 206; Madsen et al., 2019). The find of a small Nordic rune stone at approximately 72°N, on the island of Kingittorsuaq located near the modern settlement of Upernavik, also highlights just how far north the hunting expeditions reached. The rune stone probably dates to around 1200–1250 CE (Imer, 2017, p. 243) (Fig. 7.2).

It would appear that the walrus populations of Greenland were large enough to absorb the pressures of large-scale commercially driven hunting expeditions, but further palaeo-genomic research is needed to clarify the precise nature of these impacts. The cultural and geographic context was also very different in Greenland. For example, Ohthere describes his journey to the Arctic coasts of Europe as both an opportunity to hunt and also to explore the land and local peoples. In fact, it was these northern communities that provided access to ivory and furs.

By contrast, both Iceland and West Greenland south of Melville Bay were devoid of local inhabitants when the Norse colonists first arrived, forcing them to hunt walruses themselves. In Iceland, the prime walrus hunting areas were located close to the farmsteads, generating few logistical problems. In Greenland, however, a more complex system of settlement and mobility was required because permanent agricultural settlements could only be established in the warmer fjord areas in subarctic southern Greenland, as well as in inland areas close to the modern settlement of Nuuk. The rich walrus hunting grounds were found in more northerly latitudes, requiring long and dangerous voyages.

Over time, the rich Greenlandic hunting grounds came to dominate the supply of walrus ivory to the markets of western Europe, a pattern that was established already when the Icelandic arrived in Greenland. This trend intensified in the beginning of the 1100's and persisted until the mid-1400's (Star et al., 2018; Barrett et al., 2020). The increased volume of export from Greenland also coincided with the establishment of the Greenlandic bishopric in 1124 CE (Star et al., 2018, p. 5).

The organisation of Norse walrus hunting expeditions in Greenland

The Norse hunting expeditions required special vessels, which were robust enough to navigate ice-filled northern waters. Records dating to the beginning of the 1300's note that the wealthiest Greenlandic farmers commissioned large ships, which were explicitly built for hunting expeditions to northern areas, and that they also owned all the hunting tools and other equipment (GHM III, 242–243). It also seems likely that the bishop at *Garðar* was involved in sponsoring these northern expeditions (GHM III, 240–241) (Fig. 7.3).

Only a few walrus bones have been found at Norse archaeological sites, most notably in the Western Settlement. This may indicate that there was a small local

FIGURE 7.3

The small settlement of Igaliku, once the Bishop's seat during the Norse occupation. *Photo: J. Arneborg, 2012.*

walrus population here when the Norse arrived around 1000 CE. These bones were only recovered from the earliest midden layers at the farms, suggesting that local hunting had taken place after the settlement was first established, but that this local population was either hunted to extinction or driven away by human exploitation (Enghoff, 2003, p. 39; Frei et al., 2015, p. 9; McGovern, 1985; McGovern et al., 1996, p. 114; Nyegaard, 2018).

Norse hunting methods

Unfortunately, the details of how the Greenlandic Norse equipped their walrus-hunting expeditions remain unclear; for example, we do not know whether walruses were hunted in open water or at haul-out sites on the land or ice. If open-water hunting was practised, then a barbed iron harpoon, possibly attached to a long line, would have been required to both kill and retain the carcass. If walruses were instead hunted at haul-out sites they could have been dispatched with a simple spear. Both spears and harpoons are mentioned in legal texts in connection with walrus hunting in Iceland, so it is clear that both these technologies would have been available (Benediktsson, 1981a, p. 552). However, the texts also indicate that the hunt mostly took place at sea and involved harpoons, with the impaled animals then being driven ashore. In Iceland, hunters appear to have operated individually but also in groups, with each hunter gaining a personal share of any proceeds (Benediktsson, 1981a, p. 551f.; Benediktsson, 1981b, p. 669ff.; Lárusson, 1981, pp. 168–172).

The walrus hunt in Greenland was organised differently. Greenlandic walrus hunts operated on a larger scale than in Iceland and were coordinated by local elites interested in commercial gain. Additionally, unlike in Iceland, no harpoons have been found at Norse sites in Greenland, making it more likely that walruses were hunted at haul-out sites with spears. There were also no Greenlandic laws pertaining to individual ownership or rights to the catch as the expeditions were funded by the elites who then retained any profits.

The expeditions started in the spring as soon as the retreating sea ice permitted voyages to the northern hunting waters. Walruses are very effective swimmers and can dive to significant depths, but are heavy and clumsy on land. When hauled-out on land, walruses are often found in large social groups, making them vulnerable to human predation. For these reasons, Norse and later Inuit hunters appear to have used a similar approach to hunting walruses by first killing the animals closest to the water, thereby blocking the escape route for other walruses, and containing the walrus herd on land. Hunters could then pick off the remaining walruses one-by-one. However, this is no easy task when hunting with spears, as walruses are large and aggressive, and the hunter has to approach very close to ensure that the lance would be plunged deeply into the correct part of the animal.

While it is clear that the Norse engaged in active hunting, there are indications that they may also have tried to obtain walrus ivory and other Arctic goods via trade, barter or exchange with Arctic Indigenous peoples, whom they called 'Skraellings'. There is little doubt that the Norse encountered other human cultures on their voyages, though the precise nature of these relationships—and also which particular groups were involved—remains poorly understood. Evidence for such contact includes reports by homecoming Norse hunters in 1266 CE that they had sailed further north than usual, but had failed to encounter Skraellings. Upon hearing of this, members of the main church at *Garðar* immediately dispatched another boat to the north to search for them (GHM III, 240–241). This suggests that in addition to funding hunting expeditions, the local Norse elites also used these voyages to seek out and conduct trade with Arctic Indigenous peoples (Fig. 7.4).

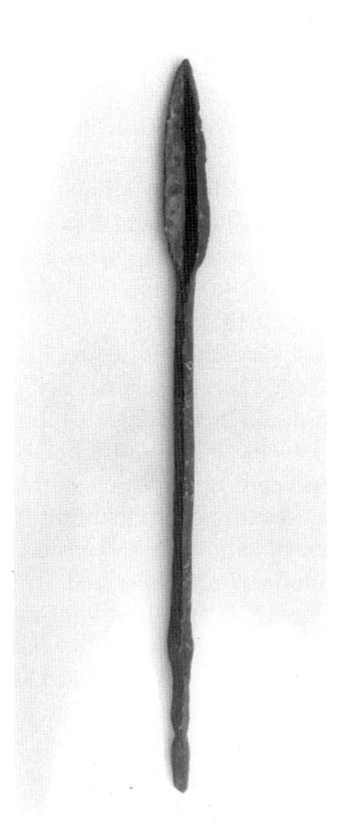

FIGURE 7.4

An iron spear, probably used for walrus hunting. It was found at the Western Settlement farm V53d. Length: 36 cm.

Courtesy of the National Museum of Denmark, licence CC-BY-SA.

The economic significance of walrus hunting to Greenlandic Norse society

The Greenlandic Norse settlements were controlled by a group of elite farmers, whose authority rested upon their ownership of large landholdings, control of the churches located on their farms (Vésteinsson et al., 2019), and control over the harvesting of Arctic trade goods, especially walrus ivory. Greenlandic Norse were able to exert control over the hunting expeditions, but monopolising the export trade was more challenging because most Greenlanders did not own large seagoing vessels suited for long open-water voyages to Iceland and Scandinavia. Instead, Norwegian merchants are likely to have controlled the trade, sailing on an unpredictable basis to engage with the Greenlanders (Magerøy, 1993, p. 57ff.).

With the establishment of the Bishop's seat at *Garðar* in 1124 CE, and its subsequent integration into the Kingdom of Norway in 1262 CE, a set of powerful new actors became involved in mediating the trade and its profits, namely the Church and the Norwegian Crown. The Norwegian kings enforced a monopoly on trade with Greenland, which meant that the merchants had to obtain royal permission to trade with Greenland, and then pay duties to the king. Central to the payment of duties was walrus ivory, and the tithes demanded by the Catholic Church (and ultimately, the Pope) also had to be paid in walrus tusks. In 1328 CE, the Flemish merchant Johannes Dipre bought 127 'lispund' of walrus ivory, representing the Greenlanders' tithe payments to the Pope (this probably equated to around 500 tusks, representing at least 250 individual walruses) (Keller, 2010, p. 3).

Greenlandic trade goods were shipped to the European markets via Bergen, Norway. This made the Norse Greenlanders heavily dependent upon the interests of both the kings and also the Church. These institutions also relied in turn upon the Norse outposts to supply the lucrative European markets, meaning that the Norwegian elite were dependent upon maintaining good connections with the Greenlandic elite given their control over the walrus hunt. Church-building activities in Greenland in the second half of the 13th century CE and afterwards testifies to the growing Norwegian influence in Greenland.

Prior to the establishment of connections with the Kingdom of Norway, the early Greenlandic churches were erected by influential local landowners using the traditional building materials of turf and unworked stone. After their integration into the Kingdom of Norway, proper churches were built of lightly-worked stone, complete with mortared walls. This suggests that the newer churches were probably erected under the supervision of foreign master builders (Bertelsen and Schnohr, 2019), who may have been dispatched to Greenland by either the Church or the king. On a political level, it may point to growing alliances between elite Greenlandic farmers, on whose land the churches were built, and the Norwegian rulers who were seeking to project their control and influence over this important part of the lucrative ivory trade.

The ships that were sent to Greenland's northern hunting waters were owned by a few elite farmers and probably also the Bishop's seat (i.e., the Church), but were manned by hunters who spent their daily lives as farmers on medium-to-small-sized farms. We do not know what the conditions were like for the hunters, but participation in the hunting and subsequent working of the tusks may have constituted payment of taxes and duties to the elite farmers and ecclesiastical tithes to the Church.

The processing of walrus ivory in Norse Greenland

Zooarchaeological evidence from Greenlandic Norse sites indicates that the northern hunting expeditions concentrated on obtaining valuable ivory. Walrus carcasses were left behind at the hunting grounds. Only the bacula (penis bones), tough hides (used to make rope) and the anterior part of the skull containing the tusks and jaw were transported back to the main settlements for further processing (McGovern, 1985, p. 89).

Exactly how the working of the skulls and the extraction of ivory was conducted is uncertain. Fragments of walrus skulls, bacula and teeth have been found at nearly all archaeologically-investigated Norse Greenlandic farmsteads, but generally in very small quantities at each site. One major challenge to understanding ivory processing is that most excavations of large Greenlandic farms were undertaken in the 1920's and 1930's. In this period, only the 'finer' cultural objects were retained and transported back to museums and other collections. Production waste generated by ivory extraction was most probably not regarded as important and left behind, obscuring the degree to which different farms participated in ivory production. Another problem is that there was poor stratigraphic control in these earlier excavations, and so any finds of walrus tusks or bones can only be dated to the general occupation phase. For most farms in the Eastern Settlement this phase is from the end of the 10th until mid 15th centuries CE, and in the Western Settlement from around the 11th until end of the 14th centuries CE.

Removal of the valuable tusks from a walrus skull is best undertaken some weeks after the animal has been killed (McGovern, 2011, p. 6). Traces of working on the tusk and skull fragments indicate that the tusks and other teeth were carefully removed from the upper jaw (maxillae) by use of a narrow-bladed chopping tool, probably a special type of chisel (McGovern, 2011, p. 6). However, no examples of such implements have ever been identified among the large tool assemblages that are typically recovered from Norse farmsteads in Greenland. In fact, only small broken fragments of walrus ivory (likely resulting from tusk-removal) have been found in Greenland. It seems that 'raw' ivory was exported to Europe where specialist carvers would then produce ornamental artefacts.

More recent studies have highlighted clear imbalances between farmsteads, with the elite farmstead *Sandnes* V51 in the Western Settlement having large quantities of walrus remains (McGovern, 1985). A few more recent excavations with good stratigraphic control have further confirmed this impression, with most farmsteads appearing to undertake only very limited ivory production. Most of the work appears to have been concentrated at elite farmsteads like *Sandnes in* the Western Settlement, and at *Garðar*, which was the large Bishop's farmstead in the Eastern Settlement. Moreover, the location of *Sandnes* at the Western Settlement, which was situated midway between the northern hunting grounds and the larger Eastern Settlement, may have occupied a nodal position in the wider Norse network of walrus hunting, ivory production and export (McGovern et al., 1996).

Walruses were important from the start of Norse settlement in Greenland, with ivory production continuing until the end. Walrus bones from the settlements are consistently recovered from house floor levels and middens, and range in date from the earliest occupation up until the end of the settlements (late 14th century CE in the Western Settlement and mid 15th century CE in the Eastern Settlement). At the farm Ø17a at Narsaq, Eastern Settlement (only inhabited in the first century or so of the settlement), the earliest walrus skull fragments, bacula and a few meat-bearing bones can be dated to the end of the 10th century CE (McGovern et al., 1993). In the Western Settlement, walrus bone debitage and tooth fragments have also been found in the earliest settlement layers at *Sandnes*, and at the medium-sized farm *Gården under Sandet* ('Farm under the Sand'). At both sites, skull fragments have been found, as well as pieces of tooth, while the meat-bearing bones date only to the very earliest phase of settlement. It appears that walrus ivory was being extracted at both farmsteads right up until the final abandonment of the settlement in the late 14th century CE (Enghoff, 2003, p. 25ff., p. 39; McGovern et al., 1996, p. 106).

In some instances, walrus tusks were exported from Greenland while still embedded in the cranium, with the cranium also often carved and decorated. Whether such crafting took place in Greenland or at a later date in Europe is uncertain, though it is clear that the decorated skulls were exotic and valuable objects that had their own intrinsic value (Roesdahl, 2005, p. 187; Stoklund and Roesdahl, 2002). It is also clear that some carving work was conducted in the Norse Greenlandic settlements, with remaining walrus teeth crafted on farms into small decorative, ornamental or domestic objects, such as amulet figures, gaming pieces and buttons. In contrast, the mandibles were often carved into circular gaming pieces for board games (Nyegaard, 2018, p. 36), and the bacula used for clubs or to haft iron tools.

To date, only two finely-crafted objects carved from walrus ivory have ever been recovered from settlements in Greenland. One is the upper part of a crozier (a symbolic religious staff), which was found in a bishop's grave at the cathedral at *Garðar* during an excavation conducted in 1926 (Nørlund and Roussell, 1930, pp. 66–68). The other is a queen figure from a chess set, which was recovered from an Inuit house on the small island of Qeqertaq, near Sisimiut, located far north of the Western Norse Settlement (Inventory, Protocol in the Danish National Museum). It is uncertain whether either of these objects was actually

FIGURE 7.5

Crozier carved from walrus ivory and ring found in a bishop's grave at Igaliku. The bishop in the grave has been ^{14}C dated to the 1200s. Height: 14 cm.

Courtesy of the National Museum of Denmark, licence CC-BY-SA.

FIGURE 7.6

The rear side of a chess piece found in Inuit house near Sisimiut. Height: 8.4 cm.

Courtesy of the National Museum of Denmark, licence CC-BY-SA.

carved in Greenland or made with exported ivory in a workshop in northern Europe, and then subsequently brought back to Greenland (Roesdahl, 1995, p. 19ff.) (Figs 7.5 and 7.6).

Human–walrus interactions in Norse belief

A range of evidence suggests that the Norse both feared and respected the walrus. For example, The King's Mirror (*Konungs skuggsjá*) is a Norwegian educational text from the middle of the 13th century CE, supposedly intended for the education of the sons of Norwegian King Håkan Håkansson (reigning from 1217–1263 CE). The text takes the form of a dialogue between a wise father and his very curious son, and deals with politics, morals and ceremonial etiquette, as well as the mysterious animals and nature of the 'Northern World'. Walruses are described as being greedy, wild and ferocious, intent only on harming humans. In his description of Greenland, 'the father' also explains why Norwegian merchants would risk such dangerous voyages to the settlements. First, it brought fame, second, it was motivated by curiosity, but third, it was a means to acquire goods: "... one searches for goods everywhere, where there is something to acquire, even if it is associated with much danger." (Konungs skuggsjá, 1926, p. 14). The best strategy was to obtain good prices for common materials such as iron and perhaps also timber, which could be transported to Greenland where they were in short supply, and exchanged for Arctic commodities, especially walruses, which commanded high prices at home: "... the ropes ... that are cut off the fish called walruses and are called 'heavy ropes', as well as the teeth of these fishes" (Konungs skuggsjá, 1926, pp. 14, 23).

Of course, it was not only the merchants that exposed themselves to mortal danger in the quest for valuable Arctic goods, but also the Norse who voyaged to the northern hunting grounds, confronting the ferocious animals face-to-face while armed only with simple spears. Numerous walrus skulls that have been recovered from the chancel of the Bishop's church at *Garðar*, as well as in the churchyard near the eastern gable of the chancel. The structured nature of both of these deposits may suggest that walruses commanded a deeper respect in the Norse colonies than previously thought. The report from the 1926 archaeological excavations indicates that around 20 and 30 complete or fragmented walrus skulls were found in the churchyard, with several "... in rows, the one beside the other ..." (Nørlund and Roussell, 1930, pp. 138, 190, see also Seaver, 1996, p. 31). The skulls date to the 11th century CE (Arneborg et al., 2012, p. 16), and were found together with many other animal bones and small finds that are typical of midden deposits. These combined deposits may have either accumulated at the site prior to establishment of the churchyard, or they may have been brought to the site as fill to level out the site of the cathedral and the churchyard (proposed by Nørlund and Roussell, 1930, p. 138). If so, the deposition of walrus skulls may have simply served a practical purpose, or it may point to more symbolic or ritualised practices. Either way, their presence confirms that the trade in walrus tusks was already important in the 11th century CE.

Other finds also hint at the respect that the local Norse community had for walruses. Such finds include a small walrus amulet carved from a walrus molar,

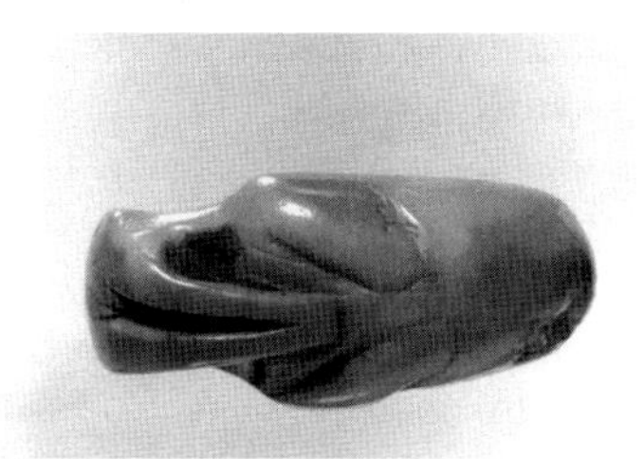

FIGURE 7.7

Walrus amulet carved from walrus tooth (molar) found at the Western Settlement farm V52a. Dated 1300s. Length 4 cm.

which was found inside the house at farm V52a at the Western Settlement (Roussell, 1936, p. 184). Wearing this amulet around the neck may have reassured a nervous hunter as he stepped up to confront his first walrus with a flimsy lance (Fig. 7.7).

The final ivory exports from Greenland

The numbers of ships sailing between Norway and Iceland declined significantly around the 15th century CE (Magerøy, 1993, p. 151), and the official royal approvals for sailing to Greenland apparently ceased completely around 1410 CE (Magerøy, 1993, p. 131). This appears to have signalled the end of Greenlandic ivory exports to European markets, but also the end of European exports to these remote settlements.

What triggered these developments is still open for debate. It has been suggested that Norse Greenlanders deliberately stopped exporting walrus tusks because they were dissatisfied with Norwegian and Danish influence over them (Seaver, 2009). However, the ceasing of trade caused great damage to the local communities who were still heavily dependent economically and culturally with their Scandinavian 'homelands'. Other academics have highlighted the small and vulnerable nature of the isolated Norse societies in Greenland, which may only have numbered 2000–2500 people at its peak in the mid-13th century CE (Lynnerup, 2011), and then went into steady decline until the eventual abandonment of the settlements in the 15th century CE. The end came first to the Western Settlement around the same time that remoter farms on marginal lands were already being depopulated in the larger Eastern Settlements (Madsen, 2014).

A number of converging problems may have ultimately led to the collapse of Norse Greenlandic settlement. Firstly, with declining populations, there may simply have been an insufficient labour force to support the long-range hunting expeditions that were at considerable distance and required a significant investment of time during the summer. Secondly, Norwegian society was devastated by the terrible plague years of 1349–1350 CE, which exacerbated the effects of an earlier agricultural crisis that had played out in the early 14th century CE (Bagge and Mykland, 1987, p. 20ff.), both reducing the number of sailings to Iceland and especially to Greenland. Thirdly, in 1380 CE Norway became part of the Danish-Norwegian Realm, resulting in centralisation of power in Denmark, whose political and economic interests were more strongly focused southwards. The entire character of European trade had also shifted, and new players were entering this field. In particular, the German Hanseatic League was far better positioned to dominate the bulk trade in mass-produced articles to meet the needs of the growing network of expanding towns across northern Europe (Enemark, 2009). Finally, the demand for Arctic walrus ivory in Europe may have declined due to the increased availability of elephant ivory (Roesdahl, 2005).

Despite these developments, walrus tusks appear to have retained some value, as confirmed by a receipt issued in 1387 CE by Vinald, Archbishop of Nidaros (Tronheim), at the Danish castle of Vordingborg in southern Sjælland. The receipt details the treasures and prestigious artfacts that Vinald's predecessor had taken to Denmark from the treasury of the Norwegian Archbishop's Seat, and no less than 97 walrus tusks, both "klene vnde grot" ("small and large") are listed (Dipl. Norv. pp. 1, 507).

FIGURE 7.8

Hvalsey Church, erected after 1250–1300 CE. The church was most probably built by foreign masons. The last written evidence of life in the Eastern Settlement was a letter written at Garðar in 1409 CE confirming the wedding between an Icelandic couple. The wedding had taken place in the Hvalsey church in September 1408 CE.
Photo: J. Arneborg, 2015.

Archaeological evidence suggests that life did continue at some Norse farms in the Eastern Settlement in Greenland until the middle of the 15th century (Arneborg, 1996), and that these communities may have kept in contact with the wider world, and potentially exported some walrus ivory. However, the scale of these operations could in no way match the organised commercial trade in ivory that had been coordinated by Norwegian authorities in Bergen and Trondheim, and it is perhaps not surprising that the last remote Greenlandic outpost was eventually abandoned by the mid 15th century CE (Fig. 7.8).

References

Arneborg, J., 1996. Burgunderhuer, baskere og døde øde nordboer i Herjolfsnæs, Grønland. Nationalmuseets Arbejdsmark. København, pp. 196–199.

Arneborg, J., Lynnerup, N., Heinemeier, J., Møhl, J., Rud, N., Sveinbjörnsdóttir, Á.E., 2012. Norse Greenland Dietary Economy ca. AD 980 – ca. AD 1450, Introduction. In, Greenland Isotope Project, Diet in Norse Greenland AD 1000 – AD 1450. Journal of the North Atlantic 3, 1–39.

Bagge, S., Mykland, K., 1987. Norge i dansketiden. Politikens Danmarks historie. Politikens Forlag, p. 334.

Barrett, J.H., Boessenkool, S., Kneale, C.J., O'Connell, T.C., Star, B., 2020. Ecological globalisation, serial depletion and the medieval trade of walrus rostra. Quaternary Science Reviews 229, 106122. Available from: https://doi.org/10.1016/j.quascirev.2019.106122.

Bately, J., 2007b. Text and translation, the three parts of the known world and the geography of Europe north of the Danube according to Orosius' *Historiae* and its Old English version. In: Bately, J., Englert, A. (Eds.), Ohthere's Voyages. Maritime Culture of the North 1. The Viking Ship Museum in Roskilde, Roskilde, pp. 40–58.

Benediktsson, J., 1981a. Jakt, Island. Kulturhistorisk Leksikon Nordisk Middelalder 7, 551–552.

Benediktsson, J., 1981b. Säljakt, Island. Kulturhistorisk Leksikon Nordisk Middelalder 17, 699–701.

Bertelsen, T., Schnohr, J., 2019. Muslingemørtel og megalitter. Skalk 3, 6–9.

Brisbane, M.A., Makarov, N.A., Nosov, E.N., 2012. Medieval Novgorod in its wider contect. In: Brisbane, M.A., Makarov, N.A., Nosov, E.N. (Eds.), The Archaeology of Medieval Novgorod in Context. Oxbow Books, pp. 1–9.

Curnow, K., 2000. In: Friedmann, J.B., Figg, K.M. (Eds.), Trade, Travel and Exploration in the Middle Ages, An Encyclopedia. Routledge, New York London, pp. 293–295.

Dipl.Norv. Diplomartarium Norvegicum vol 1, letter 508. Available from: https://www.dokpro.uio.no/perl/middelalder/diplom_vise_tekst_2016.prl?b = 509&s = n&str = .

Enemark, P. 2009 Hansestæder. Den Store Danske, Gyldendal. http//denstoredanske.dk/index.php?sideId = 88854.

Enghoff, I.B., 2003. Hunting, Fishing and Animal Husbandry at The Farm Beneath the Sand, Western Greenland, 28. Meddelelser om Grønland, Man & Society.

Frei, K., Coutu, A.K., Smiarowski, K., Harrison, R., Madsen, C.K., Arneborg, J., et al., 2015. Was it for walrus? Viking age settlement and medieval walrus ivory trade in

Iceland and Greenland. World Archaeology . Available from: https://doi.org/10.1080/00438243.2015.1025912.
GHM, 1838–1845. Grønlands Historiske Mindesmærker vol. I–III. Det kongelige nordiske Oldskrift-Selskab, Kjøbenhavn.
Hofstra, T., Samplonius, K., 1995. Viking expansion northwards, mediaeval sources. Arctic 48 (3), 235–247.
Imer, L.M., 2017. Peasants and prayers, The Inscriptions of Norse Greenland, 25. Publications from the National Museum.
Jónsson, F., 1930. Det gamle Grønlands Beskrivelse af Ívar Bárðarson. Levin & Munksgaards Forlag, København.
Keighley, X., Pálsson, S., Einarsson, B.F., Petersen, A., Fernández-Coll, M., Jordan, P., et al., 2019. Disappearance of Icelandic Walruses coincided with Norse Settlement. Molecular Biology and Evolution 36 (12), 2656–2667. Available from: https://doi.org/10.1093/molbev/msz196.
Keller, C., 2010. Furs, fish, and ivory: medieval Norsemen at the Arctic Fringe. Journal of the North Atlantic 3, 1–23.
Konungs skuggsjá, 1926. Dansk oversættelse ved Finnur Jónsson. Udgivet af Det kongelige nordiske Oldskriftselskab, Gyldendalske Boghandel. Nordisk Forlag, Kjøbenhavn. Available from: http//heimskringla.no/wiki/Kongespejlet.
Lárusson, M.M., 1981. Hvalfangst, island. Kulturhistorisk Leksikon Nordisk Middelalder 7, 168–172.
Lebecq, S., 2007. Communication and exchange in nordwest Europe. In: Bately, J., Englert, A. (Eds.), Ohthere's Voyages. Maritime Culture of the North 1. The Viking Ship Museum in Roskilde, Roskilde, pp. 170–179.
Lynnerup, N., 2011. When populations decline. Endperiod demographics and economics of the Greenland Norse. In: Meier, T., Tillesen, T. (Eds.), Ûber die Grenzen und zwischen den Disziplinen. Archaeolingua Alapitvany, Budapest, pp. 335–345.
Madsen, C.K., 2014. Pastoral Settlement, Farming and Hierarchy in Norse Vatnahverfi, South Greenland (PhD Dissertation). Faculty of Humanities, University of Copenhagen, p. 442.
Madsen, C.K., Ravn, M., Sand, O., 2019. Grønlandstogtet. Arkæologisk Forum 40, 8–17.
Magerøy, H., 1993. Soga om austmenn. Nordmenn som siglde til Island og Grønland I mollomalderen. Det Norske Videnskaps-Akademi II. Hist.-Filos. Klasse Skrifter Ny Serie no. 19.
Makarov, N.A., 2007. The land of the *Beormas*. In: Bately, J., Englert, A. (Eds.), Ohthere's Voyages. Maritime Culture of the North 1. The Viking Ship Museum in Roskilde, Roskilde, pp. 140–149.
Makarov, N.A., 2012. The fur trade in the economy of the northern borderlands of medieval Russia. In: Brisbane, M.A., Makarov, N.A., Nosov, E.N. (Eds.), The Archaeology of Medieval Novgorod in Context. Oxbow Books, pp. 381–390.
McGovern, T.H., 1985. Contributions to the Paleoeconomy of Norse Greenland 1985. Acta Archaeologica 54. København, pp. 73–122.
McGovern, T.H., 2011. Walrus Tusks & Bone from Aðalstræti 14-18, Reykjavík, Iceland. CUNY Northern Science & Education Center NORSEC Report 55. Available from https//www.nabohome.org/uploads/nabo/rvk_walrus_2011_report_2.pdf.
McGovern, T.H., Bigelow, G.F., Amorosi, T., Woollett, J., Perdikaris, S., 1993. Narsaq—a Norse *landnáma* farm. In: Vebæk, C.L. (Ed.), Meddelelser om Grønland, 18. Man & Society, pp. 58–74.

McGovern, T.H., Amorosi, T., Perdikaris, S., Woollett, J., 1996. Verterate Zooarchaeology of Sandnes V51, Economic Change at a Chieftain's Farm in West Greenland. Arctic Anthropology 33 (2), 94–121.
Meldgaard, J., 1995. Eskimoer og Nordboer i Det yderste Nord. Nationalmuseets Arbejdsmark. Nationalmuseet, København, pp. 199–214.
Muus, B., Salomonsen, F., Vibe, C., 1981. Grønlands Fauna. Gyldendal, København.
Nyegaard, G., 2018. Dairy farmers and seal hunters, subsistence on a Norse farm in the Eastern settlement, Greenland. Journal of the North Atlantic 37, 1–80.
Näsman, U., Roesdahl, E., 2003. Scandinavian and European perspectives—Borg I. In: Munch, G.S., Johansen, O.S., Roesdahl, E. (Eds.), Borg in Lofoten. Tapir Academic Press, Trondheim, pp. 283–299. Lofotr. The Viking Museum at Borg, Lofoten.
Nørlund, P., Roussell, A., 1930. Norse Ruins at Gardar. Meddelelser om Grønland LXXVI. København.
Roesdahl, E., 1995. Hvalrostand elfenben og nordboerne i Grønland. Odense Universitetsforlag.
Roesdahl, E., 2005. Walrus ivory—demand, supply, workshops, and Greenland. In: Mortensen, A., Arge, S.V. (eds.) Viking and Norse in the North Atlantic. Select papers from the Proceedings of the Fourteenth Viking Congress, Tórshavn 19–30 July 2001. Annales Societatis Scientiarum Færoensis Supplementum XLIV, Tórshavn, pp. 182–191.
Roussell, Aa, 1936. Sandnes and the neighbouring farms. Meddelelser om Grønland 38 (2).
Seaver, K., 1996. The Frozen Echo. Stanford University Press, Stanford, California.
Seaver, K., 2009. Desirable teeth, the medieval trade in Arctic and African ivory. Journal of Global History 4, 271–292. Available from: https://doi.org/10.1017/S1740022809003155. London School of Economics and Political Science.
Star, B., Barrett, J.H., Gondek, A.T., Boessenkool, S., 2018. Ancient DNA reveals the chronology of walrus ivory trade from Norse Greenland. Proceedings Royal Society B 285, 20180978. Available from: http//doi.org/10.1098/rspb.2018.0978.
Stoklund, M., Roesdahl, E., 2002. En dekoreret hvalrosskalle med tænder og runer i Le Mans—og om runeindskrifter på hvalros-og narhvaltand. Aarbøger for nordisk Oldkyndighed og Historie, pp. 163–184.
Stylegar, F.-A., 2015. Eastern imports in the Arctic. In: Minaeva, O., Blomquist, L. (Eds.), Scandinavia and the Balkans. Cambridge Scholars Publishing, pp. 204–214.
Tegengren, H., 1962. Valrosstanden i Världshandeln. Nordenskiöld-samfundets Tidsskrift XXII, 3–37. Helsingfors.
Urbanczyk, P., 1992. Medieval Arctic Norway. Institute of the History of Material Culture, Polish Academy of Sciences, Warszawa.
Valtonen, I., 2007. Who were the *Finnas*? In: Bately, J., Englert, A. (Eds.), Ohthere's Voyages. Maritime Culture of the North 1. The Viking Ship Museum in Roskilde, Roskilde, p. 92.
Vésteinsson, O., Hegmon, M., Arneborg, J., Ride, G., Russel, W.G., 2019. Dimensions of inequality. Comparing the North Atlantic and the US Southwest. Journal of Anthropological Archaeology 54, 172–191. Available from: https//doi.org/10.1016/j.jaa.2019.04.004.
Wallerström, T., 1995. Norrbotten, Sverige och Medeltiden. Lund Studies in Medieval Archaeology 15 (1), Stockholm.

CHAPTER

The exploitation of walrus ivory in medieval Europe

James H. Barrett

McDonald Institute for Archaeological Research, Department of Archaeology, University of Cambridge, Downing Street, Cambridge, United Kingdom

Department of Archaeology and Cultural History, NTNU University Museum, Trondheim, Norway

Trinity Centre for Environmental Humanities, Arts Block, Trinity College Dublin, College Green, Dublin, Ireland

Chapter Outline

Introduction

Walrus ivory was valued for both craftwork and high art in medieval Europe, albeit with roles that varied across space and time. As a raw material, walrus ivory's size and heterogeneity made it ideally suited to objects of modest dimensions, and/or where the pearly core of a tusk could be used to great decorative effect (Fig. 8.1). Although sometimes interpreted as a poor-quality substitute for elephant ivory (see below), in certain times and places, it was clearly also a desirable exotic good in its own right.

This chapter outlines the shifting popularity and roles of walrus ivory in Europe throughout the Middle Ages. I explore the sources and networks via which walrus tusks were obtained, and the workshops where, on present

The Atlantic Walrus. DOI: https://doi.org/10.1016/B978-0-12-817430-2.00009-1

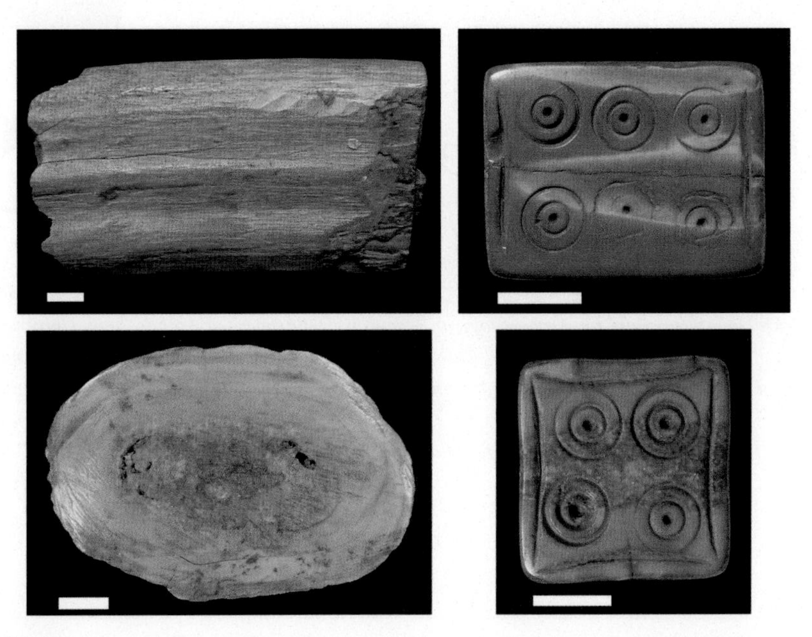

FIGURE 8.1

Finds from Dublin showing the characteristics of worked walrus ivory. Top left: cut segment of tusk (site E190, Fishamble Street III, number 4469, early-mid 11th century CE). Bottom left: cross-section of tusk showing the thin outer rim of cementum, the thick zone of primary dentine and the pearly core of secondary dentine (site E122, Christchurch Place, number 14325, undated). Top right: parallelepipedal die showing mainly primary dentine, but with remnants of lighter cementum where the tusk surface was originally undulated (site E122, Christchurch Place, number 12927, mid-late 11th century CE). Bottom right: the end of the same parallelepipedal die, showing the pearly core of the tusk sandwiched between primary dentine. Scale bars are 1 cm.

Credit: Photographs: J.H. Barrett, © National Museum of Ireland.

knowledge, ivory is known to have been finished (Fig. 8.2). I discuss how walrus ivory was paid for, and how much medieval Europeans actually knew about the natural history of the animal from which it came. Finally, the chapter concludes with a brief outline of the causes and consequences of the trade in walrus ivory for regions of supply and demand.

The medieval use of walrus ivory

The medieval trade in walrus ivory occurred as urban and ecclesiastical centres around the North and Baltic Seas were growing, and while Scandinavian voyaging—and sometimes settlement—extended ever further into the Barents Sea

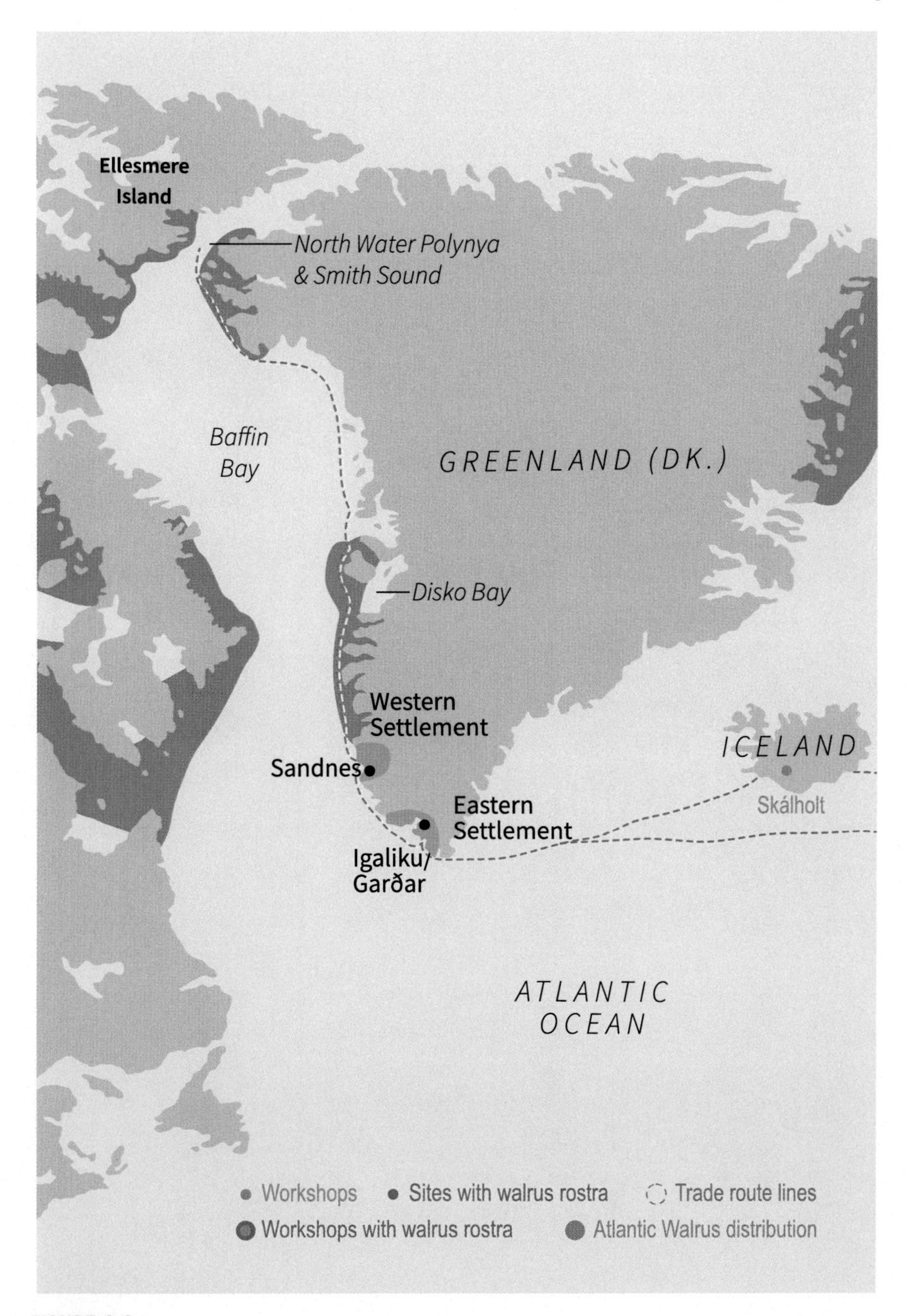

FIGURE 8.2

Map showing the spatial extent of walrus ivory trade during the Middle Ages. Probable walrus ivory workshops are shown across medieval Europe. Finds of medieval walrus rostra (the front of the skull, to which traded tusks were initially attached) are also shown. Note that the medieval range of the walrus was more extensive than the modern distribution used in this map; walruses would also have been found in northernmost Norway and (until they were hunted out in the Viking Age) Iceland.

Illustration courtesy of: Elena Kakoshina, after Fig. 8.1 in Barrett et al. 2020

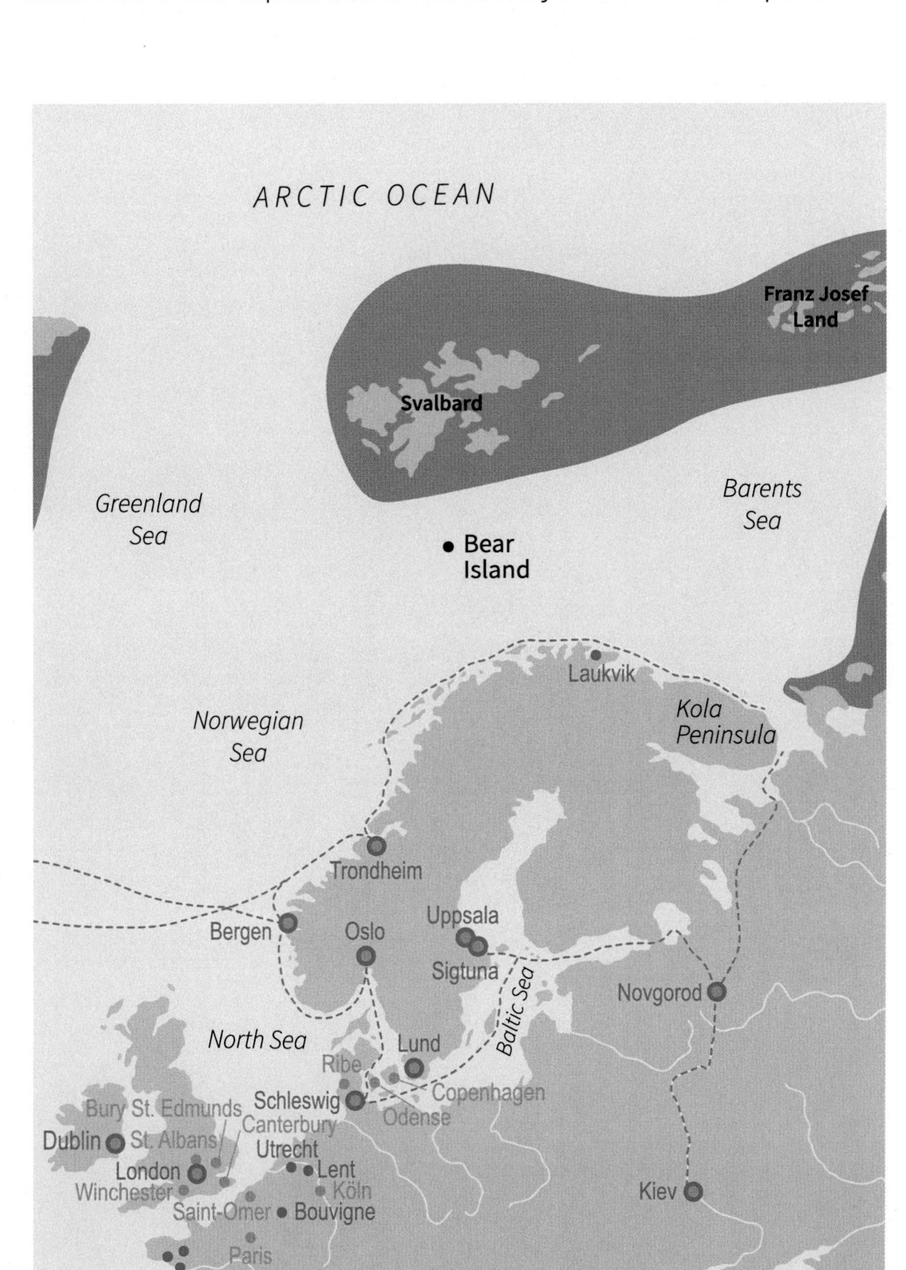

FIGURE 8.2

Continued

region and the North Atlantic. Based on surviving artefacts, European demand for walrus ivory first rose to a significant scale around 1000 CE. The author's primary survey of the existing material culture is on-going, but this chronological pattern is clear, and has long been recognised (Roesdahl, 1998: p. 18; Smith, 2016). There are isolated examples of earlier trade in walrus ivory, such as a famous Anglo-Saxon account regarding an Arctic Norwegian chieftain of the late 9th century (see below), but the decades around 1000 CE were pivotal. Two drivers of demand are evident at this time. In Scandinavian and Scandinavian-influenced centres of walrus ivory import (e.g., Trondheim and Novgorod), the material was an abundant resource and was employed for high-quality versions of commonplace objects. Simple gaming pieces are prime examples, with early examples from as far afield as Dublin (Caulfield, 1992a: p. 385), Trondheim (McLees, 1990), Sigtuna (O'Meadhra, 2001: pp. 72, 76; Karlsson, 2016: pp. 92–93) and Novgorod (Smirnova, 2001: pp. 14, 16). Knife handles were also made of walrus ivory, especially in Novgorod (Smirnova, 2001: pp. 14, 17), and plain combs of this material were produced in both Novgorod (Smirnova, 2005) and in Scandinavian towns of the Baltic Sea region (such as Sigtuna) (Wikström, 2011; Söderberg, pers. comm.).

At a greater distance from these main importing centres regions such as southern England used walrus ivory for the most precious of objects in the years around 1000 CE. Many of these objects were produced in ecclesiastical workshops. At Winchester Cathedral, for example, crucifix corpora were manufactured to a very high standard. During archaeological excavation of buildings associated with a royal monastery, both a finished (albeit broken) example and a rough-out of a crucifix corpus forearm have been recovered (Biddle, 1990: pp. 260, 262; Beckwith, 1990: pp. 760–761) (Fig. 8.3). As inferred from art history, walrus ivory objects were also produced in Continental workshops of the late 10th and 11th centuries CE, including Germany (e.g., Legner, 1985: p. 427; Williamson, 2010: p. 239). Knowledge and materials moved freely across the English Channel. For instance, a pectoral reliquary cross (Williamson, 2010: pp. 238–241) from the collection of the Victoria and Albert Museum exemplifies this milieu in time and space. The cross has a corpus of walrus ivory, carved to an exceptionally high standard in the style of the Winchester workshop, mounted (probably at Essen Abbey, Germany) on a cross of imported cedar, encased in gold sheet decorated with gold wire and enamel mounts (Fig. 8.4).

In locations such as England, distant from the main transshipment centres of walrus ivory, simple objects such as combs were rare during the 10th and 11th centuries CE. When present, moreover, they could be highly decorated—an example being a late 10th or early 11th century CE Anglo-Saxon comb with interlace and animal ornamentation currently in the British Museum's collection (Wilson, 1960).

Medieval European demand for walrus ivory peaked during the 12th century CE, at the height of Romanesque art (Gaborit-Chopin, 2003; Williamson, 2010). Concurrently, the range of uses for walrus ivory became more generalised from

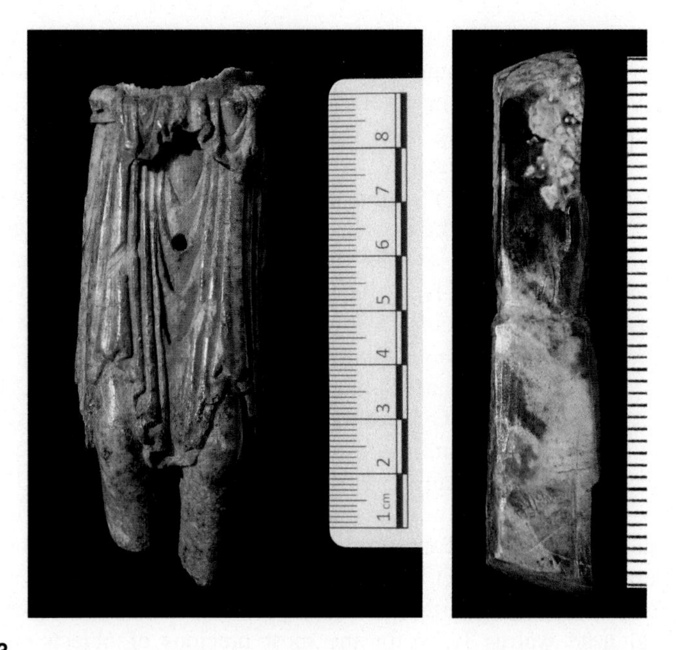

FIGURE 8.3

Crucifix corpus (number CG SF2412, late 10th or 11th century CE) and an unfinished hand with forearm (number CG SF2714, late 11th or early 12th century CE, scale in mm) from excavations at Winchester Cathedral.

Credit: Photographs: J.H. Barrett, by permission of the Winchester Excavations Committee. Not to be reproduced without written permission.

place to place. Some local trends remained (for example the frequent use for combs and knife handles in Novgorod [Smirnova, 2001, 2005]), but overall, walrus ivory became a raw material for use in both high art, and fine versions of more routine objects, throughout its European distribution. The clear distinction between centres of import and more distant workshops blurred. On the one hand, Scandinavian centres of walrus ivory trade became responsible for high-quality ivory working. Well-known examples include the Lewis Chessmen (Caldwell et al., 2009; Dectot, 2018) (Fig. 8.5) and the Munkholmen tau cross head (Blindheim, 1972: p. 47; Gundestrup, 1991: p. 188) (Fig. 8.6), all of which were probably produced in Trondheim. On the other hand, simple objects, such as decorated discs for the game of tables (related to backgammon) were by then being made in locations such as southern England (e.g., at St Albans, see Williamson, 2010: pp. 418, 420) and the Rhineland (e.g., Cologne, see Gaborit-Chopin, 2003: pp. 252–253), where the raw material had previously been treated with greater reverence.

The roles of walrus ivory diverged once again as Romanesque art was replaced by Gothic. Outside of Scandinavia and Russia, ecclesiastical ivory

FIGURE 8.4

Reliquary cross (museum number 7943-1862, late 10th century CE, length of ivory corpus 112 mm) having a walrus ivory corpus in the style of the Winchester workshop, mounted in a contemporary purpose-made cross of cedar and gold that was probably crafted at Essen Abbey near Cologne, Germany. This object shows the high value of walrus ivory outside Scandinavia in the years around 1000 CE, and the exchange of materials and knowledge between European workshops.

Credit: © Victoria and Albert Museum, London.

carving shifted to using elephant tusk in the 13th century CE (Roesdahl, 1995: pp. 30–31; Guérin, 2010; Williamson and Davies, 2014). This transition was influenced by an increase in the availability of elephant ivory, thought to result from increased trans-Saharan trade with West Africa, and then transshipment by sea to western European centres such as Paris (Guérin, 2010, 2017). Elephant ivory tusks are larger and the ivory more homogeneous, affordances that facilitated the large sculptures for which the Gothic period is well known. Elephant ivory also became a preferred medium for some secular objects outside Scandinavia and Russia, most notably mirror-backs and combs (Williamson and Davies, 2014: pp. 562–631). Nevertheless, walrus tusk did retain a niche role—in the manufacture of some chess pieces in England and Germany (Williamson and Davies, 2014: pp. 718–723).

Within Scandinavia and at Novgorod, conversely, walrus ivory continued to serve a wide variety of uses beyond the 13th century CE, even until the beginning

FIGURE 8.5

Select chess pieces (the Lewis Chessmen) from the hoard of over 93 walrus ivory objects (late 12th or early 13th century CE) found at Uig on the Isle of Lewis, Scotland.

Credit: Image © National Museums Scotland.

FIGURE 8.6

Highly-decorated object of walrus ivory (museum number 9101, c.1175–1200 CE, 120 mm long), probably from a crozier in the shape of a tau cross, found on Munkholmen (an island monastery near Trondheim).

Credit: Photograph: Lennart Larsen, National Museum of Denmark, CC-BY-SA.

of the 15th century CE (e.g., see Haug, 1996; Smirnova, 2001, 2005; den Hartog, 2012: pp. 9, 23; Barrett and Hansen, in prep). For example, in Norway (probably Trondheim), walrus ivory was still sufficiently valued to serve as raw material for

FIGURE 8.7

The ornate Wingfield-Digby crozier (museum number A.1–2002, 258 mm tall) made in Norway late in the 14th century CE. It illustrates the continued use of walrus ivory within Scandinavia after workshops in England and Continental Europe had switched to using elephant tusk for the sculpting of ecclesiastical objects.

Credit: © Victoria and Albert Museum, London.

the exquisite Wingfield-Digby crozier of the late 14th century CE (Williamson and Davies, 2014: pp. 442–445) (Fig. 8.7). By the mid 15th century CE, however, use of walrus ivory had almost completely ceased everywhere in Europe (Barrett, unpublished data), before reappearing in the postmedieval period concurrent with the rise of Arctic whaling (Rijkelijkhuizen, 2009).

Sources and networks

Sources

Having sketched an outline of the uses to which walrus ivory was put through the Middle Ages, new questions emerge. From where did it come, through what networks did it flow, and did these networks change over time? These are questions about which long-standing assumptions hold true, in part, but where recent research is also changing and clarifying our understanding.

Prior to the Norse settlement of Greenland (c.985 CE), it has long been thought that the main source of walrus ivory imported into western Europe must have been Arctic Russia and/or Finnmark in northernmost Norway—reached via coastal Norway (Tegengren, 1962; Roesdahl, 2001, 2007). We are told as much in the remarkable late 9th century Anglo-Saxon account of Ohthere, an Arctic Norwegian chieftain who brought walrus tusks to Alfred the Great of England (Bately, 2007). Ohthere had acquired the tusks, along with walrus skin ropes (see below), from Sámi hunters (called *Finnas* in the primary source) (Bately, 2007). The natural distribution of walruses makes it likely that they were hunted on the Kola Peninsula or further east (e.g., at the Kanin Peninsula). However, prior to the 19th century extirpation of walruses from Bear Island (in the middle of the Barents Sea), they were more frequent visitors to the coast of Finnmark (Munthe-Kuas Lund, 1954; Lønø, 1972). Thus, walruses could conceivably also have been harvested in the northernmost parts of Norway. Ohthere probably also entered the White Sea (Englert, 2007), but contrary to popular belief, it is unlikely to have been a major walrus hunting ground (Barrett et al., 2020 and references therein).

In Russia, it is also probable that Novgorod acquired walrus ivory through trade networks with Indigenous hunters, ancestral to modern Nenets peoples, *east* of the White Sea. Certainly the town's fur-trading activities in this direction became extensive over the course of the Middle Ages (Martin, 1986; Makarov, 2012; Brisbane, 2020: p. 12). The degree to which Novgorod's economic catchment extended to the Arctic coast of the Pechora Sea (the south-eastern subdivision of the Barents Sea) in the 10th and 11th centuries CE is unclear, but indirect evidence suggests this may have been the case. 'Fish teeth' (assumed to be walrus tusks) from what is now Russia, are referred to in Islamic sources of this date (e.g., see Ettinghausen, 1950: pp. 121–122; Storå, 1987: p. 122; see also Gillman, 2017). Building on these accounts, the potential importance of Viking-age walrus hunting between the Kanin and Yugorsky peninsulas has recently been popularised by Birgisson (2014), although this possibility has yet to be confirmed by archaeological evidence.

One might speculate—very tentatively—that the complete walrus tusk from an early 10th century CE context at the elite centre of Ryric (also known as Rurikovo) Gorodishche near Novgorod (Giria and Dorofeeva, 2010; Носов et al., 2012: pp. 62–63; Nosov, pers. comm.) derived from western trade routes. It is an interesting outlier, predating most discoveries of walrus ivory in Russia, and has an incised triquetra-a popular Carolingian/Ottonian and Insular (British and Irish) design that was adopted on Viking-age Scandinavian coins (Giria and Dorofeeva, 2010: p. 73; Garipzanov, 2011).

Iceland, like Finnmark and/or Arctic Russia, was another source of walrus ivory prior to the settlement of Greenland. A local population of walruses existed at the time of Landnám (initial settlement) c.870 CE. Archaeological and historical evidence suggests this population was quickly extirpated in the Viking Age, with only occasional visiting animals hunted thereafter (Frei et al., 2015; Murray-Bergquist, 2017; Keighley et al., 2019). Nevertheless, some 10th and 11th century CE walrus ivory finds from distant centres may be from Iceland. For example, based on ancient DNA analysis, the mitochondrial genome of one early 11th century tusk offcut from Sigtuna (Star et al., 2018) appears most closely related to Icelandic samples (Keighley et al., 2019: p. 2660).

Several sources of walrus ivory were thus available to European carvers prior to the settlement of Greenland, but the volume of material seems to have been very small. Excluding Iceland and northern Norway (which had local sources), and artefacts with date ranges that include the 11th century CE (when Greenland had been settled), very few medieval European finds of walrus ivory predate 1000 CE. Key examples of these rare finds include a set of five plaques from a composite Carolingian object (perhaps a portable altar) dated by non-absolute means to the late ninth century CE (Gaborit-Chopin, 2003: pp. 165–166), the early 10th century CE tusk from Ryric Gorodishche noted above, a 10th century CE offcut from Fishamble Street in Dublin (Barrett, unpublished data) and a 10th century CE semimanufacture of a gaming piece from the Kvarteret Urmakaren excavation in Sigtuna (Söderberg, pers. comm.).

It is important to highlight that not all reported examples of walrus ivory in the published literature and museum catalogues are, in fact, ivory, and initial identifications have sometimes been revised in subsequent reporting. For example, hemispherical gaming pieces from the 8th century CE Salme boat burial in Estonia have been reported as walrus ivory (Price et al., 2016: pp. 1025, 1028), but the objects were shown to be whale bone (Peets and Maldre, 2010; Maldre, pers. comm.). Similarly, reference in the popular literature to walrus ivory at the early Viking-age town of Kaupang in southern Norway (Brown, 2015: p. 51) is based on preliminary reporting of a single object (Skre and Stylegar, 2004: p. 25) subsequently confirmed to be either burnt bone or antler. These are two of many examples internationally, and should not be taken as implied criticism. Walrus ivory can usually be identified definitively under low magnification given an experienced observer, but inaccurate preliminary identifications, often during initial artefact cataloguing, are sometimes recorded and perpetuated.

As noted above, walrus ivory first became a common medium for the arts and crafts of medieval Europe in the decades around 1000 CE. This was true in both western Europe (e.g., Dublin, where the author has documented 12 finds of the late 10th and early 11th centuries CE) and eastern Europe (e.g., Novgorod, see Smirnova, 2001, 2005). The temporal correlation of walrus finds with the settlement date of Norse Greenland (c.985 CE) suggests the impact of this abundant new source. If some of Novgorod's raw material was instead coming from northern Russia at this date, then a widespread change in fashion, perhaps itself inspired by the new availability of Greenlandic ivory, may also have played a role in the rising popularity of the material.

It is important to determine where the increasing abundance of ivory during the 11th-12th centuries CE was from, given the possibility of Arctic Russian walrus ivory being sourced in Novgorod, and the information regarding trade in Finnmark or the Kola peninsula in Ohthere's 9th century account. Assuming Iceland's walruses were extirpated relatively quickly (Frei et al., 2015; Keighley et al., 2019), the questions arises as to whether ivory originated from new sources in Greenland, from the Barents Sea Region, or both? This question has been addressed by recent ancient DNA (aDNA) and stable isotope analyses. The archaeological finds making this research possible are pieces of medieval walrus skull (rostra) discovered across northern Europe. These bone objects can be sampled for analysis without the need to damage carved ivory artefacts (Star et al., 2018; Barrett et al., 2020).

In the 10th century CE, the small quantity of walrus ivory being traded was likely transported in the form of extracted loose tusks. This can be inferred based on the Ryric Gorodishche example discussed above (Носов et al., 2012: pp. 62–63), and also a cache of three tusks buried under a house floor at Aðalstræti 14–18 in Iceland (McGovern, 2011; Frei et al., 2015: pp. 4–5). By the 11th century CE, however, the raw material was at least sometimes transported as pairs of tusks still attached to the front of the walrus skull (the rostrum) (Barrett et al., 2020). This practice protected the hollow proximal ends of the tusks and provided a convenient package for storage and transportation. Through time, from the 11th century to the 14th century CE, the rostra themselves were increasingly modified by deliberate carving. Rostra were carved for decorative effect and to prethin the bone around the tusk sockets to ease extraction of the ivory on delivery (Barrett et al., 2020; Fig. 8.8). In a very few instances the resulting packages, complete with tusks, survive intact, having been kept as curiosities (Roesdahl and Stoklund, 2006; Imer, 2017: p. 244). However, in most cases rostra were broken up during the Middle Ages, to extract the ivory as intended, and pieces of rostra have been discovered during archaeological excavations (Barrett et al., 2020).

European finds of medieval rostra (67 of which have now been catalogued [Barrett et al., 2020]) are concentrated in towns that had walrus ivory workshops, and probably also served as network nodes for transshipment and onward trade. The earliest such nodes included Dublin, Trondheim and Schleswig. Subsequently, Dublin may have ceased its involvement in the trade and Bergen adopted a primary role (Barrett et al., 2020). From these main centres of redistribution, smaller numbers of rostra with tusks were transported to ivory workshops in England, France, the Netherlands, Sweden, Russia (Novgorod) and Ukraine (Kiev). Extracted tusks were probably also traded, to these and other destinations (see below).

During initial processing, the rostra were removed and modified in a consistent series of steps. This operational sequence (*chaîne opératoire*) implies that they derived from a community of practice sharing clear ways of doing things. One decorated rostrum was discovered at Igaliku (Garðar), the former bishop's farm in Greenland, suggesting that Greenland was part of this community. Moreover, based on aDNA, many of the European rostra finds were from walruses of a 'western' clade that is exclusively distributed in western Greenland and the Canadian Arctic (Star et al., 2018). Stable isotope analysis of the minority of medieval rostra belonging to a more widely-distributed DNA clade (that is present

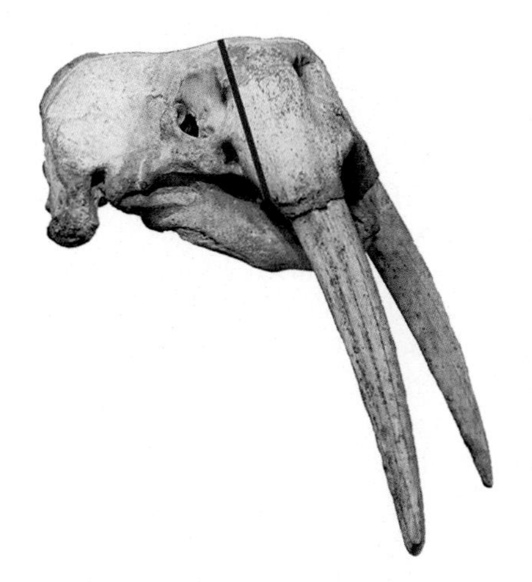
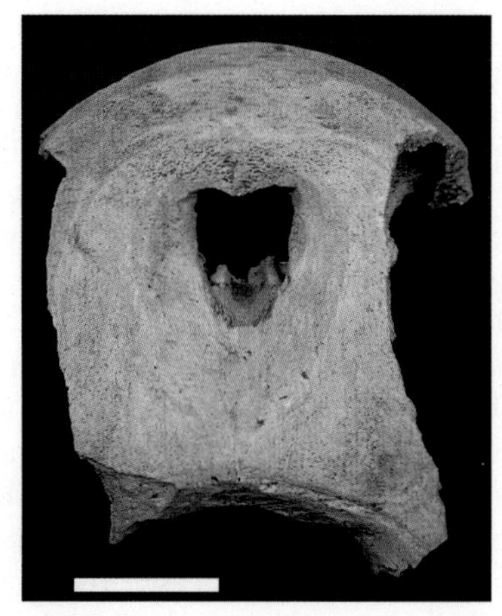

FIGURE 8.8

Left: Modern walrus skull showing the plane along which the rostrum (complete with tusks) was typically removed for trade in the Middle Ages. Right: A modified walrus rostrum (museum number BRM4/4176/1, late 12th to early 13th century CE, scale bar 5 cm) from archaeological excavations in Bergen, showing carved surfaces and breakage of the tusk sockets to remove the ivory. This rostrum was probably imported from Greenland based on ancient DNA and stable isotope evidence (Barrett et al., 2020).

Credit: Photographs: J.H. Barrett, courtesy of the University Museum of Bergen.

throughout the range of the Atlantic walrus, including in Arctic Europe) showed that most were also consistent with a Greenlandic origin (Barrett et al., 2020).

Thus, present evidence suggests Norse Greenland was the main source of walrus ivory imported into Europe between the 11th and 15th centuries CE. At least, this was so of ivory transported together with rostra. Thus far, given the need for destructive sampling, ivory artefacts themselves have only been the subject of pilot studies using aDNA (Star et al., 2018) or stable isotopes (Frei et al., 2015). It thus remains to be seen if fully-extracted tusks, lacking attached rostra, were regularly traded from other locations, such as Arctic Russia.

Workshops

The main centres where Greenlandic rostra were imported and presumably redistributed also had ivory workshops; Dublin (e.g. Caulfield, 1992a, 1992b), Trondheim (McLees, 1990), Schleswig (Ulbricht, 1984: p. Taf. 54.7) and Bergen (Barrett and Hansen, forthcoming) have all produced offcuts and/or unfinished objects. Conversely, other major centres of walrus ivory working have yielded few or no known rostra fragments. Within Scandinavia, Sigtuna is a prime example, having many tusk offcuts and finished artefacts (Karlsson, 2016: pp. 91–93,

95–99) but only two rostra (Barrett et al., 2020). Historical evidence indicates that the ecclesiastical see of Skalholt in Iceland was also known for its ivory carving. There is a unique reference to a named artisan, Margrét *in haga* ('the adroit'), who worked under the patronage of Bishop Páll Jónsson early in the 13th century CE (Egilsdóttir, 2002: p. 325; Brown, 2015). The reference to Margrét is also notable as evidence for a female ivory worker. Winchester and St Albans in England, other known workshops, have also not produced rostra. Furthermore, Cologne in Germany was a major 12th century CE centre of walrus ivory carving, from which ivory offcuts have been found (Berke, 1997; Gechter and Schütte, 1995: pp. 133–134), but no known rostra. Perhaps the most striking example is Novgorod. It has yielded the largest assemblage of walrus ivory known from any medieval European settlement, and includes offcuts (e.g., Smirnova, 2001, 2005). However, only a single rostrum (late 13th century CE) is presently known from this site (Smirnova, 2001: pp. 10–11, 17; Barrett et al., 2020).

Skalholt may have had access to ivory from the occasional walruses that still visited Iceland's shores after a local population was extirpated in the Viking Age. In other cases—such as Winchester and Cologne—it seems likely that Greenlandic ivory was typically acquired as extracted tusk material from the main nodes of the rostra trade. In Novgorod, this may also have been the case. The single rostrum found there has not yet been analysed using aDNA or stable isotopes, but is modified in the same way as other European finds now known to be from Greenland (Barrett et al., 2020). As considered above, however, a European Arctic source of extracted tusks (traded without attached rostra) is also probable for this town.

It is not yet possible to compile a comprehensive list of medieval walrus ivory workshops, but based on archaeological evidence of broken-up rostra and/or ivory offcuts, other examples include: Oslo (Barrett et al., 2020), Lund (Bergquist and Lepiksaar, 1957: pp. 46, 48–49, 75; Wahløø, 1992: p. 390; Carelli, 1998; Kulturen, 1999: pp. 363, 1555, 1563), Uppsala (Karlsson, 2016: pp. 176–177), Ribe (in its 13th century phase) (Sydvestjyske Museer, n.d.: pp. 200003570; Søvsø, pers. comm.), Odense (Brandt et al., 2018), Copenhagen (Lyne et al., 2015: pp. 80–81), London (Bowsher et al., 2007: p. 343), Canterbury (Riddler, pers. comm.) and Kiev (Сагайдак et al., 2008: pp. 143–144; Khamaiko, 2018: pp. 150–152). In the case of Canterbury, the archaeology is supported by a 13th century CE reference in *Hrafns saga Sveinbjarnarsonar* describing the donation of tusks from a (by that time, rare) Icelandic walrus kill to the shrine of St Thomas Becket (Helgadóttir, 1987: p. 3; Murray-Bergquist, 2017: p. 41). On art-historical grounds, walrus ivory workshops are also thought to have existed in St Albans, Bury St Edmunds, Paris, in Saint-Omer, and along the River Meuse, perhaps in Liège (Parker and Little, 1994; Gaborit-Chopin, 2003: pp. 230–232, 243–245). Some workshops were in ecclesiastical settings (e.g., Canterbury and Winchester as discussed above). Others were in secular contexts, such as at Sigtuna where ivory offcuts were found near the early 11th century CE mint of Olof Skötkonung (Ros, 2009: p. 84; Söderberg, pers. comm.).

Trade routes

The routes connecting inferred major nodes of ivory redistribution (e.g., Trondheim, Bergen, Schleswig and Dublin) with each other and more distant workshops cannot be reconstructed with any level of detail. It is very rare that specific journeys are recounted in the historical sources, and walrus tusks are not among those products routinely listed in the detailed customs accounts regarding northern European trade that become available starting in the 14th century CE (e.g., see Nedkvitne, 2014, 2019: p. 171). What information exists is instead anecdotal or indirect, such as the reference in an early 13th century Old Norse translation of *Tristan and Iseult*, which refers to walrus tusks as among the trade goods imagined to be on a Norwegian merchant ship in France (Nedkvitne, 2019: p. 171).

We might speculate that trade from Norway to Schleswig followed the coastal route to Hedeby, as described by Ohthere in the late 9th century (Englert, 2007). In fact, Hedeby, like Schleswig, has yielded tusk offcuts (Reichstein, 1991: pp. 65–66; Schietzel, 2014: p. 352). For trade to the Rhineland, including Cologne, one important waypoint may have been Utrecht, through which passed a 12th century pilgrimage route extending from Iceland and Norway to Rome and Jerusalem (Marani, 2012: p. 21). Two modified walrus rostra, now recognised as probable medieval finds, were discovered in Utrecht in 1941 (van Deinse, 1944; van Bree, 1968; Barrett et al., 2020, numbers R44–R45). Trade to Continental workshops via Flanders is similarly probable. In the latest known medieval record of the sale of Greenlandic walrus ivory, which occurred in Bergen in 1327, a merchant from Ypres purchased over 500 tusks originally paid as tithe (Keller, 2010: p. 3; Nedkvitne, 2019: pp. 126–127).

Trade from Trondheim and Bergen to England was commonplace in the Middle Ages. The main destinations, including both seasonal fairs and towns (Moore, 1985; Nedkvitne, 2014: pp. 25–33, 53–79, 146–186), varied through time. In the north, 'Danelaw' towns such as York and Lincoln had Scandinavian links from the Viking Age onward and both have produced objects of walrus ivory (e.g. MacGregor et al., 1999: pp. 1981–1982, 2047; Steane et al., 2016: p. 290). In the south, one might envision that Winchester (served by its coastal port of Southampton) would have been particularly significant in the early 11th century CE, when it served as a *de facto* capital of Knut's empire (which spanned England, Denmark and Norway). Winchester's role as a centre of walrus ivory carving has been discussed above. London's trade with Norway was especially important in the late 12th and early 13th centuries CE, based on the evidence of English pottery finds in Bergen and Trondheim (Reed, 1990; Blackmore and Pearce, 2010: pp. 76–77). Two fragmentary walrus skulls from the London Guildhall excavation provide evidence of a workshop operating around this time (Bowsher et al., 2007: p. 343). Later in the Middle Ages, east coast ports such as King's Lynn, Boston, Grimsby, Ravenser Odd and Hull are known to have taken on greater importance in Anglo-Norwegian trade (Carus-Wilson, 1963; Rigby, 1993; Evans, 1999). None has yet produced evidence for the working of walrus ivory, but these ports served major centres, such as York, Lincoln, Norwich and, in the case of King's Lynn, London.

The trade connections of Viking Age and medieval Dublin stretched from the Scandinavian North Atlantic (including Iceland and Norway) to Britain, Denmark and France (Wallace, 1987, 2015; McCutcheon, 2006; Woods, 2013). Although Dublin fell to Anglo-Norman colonists in the 1170's, Norwegian interest in Ireland and the Irish Sea province continued into the 13th century CE. In fact, the mid-13th century merchant's text from which we know much about the Greenlandic trade, *The King's Mirror*, also discusses Ireland at length (Holm-Olsen, 1983: pp. 21–26). It is not possible to reconstruct medieval Dublin's many trade routes in this brief chapter. Nevertheless, it is relevant that the extraordinary late 12th century CE hoard of walrus ivory objects found on the Isle of Lewis in Scotland (which included the Lewis Chessmen) lies on the sea lanes between Trondheim (where they were likely manufactured) and Ireland (see Caldwell et al., 2009: pp. 165–166).

Trade from source regions to the main centres of transshipment (and sometimes also direct to other workshops) can be reconstructed to varying degrees. The earliest small-scale trade—of the kind included in the Ohthere account—first entailed Norwegian voyages to Finnmark and the Kola Peninsula to acquire walrus tusks from Sámi hunters (Bately, 2007; Englert, 2007). Transport south may then have followed the coastal voyage described for Ohthere's journey from his home in northern Norway to Hedeby. Ohthere also visited England, with walrus tusks on board. This trip could have continued from Hedeby, along the southern shore of the North Sea to London or Winchester, or entailed a separate, open-water crossing (Sawyer, 2007; Storli, 2018: pp. 113, 124–126).

Viking-age and medieval Iceland lacked proper towns. Trade was conducted when Icelanders sailed abroad on their own vessels (until c.1200 CE) and when merchants (mostly Norwegians) visited wealthy Icelandic farms or a small number of seasonal stations such as Kolkuós and Gásir (Vésteinsson, 2009; Þorláksson, 2010; Harrison, 2014; Helle, 2019). The trade with seasonal stations was established by the 12th century, and evidenced by a walrus tusk offcut found at Gásir in North-East Iceland, probably dating to the 14th century CE (Harrison et al., 2008: p. 112; pers. comm.; Pierce, 2014: p. 175). Iceland's trade routes may have been diverse between Landnám (Settlement, approximately 870 CE) and the 12th century CE. It has been suggested, for example, that there were connections between the island and Hedeby (Hilberg and Kalmring, 2014). By the 13th and 14th centuries CE, however, most Icelandic trade was conducted by Norwegians and funneled through Bergen (Helle, 2019). Direct trade with Iceland by merchants from elsewhere, for example, England and Germany, did not occur until the 15th century.

Trade with Greenland was sometimes conducted via Iceland (occasionally, by Greenlanders themselves), but direct sailing from Norway also occurred (Pierce, 2014: pp. 175–176; Nedkvitne, 2019: pp. 203–205). After rounding Cape Farewell (Greenland), landfall might have been made at Herjolfsnes (the most southerly-known Norse site) or its neighbouring harbour Sandhavn (Golding et al., 2011), before onward travel elsewhere within the Eastern and/or Western settlements. Walrus products were prepared for trade in both settlements (McGovern, 1985; McGovern et al., 1996; Enghoff, 2003; Smiarowski et al., 2017), having been acquired by the Greenlanders during summer expeditions to the northern hunting grounds (Roesdahl, 1995; Ljungqvist, 2005; Gulløv, 2016) (Fig. 8.2).

The organisation of the northern hunt within Greenland is discussed thoroughly by Arneborg in chapter 7 of this volume. It is pertinent only to add that recent evidence raises the possibility that the western Greenlandic walrus population became overexploited in the Middle Ages (Barrett et al., 2020), similar to Icelandic walruses around the 11th–12th centuries CE (Keighley et al., 2019). By the 13th–14th centuries CE, Norse walrus hunting had switched from exclusively large males to smaller, often female, animals. Concurrently, there was a focus on animals of the western genetic clade, which is most represented in the northernmost walrus populations of western Greenland and Arctic Canada (Star et al., 2018; Barrett et al., 2020). Moreover, Norse finds of this date occur in Thule Inuit settlements on either side of Smith Sound, in North-West Greenland and Ellesmere Island (Holtved, 1944: pp. 298–302; Schledermann and McCullough, 2003; Appelt and Gulløv, 2009; Gulløv, 2016).

Combining all of this evidence, it seems likely that walrus hunting, and probably also trade with Inuit for walrus ivory, had reached extreme and unsustainable limits (Barrett et al., 2020). Iceland's walruses had previously been extirpated (Keighley et al., 2019), and the search for Greenland's walrus ivory probably extended further and further north, perhaps as far as the Indigenous Canadian Arctic on Ellesmere Island. Subsequently, in the 15th century CE, the use of walrus ivory evaporated in Europe (Barrett, unpublished data), and Norse settlement in Greenland ended. It is unlikely to be a coincidence that Greenland's more northerly Western Settlement, which was probably more involved in walrus hunting, was the first to be abandoned in the mid-to-late 14th century CE. The Eastern Settlement held out longer with a more diversified economic base including more pastoral agriculture though it, too, was abandoned by the mid 15th century. The late-medieval decline in the trade of walrus ivory is considered further below.

The trade in walrus tusks allowed Norse Greenlandic settlements to import goods that were not available locally, and to pay taxes or tithes imposed upon them. The mid-13th century text *King's Mirror* suggests that Greenland's most important imports were iron and timber (Holm-Olsen 1983: p. 29). Ivory also paid for religious provision in the island's Norse settlements. Tithes were paid in tusks (Nedkvitne, 2019: pp. 124–128), potentially providing the initial incentive for the creation of a Greenlandic bishopric in 1124 CE (Seaver, 2009: p. 282; see Arneborg, this volume) according to the Icelandic text *Grænlendinga tháttr* (Sveinsson and Þórðarson, 1935).

It is unfortunate that we know little about exchange rates, or the price of walrus ivory. There is one definite exchange rate, and it comes from the 1327 transaction in Bergen noted above. Nedkvitne (2019: pp. 126, 178) infers from the transaction that two tusks were worth one Norwegian mark, or three English shillings, equivalent to approximately 100 dried cod. This valuation places walrus ivory as a valuable, but not precious commodity, at the time of its declining demand in Europe. The exchange rate must have been higher during the 11th and 12th centuries CE, when demand was rising, but we cannot quantify this with available evidence.

Medieval European knowledge of the walrus

How much did European consumers know of the source of the walrus ivory they carved? What knowledge was in circulation regarding its geographical origin and the natural history of the animal from which it derived? In asking this question it is worth raising the example of narwhal tusks. Their popularity rose with that of walrus ivory in England and Continental Europe, and continued thereafter (Pluskowski, 2004). While the Norse settlements across the North Atlantic existed, western Greenland (especially Disko Bay) must have been the main source for both walrus and narwhal materials. Narwhals, including beached remains from mass mortalities through entrapment in sea ice, would have been more accessible to medieval Europeans here than in any other part of their natural range (see Siegstad and Heide-Jørgensen, 1994; Heide-Jørgensen, 2018).

Yet narwhal tusk was widely regarded in medieval Europe as originating from unicorns (Ettinghausen, 1950; Pluskowski, 2004). Knowledge of the real whale did spread within some circles. Realistic narwhals are described in the Norwegian *King's Mirror* (Holm-Olsen, 1983: p. 16), and Albertus Magnus (the German author of an encyclopaedic 13th century natural history, *De animalibus*) shows vague awareness of a sea 'fish' having a horn on its forehead (Kitchell and Resnick, 1999: p. 1693). For many in England and Continental Europe, however, the narwhal's geographic origin and natural history were overridden by popular belief and the ecclesiastical symbolism of unicorns (Pluskowski, 2004).

What, then, of walrus ivory? Within Scandinavian nodes of transshipment, there is artefactual evidence that the real animal was understood and appreciated. In both Trondheim (Christophersen and Nordeide, 1994: pp. 35; 248–249; Fig. 8.9) and Bergen (Grieg, 1933: pp. 384–385), figurines of walruses have been found that are relatively anatomically correct. These figurines are carved from the small cheek teeth of walruses, rather than the tusks, and are examples of folk art rather than high art. Nevertheless, they accurately depict key anatomical characteristics, such as the tusks, flippers and tapered torso. Both examples attempt to represent the characteristic vibrissae (whiskers) of walruses, and the Bergen find has incised notches that may represent the folded skin of the neck. The Trondheim object is from the mid-13th century (Christophersen and Nordeide, 1994: pp. 35, 248–249); the Bergen find is less well stratified, but probably also of medieval date (Barrett and Hansen, in prep).

FIGURE 8.9

Figurine of a walrus (Library site, number N28915, c.1225–1275 CE, 41 mm long) found during archaeological excavation in Trondheim. It was carved from a walrus cheek tooth (rather than a tusk), perhaps by a walrus hunter, as it represents anatomically-accurate features of the animal.

Credit: Photograph: Åge Hojem, NTNU University Museum, Trondheim.

It seems likely that these objects were carved in Greenland (another example was excavated near the Western Settlement farm of Sandnes [Roussell, 1936: pp. 123 – 124]), or by Greenlanders resident in Norway (Christophersen and Nordeide, 1994: p. 248). The townspeople of medieval Trondheim and Bergen may have been only one degree of separation away from a walrus hunt. Thus, we might ask how widespread knowledge of the walrus was beyond Greenland—both in Scandinavia and further afield within Europe?

Icelandic and Norwegian written sources, such as *Hrafns saga Sveinbjarnarsonar* (Helgadóttir, 1987: p. 3) and *The King's Mirror* (Holm-Olsen, 1983: p. 29), show that, by the mid-13th century CE, literate as well as mercantile and hunting circles, did have familiarity with the real animal. In Continental Europe, conversely, even the most informed commentator had only an imperfect understanding. Albertus Magnus combines folklore and fact in his description:

"... *the hairy whales, and others as well, have the longest tusks. They use these to hang from the rocks of cliffs when they sleep. At such a time a fisherman, coming close, separates as much of its skin as he can from the blubber near its tail. He passes a strong rope through the part he has loosened and he then ties the ropes to rings fixed into a mountain or to very strong stakes or trees. Then, by throwing rocks at the head of the fish from a large sling, he wakes it up. Aroused, the fish, as it tries to escape, draws off its skin from its tail down along its back and head and leaves it behind. Later, not far from that place, it is captured in a weakened state, either swimming bloodless in the water or lying half-alive on the shore. Strips of this animal's skin are very strong and are used for lifting huge weights on pulleys. They are always displayed for sale in the marketplace at Cologne.*" (Kitchell and Resnick, 1999: p. 1671)

In summary, the animal was known in medieval Europe, but only vague details regarding its natural history accompanied traded walrus products beyond Scandinavia.

Discussion: causes and consequences

The utilisation of walrus ivory in medieval Europe was influenced by the shifting interplay of supply and demand, as determined by global changes in artistic fashion, trade routes and wildlife abundance. It is probable that demand greatly exceeded supply until the end of the 12th century CE, at least amongst the most distant consumers. In the ninth century CE, when only rare imports via Arctic Norway are likely to have reached western Europe, the Ohthere account informs us that tusks were a gift worthy of royalty (Bately, 2007: p. 45). In the decades around 1000 CE, as walruses were being extirpated from Iceland (Keighley et al., 2019) and Greenland was first being settled, only precious objects were made of this material in southern England and Continental Europe.

Both demand and supply boomed in the 12th century CE, when walrus ivory was used for everything from book mounts and tabernacles, to chess and tables pieces (Williamson, 2010). The increased supply may be attributed in part to the creation of a Greenlandic bishopric in 1124 CE, and the role of the church in organising the trade (Star et al., 2018; see Arneborg chapter 7). Historical sources (e.g., Sveinsson and Þórðarson, 1935) and finds of multiple walrus skulls at the bishop's farm of Garðar (Igaliku) (Degerbøl, 1929) are both consistent with this interpretation.

However, this extensive use of walrus ivory may have begun to reduce the walrus population of western Greenland. There is archaeological evidence for possible over-hunting by the 13th–14th centuries CE, as smaller, female walruses of the 'western' genetic clade (most exclusively distributed among Greenland's northernmost populations) were increasingly likely to be imported to Europe (Barrett et al., 2020).

Surprisingly, this evidence for possible resource depletion coincided with a major decrease in European *demand* for walrus ivory, due at least in part to increased availability of elephant ivory. It has thus been proposed that a decrease in the unit value of walrus ivory meant that Norse Greenlanders needed to procure more of it to maintain their balance of trade with Europe (Barrett et al., 2020). It is in this context that we can best understand papal correspondence of 1282 CE, which requested that Greenlandic tithes of walrus ivory be converted into gold and silver in Norway before payment to Rome (DN Diplomatarium Norvegicum, 2011: p. 71). Walrus ivory was no longer as highly valued, but remained what Greenlanders had to offer. Rather than an influx of elephant ivory to Europe leading to reduced walrus hunting, the opposite may instead have occurred. As argued above, the hunt may ultimately have extended as far north as Smith Sound in the 13th and 14th centuries CE, involving voyages of unsustainable distance and risk. Facing these and other pressures, not least the climate change of the Little Ice Age (Jackson et al., 2018; Lasher and Axford, 2019), Norse Greenland was soon abandoned. In the 15th century CE, walrus ivory became rare even in Scandinavia, perhaps due to supply constraints. It was in postmedieval times, for example, with the discovery of the Svalbard archipelago in the late 16th century CE, that walrus products once again re-entered European trade networks (e.g., Rijkelijkhuizen, 2009).

Acknowledgements

The following organisations and individuals kindly provided access to collections, contextual information and/or other advice: National Museum of Denmark (J. Arneborg), Natural History Museum of Denmark (K. M. Gregersen, K.H. Kjær), National Museum of Iceland (L. Árnadóttir and Á. Guðmundsson), The Agricultural University of Iceland (A. H. Pálsdóttír), City University of New York (T. McGovern), NTNU University Museum (A. Christophersen, J. A. Risvaag, J. Rosvold), University Museum, University of Bergen (G. Hansen, A. K. Hufthammer), AHKR, University of Bergen (S. Nordeide), Museum of Cultural History, University of Oslo (M. Vedeler), NIKU (L. M. Fuglevik), National Museum of Ireland (M. Sikora, A. Halpin), York Archaeological Trust (L. Carter), The Collection, Lincoln (A. Lee), Norwich Castle Museum and Art Gallery (T. Pestell), The British Museum (S. Brunning, V. Smithson), MOLA (L. Blackmore, A. Pipe), Westminster Abbey (S. Jenkins), Winchester Excavations Committee (K. Barclay, M. Biddle, M. Kirkpatrick and H. Rees), Sigtuna Museum (A. Söderberg), Uppsala University (A. Hennius), Kulturen, Lund (C. Johansson Hervén), Schleswig-Holsteinische Landesmuseen Schloss Gottorf (V. Hilberg, U. Schmölcke), Art & History Museum, Bussels (E. van Binnebeke), Le musée Vert, Le Mans (N. Morel), Tromsø Museum (T. Larssen, I. Storli, S. Wickler), the Novgorod State Museum (A. Andrienko), the Institute of Archaeology, National Academy of Sciences of Ukraine (N. Khamaiko), The Kyiv Centre of Archaeology (M. Sagaydak). Alena Kakoshina drew Fig. 8.2 based on information provided by Xénia Keighley and the author. Gitte Hansen kindly agreed the mention of Bergen finds prior to publication of our coauthored catalogue. H. C. Küchelmann and L. Smirnova shared knowledge of western European and Russian walrus finds respectively. Bastiaan Star kindly commented on the text. This research was supported by the Leverhulme Trust [MRF-2013-065] and the Research Council of Norway [projects 262777 and 230821].

References

Appelt, M., Gulløv, H.C., 2009. Tunit, Norsemen, and Inuit in thirteenth-century Northwest Greenland: Dorset between the devil and the deep sea. In: Maschner, H., Mason, O., McGhee, R. (Eds.), The Northern World, AD 900–1400. The University of Utah Press, Salt Lake City, pp. 300–320.

Barrett, J.H., Hansen, G., in prep. *Bergen as a Hub for Medieval Arctic Trade: The Walrus Ivory*, The Bryggen Papers, Bergen.

Barrett, J.H., Boessenkool, S., Kneale, C.J., O'Connell, T.C., Star, B., 2020. Ecological globalisation, serial depletion and the medieval trade of walrus rostra. Quaternary Science Reviews 229, 106122.

Bately, J., 2007. Text and translation. In: Bately, J., Englert, A. (Eds.), Ohthere's Voyages: A Late 9th-Century Account of Voyages Along the Coasts of Norway and Denmark and its Cultural Context. The Viking Ships Museum, Roskilde, pp. 40–50.

Beckwith, J., 1990. Ivory corpus from a cross. In: Biddle, M. (Ed.), Object and Economy in Medieval Winchester. The Clarendon Press, Oxford, pp. 760–761.

Bergquist, H., Lepiksaar, J., 1957. Medieval animal bones found in Lund. Archaeology of Lund. 1, 11–84.

Berke, H., 1997. Haustiere, handwerker und händler. Vorläufige archäozoologische ergebnisse der ausgrabungen am Heumarkt in Köln, *Kölner Jahrbuch* 30, 405–414.

Biddle, M., 1990. The nature and chronology of bone, antler and horn working in Winchester. In: Biddle, M. (Ed.), Object and Economy in Medieval Winchester. The Clarendon Press, Oxford, pp. 252–263.

Birgisson, B., 2014. *Den Svarte Vikingen*, Spartacus Forlag, Oslo.

Blackmore, L., Pearce, J., 2010. *A Dated Type Series of London Medieval Pottery: Part 5, Shelly-Sandy Ware and the Greyware Industries*. Museum of London Archaeology, London.

Blindheim, M., 1972. *Middelalderkunst fra Norge i andre land Norwegian Medieval Art Abroad*, Universitetets Oldsaksamling, Oslo.

Bowsher, D., Dyson, T., Holder, N., Howell, I., 2007. *The London Guildhall: An Archaeological History of a Neighbourhood From Early Medieval to Modern Times Part II*. Museum of London Archaeology Service, London.

Brandt, L.Ø., Haase, K., Collins, M.J., 2018. Species identification using ZooMS, with reference to the exploitation of animal resources in the medieval town of Odense. Danish Journal of Archaeology 7, 139–153.

Brisbane, M., 2020. The INTAS projects and the archaeological background. In: Maltby, M., Brisbane, M. (Eds.), Animals and Archaeology in Northern Medieval Russia: Zooarchaeological Studies in Novgorod and its Region. Oxbow Books, Oxford, pp. 1–14.

Brown, N.M., 2015. *Ivory Vikings*. St Martin's Press, New York.

Caldwell, D.H., Hall, M.A., Wilkinson, C.M., 2009. The Lewis Hoard of gaming pieces: a re-examination of their context, meanings, discovery and manufacture. Medieval Archaeology 53, 155–203.

Сагайдак, М.А., Хамайко, Н.В., Вергун, О.Л., 2008. *Новые находки древнерусских игральных фигурок из Киева*, in: Моця, О.П. (Ed.), Стародавній Іско-ростень і слов'янські гради II, НАН Укра'іни, Коростень (Korosten), pp. 137-145.

Carelli, P., 1998. Schack: Det medeltida feodalsamhället i miniatyr. In: Wahlöö, C. (Ed.), Kulturen 1998: Metropolis Daniae Ett stycke Europa. Kulturhistoriska Föreningen för Södra Sverige, Lund, pp. 135–148.

Carus-Wilson, E.M., 1963. The medieval trade of the ports of the Wash. Medieval Archaeology 6–7, 182–201.

Caulfield, D., 1992a. Gaming-piece and rough-out. In: Roesdahl, E., Wilson, D.W. (Eds.), From Viking to Crusader: Scandinavia and Europe 800–1200. Nordic Council, Uddevalla, Sweden, p. 385.

Caulfield, D., 1992b. Walrus skull and tusk fragment. In: Roesdahl, E., Wilson, D.W. (Eds.), From Viking to Crusader: Scandinavia and Europe 800–1200. Nordic Council, Uddevalla, Sweden, p. 385.

Christophersen, A., Nordeide, S.W., 1994. *Kaupangen ved Nidelva: 1000 års byhistorie belyst gjennom de arkeologiske undersøkelsene på Folkebibliotekstomten i Trondheim 1973–1985*, Riksantikvaren, Trondheim.

Dectot, X., 2018. When ivory came from the seas. On some traits of the trade of raw and carved sea-mammal ivories in the Middle Ages. Anthropozoologica 53, 159–174.

Degerbøl, M., 1929. Animal bones from the Norse Ruins at Gardar. In: Nørlund, P., Roussell, A. (Eds.), Norse Ruins at Gardar: The Episcopal Seat of Medieval Greenland. The Commission for Scientific Research in Greenland, Copenhagen, pp. 183–192.

den Hartog, E., 2012. On six Danish knife handles or hair parters shaped like falconers, *By, marsk og. geest: Kulturhistorisk årbog Sydvestjylland* 24, 5–27.

DN (Diplomatarium Norvegicum), 2011. *Diplomatarium Norvegicum* Vols. 1 to 21. Dokumentasjonsprosjektet, Oslo. https://www.dokpro.uio.no/perl/middelalder/diplom_-vise_tekst.prl?b = 72&s = n&str = . Consulted 25 May 2020.

Egilsdóttir, Á., 2002. *Biskupa Sögur II*, Íslenzk Fornrit XVI. Hið Íslenska Fornritafélag, Reykjavík.

Enghoff, I.B., 2003. Hunting, fishing and animal husbandry at the Farm Beneath the Sand, Western Greenland. An archaeozoological analysis of a Norse farm in the Western Settlement., Danish Polar Centre, *Meddelelser om Grønland: Man and Society*, 28, Copenhagen.

Englert, A., 2007. Ohthere's voyages seen from a nautical angle. In: Bately, J., Englert, A. (Eds.), Ohthere's Voyages: A Late 9th-Century Account of Voyages Along the Coasts of Norway and Denmark and its Cultural Context. The Viking Ships Museum, Roskilde, pp. 117–129.

Ettinghausen, R., 1950. *The Unicorn*, Studies in Muslim Iconography I. Freer Gallery of Art, Washington.

Evans, D.H., 1999. The trade of Hull between 1200 and 1700. In: Gläser, M. (Ed.), Lübecker Kolloquium zur Stadtarchäologie im Hanseraum, II: Der Handel. Verlag Schmidt-Römhild, Lübeck, pp. 59–97.

Frei, K.M., Coutu, A.N., Smiarowski, K., Harrison, R., Madsen, C.K., Arneborg, J., et al., 2015. Was it for walrus? Viking Age settlement and medieval walrus ivory trade in Iceland and Greenland. World Archaeology 47, 439–466.

Gaborit-Chopin, D., 2003. *Ivoires medievaux Ve–XVe siecle*. Réunion des musées Nationaux, Paris.

Garipzanov, I., 2011. Religious symbols on early Christian Scandinavian coins (ca. 995–1050): From imitation to adaptation. Viator 42, 35–54.

Gechter, M., Schütte, S., 1995. Der Heumarkt in Köln: Ergebnisse und Perspektiven einer archäologischen Untersuchung. Geschichte in Köln 38, 129–139.

Gillman, M.E., 2017. A tale of two ivories: elephant and walrus. In: Shalem, A. (Ed.), Treasures of the Sea: Art Before Craft? Universidad Nacional de Educación a Distancia, Madrid, pp. 81–105.

Giria, Y.Y., Dorofeeva, T.S., 2010. Unique walrus tusk from Rurik's fortress: traceological analysis. Российская Археология 2010 (1), 64–73.

Golding, K.A., Simpson, I.A., Schofield, J.E., Edwards, K.J., 2011. Norse-Inuit interaction and landscape change in southern Greenland? A geochronological, pedological, and palynological investigation. Geoarchaeology 26, 315–345.

Grieg, S., 1933. *Middelalderske Byfund fra Bergen og Oslo*. A.W. Brøggers Boktrykkeri A/S, Oslo.

Guérin, S.M., 2010. Avorio d'ogni Ragione: the supply of elephant ivory to Northern Europe in the Gothic Era. Journal of Medieval History 36, 156–174.

Guérin, S.M., 2017. Exchange of sacrifices: West Africa in the medieval world of goods. The Medieval Globe 3 (2), 97–124.

Gulløv, H.-C., 2016. Inuit-European interactions in Greenland. In: Friesen, M., Mason, O. (Eds.), The Oxford Handbook of the Prehistoric Arctic. Oxford University Press, Oxford.

Gundestrup, B., 1991. *Det Kongelige danske Kunstkammer 1737*, Nationalmuseet, Copenhagen.

Harrison, R., 2014. Connecting the land to the sea at Gásir: International exchange and long-term Eyjafjörður ecodynamics in medieval Iceland. In: Harrison, R., Maher, R.A. (Eds.), Human Ecodynamics in the North Atlantic: A Collaborative Model of Humans and Nature Through Space and Time. Lexington Books, London, pp. 117–136.

Harrison, R., Roberts, H.M., Adderley, W.P., 2008. Gásir in Eyjafjörður: International exchange and local economy in medieval Iceland. Journal of the North Atlantic 1, 99–119.

Harrison, R. Personal communication, email of 26 April 2020.

Haug, A., 1996. Kristus-figuren fra Kjøpmannsgaten. Spor 11, 42–43.

Heide-Jørgensen, M.P., 2018. Narwhal *Monodon monoceros*. In: Würsig, B., Thewissen, J. G.M., Kovacs, K.M. (Eds.), Encyclopedia of Marine Mammals, 3rd Edition Academic Press, London, pp. 627–631.

Helgadóttir, G.P., 1987. *Hrafns saga Sveinbjarnarsonar*. Clarendon Press, Oxford.

Helle, K., 2019. Bergen's role in the medieval North Atlantic trade. In: Mehler, N., Gardiner, M., Elvestad, E. (Eds.), German Trade in the North Atlantic c. 1400–1700. Interdisciplinary Perspectives. Museum of Archaeology, University of Stavanger, Stavanger, pp. 43–51.

Hilberg, V., Kalmring, S., 2014. Viking age Hedeby and its relations with Iceland and the North Atlantic: Communications, long-distance trade, and production. In: Zori, D., Byock, J. (Eds.), Viking Archaeology in Iceland: Mosfell Archaeological Project. Brepols, Turnhout, pp. 221–241.

Носов, Е.Н., Хвощинская, Н.В., Медведева, М.В., 2012. *Новгородская: Рождение, державы, Свидетельства из глубины столетий*, Лнформационно-издательское агентство «ЛЛК», St Petersburg.

Holm-Olsen, L., 1983. *Konungs skuggsiá*, 2nd ed. Norsk historisk kjeldeskrift-institutt, Oslo.

Holtved, E., 1944. Archaeological investigations in the Thule district I. Descriptive part. Meddelelser om Grønland 141, 1–308.

Imer, L.M., 2017. *Peasants and Prayers: The Inscriptions of Norse Greenland*. University Press of Southern Denmark, Copenhagen.

Jackson, R., Arneborg, J., Dugmore, A., Madsen, C., McGovern, T., Smiarowski, K., et al., 2018. Disequilibrium, adaptation, and the Norse settlement of Greenland. Human Ecology 46, 665–684.

Karlsson, J., 2016. *Spill om Djur, Hantverk och Nätverk i Mälarområdet under Vikingatid och Medeltid*, Institutionen för arkeologi och antikens kultur, Stockholms universitet, Stockholm University, Stockholm.

Keighley, X., Pálsson, S., Einarsson, B.F., Petersen, A., Fernández-Coll, M., Jordan, P., et al., 2019. Disappearance of icelandic walruses coincided with Norse settlement. Molecular Biology and Evolution 36, 2656–2667.

Keller, C., 2010. Furs, fish and ivory: medieval Norsemen at the Arctic fringe. Journal of the North Atlantic 3, 1–23.

Khamaiko, N., 2018. Gaming pieces from recent excavations of the Kyiv Podil. In: Stempin, A. (Ed.), The Cultural Role of Chess in Medieval and Modern Times: 50th Anniversary Jubilee of the Sandomierz Chess Discovery. Muzeum Archeologiczne Poznaniu, Poznań, pp. 149–156.

Kitchell Jr., K.F., Resnick, I.M., 1999. *Albertus Magnus On Animals: A Medieval Summa Zoologica Volume II*. Johns Hopkins University Press, Baltimore.

Kulturen, 1999. *Metropolis fyndkatalog*. Kulturhistoriska föreningen för södra Sverige (Kulturen), Lund.

Lasher, G.E., Axford, Y., 2019. Medieval warmth confirmed at the Norse Eastern Settlement in Greenland. Geology 47, 267–270.

Legner, A., 1985. *Ornamenta ecclesiae: Kunst und Kunstler der Romanik: 2 Katalog zur Ausstellung des Schnutgen-Museums in der Josef-Haubrich-Kunsthalle*, Schnütgen-Museum, Köln

Ljungqvist, F.C., 2005. The significance of remote resource regions for Norse Greenland. Scripta Islandica 56, 13–54.

Lønø, O., 1972. The catch of walrus (*Odobenus rosmarus*) in the areas of Svalbard, Novaja Zemlja and Franz Josef Land, *Norsk Polarinstitutt Årbok* 1970, 199 – 212.

Lyne, E., Dahlström, H., Hansen, C.H., 2015. Rådhuspladsen, Metro Cityring Project. KBM 3827, Vestervold Kvarter, Københavns Sogn Sokkelund Herred, Københavns Amt. Kurturstyrelsen j.nr. 2010-7.24.02/KBH 0015, Museum Copenhagen, Copenhagen.

MacGregor, A., Mainman, A.J., Rogers, N.S.H., 1999. Craft, industry and everyday life: bone, antler, ivory and horn from Anglo-Scandinavian and medieval York. The Archaeology of York 17, 1869–2072.

Makarov, N., 2012. The fur trade in the economy of the northern borderlands of medieval Russia. In: Brisbane, M., Makarov, N., Nosov, E. (Eds.), The Archaeology of Medieval Novgorod in Context: Studies in Centre/Periphery Relations. Oxbow Books, Oxford, pp. 381–390.

Maldre, L. Personal communication, email of 20 February 2017.

Marani, T., 2012. Leiðarvísir. Its genre and sources, with particular reference to the description of Rome. Durham theses, Durham University. Available at Durham E-Theses Online: http://etheses.dur.ac.uk/6397/, Durham.

Martin, J., 1986. *Treasure of the Land of Darkness: The Fur Trade and its Significance for Medieval Russia*. Cambridge University Press, Cambridge.

McCutcheon, C., 2006. *Medieval Pottery from Wood Quay*, Dublin, Royal Irish Academy, Dublin.

McGovern, T.H., 1985. Contributions to the paleoeconomy of Norse Greenland. Acta Archaeologica 54, 73–122.

McGovern, T.H., 2011. Walrus tusks & bone from Aðalstræti 14-18, Reykjavík Iceland, CUNY Northern Science and Education Center Zooarchaeology Laboratory report 55. Unpublished report on file Archaeological Institute Iceland and download from https://www.nabohome.org/publications/labreports/labreports.html.

McGovern, T.H., Amorisi, T., Perdikaris, S., Woollett, J., 1996. Vertebrate zooarchaeology of Sandnes V51: Economic change at a Chieftain's farm in West Greenland. Arctic Anthropology 33, 94–121.

McLees, C., 1990. *Games People Play: Gaming-pieces, Boards and Dice from Excavations in the Medieval Town of Trondheim, Norway*, Riksantikvaren, Utgravningskontoret for Trondheim, Trondheim.

Moore, E.W., 1985. *The Fairs of Medieval England: An Introductory Study*. Pontifical Institute of Mediaeval Studies, Toronto.

Munthe-Kuas Lund, H., 1954. Walrus (*Odobenus rosmarus* L.) off the coast of Norway in the past and after the year 1900, together with some observations on its migration and cruising speed. Astarte 8, 1–12.

Murray-Bergquist, K., 2017. 'To talk of many things': Whales, walrus, and seals in medieval Icelandic literature. University of Iceland., Reykjavík. Available at: https://skemman.is/bitstream/1946/27089/3/Thesis.pdf.

Nedkvitne, A., 2014. *The German Hansa and Bergen 1100-1600*. Böhlau Verlag, Köln.

Nedkvitne, A., 2019. *Norse Greenland: Viking Peasants in the Arctic*. Routledge, London.

Nosov, E. Personal communication, email of 3 January 2017.

O'Meadhra, U., 2001. Kräklan: ett unikt och viktigt fynd, in: Tesch, S., Edberg, R. (Eds.), *Biskopen i Museets Trädgård: En arkeologisk gåta*, Sigtuna Museers Skriftserie 9, Stockholm, pp. 67–90.

Parker, E.C., Little, C.T., 1994. *The Cloisters Cross: Its Art and Meaning*. The Metropolitan Museum of Art, New York.

Peets, J., Maldre, L., 2010. Salme paadijäänused ja luunupud, In: Tamla Ü., (Ed.), *Ilusad asjad: Tähelepanuväärseid leide Eesti arheoloogiakogudest*, 2010, Ajaloo Instituut, Tallinn, 2–42.

Pierce, E., 2014. Walrus hunting and the ivory trade in the North Atlantic. In: Caldwell, D. H., Hall, M.A. (Eds.), The Lewis Chessmen: New Perspectives. National Museums Scotland, Edinburgh, pp. 169–183.

Pluskowski, A., 2004. Narwhals or unicorns? Exotic animals as material culture in medieval Europe. European Journal of Archaeology 7, 291–313.

Price, T.D., Peets, J., Allmäe, R., Maldre, L., Oras, E., 2016. Isotopic provenancing of the Salme ship burials in pre-Viking Age Estonia. Antiquity 90, 1022–1037.

Reed, I.W., 1990. *1000 Years of Pottery: An analysis of pottery trade and use*, Riksantikvaren, Utgravningskontoret for Trondheim, Trondheim.

Reichstein, H., 1991. *Die wildlebenden Säugetieren von Haithabu*. Karl Wachholtz Verlag, Neumünster.

Rigby, S.H., 1993. *Medieval Grimsby: Growth and Decline*. University of Hull Press, Hull.

Rijkelijkhuizen, M., 2009. Whales, walruses, and elephants: Artisans in ivory, baleen, and other skeletal materials in seventeenth- and eighteenth-century Amsterdam. International Journal of Historical Archaeology 13, 409–429.

Riddler, I. Personal communication, email of 2 March 2017.

Roesdahl, E., 1995. *Hvalrostand elfenben og nordboerne i Grønland*. Odense Universitetsforlag, Odense.

Roesdahl, E., 1998. L'ivoire de morse et les colonies norroises du Groenland. Proxima Thule 3, 9–48.

Roesdahl, E., 2001. Walrus ivory in the Viking Age - and Ohthere (Ottar). Offa 58, 33–37.

Roesdahl, E., 2007. Walrus ivory. In: Bately, J., Englert, A. (Eds.), Ohthere's Voyages: A Late 9th-Century Account of Voyages Along the Coasts of Norway and Denmark and its Cultural Context. The Viking Ships Museum, Roskilde, pp. 92–93.

Roesdahl, E., Stoklund, M., 2006. Un crâne de morse décoré et gravé de runes: Á propos d'une découverte récente dans un musée du Mans. Proxima Thulé 5, 9–38.

Ros, J., 2009. *Stad och gård: Sigtuna under sen vikingatid och tidig medeltid*. Uppsala universitet, Uppsala.

Roussell, Aa, 1936. Sandnes and the neighbouring farms. Meddelelser om Grønland 88 (2).

Sawyer, P., 2007. Ohthere's destinations: Norway, Denmark and England. In: Bately, J., Englert, A. (Eds.), Ohthere's Voyages: A Late 9th-Century Account of Voyages Along the Coasts of Norway and Denmark and its Cultural Context. The Viking Ships Museum, Roskilde, pp. 136–139.

Schietzel, K., 2014. *Spurensuche Haithabu*. Wachholtz, Neumünster.

Schledermann, P., McCullough, K.M., 2003. Inuit-Norse contact in the Smith Sound region. In: Barrett, J.H. (Ed.), Contact, Continuity and Collapse: The Norse Colonization of the North Atlantic. Brepols, Turnhout, pp. 183–206.

Seaver, K.A., 2009. Desirable teeth: the medieval trade in Arctic and African ivory. Journal of Global History 4, 271–292.

Siegstad, H., Heide-Jørgensen, M.P., 1994. Ice entrapments of narwhals (*Monodon monoceros*) and white whales (*Delphinapterus leucas*) in Greenland. Meddelelser om Grønland, Bioscience 39, 151–160.

Skre, D., Stylegar, F.-A., 2004. *Kaupang: The Viking Town*. University of Oslo, Oslo.

Smiarowski, K., Harrison, R., Brewington, S., Hicks, M., Feeley, F.J., Dupont-Hébert, C., et al., 2017. Zooarchaeology of the Scandinavian settlements in Iceland and Greenland: Diverging pathways. In: Albarella, U., Rizzetto, M., Russ, H., Vickers, K., Viner-Daniels, S. (Eds.), The Oxford Handbook of Zooarchaeology. Oxford University Press, Oxford, pp. 147–163.

Smirnova, L., 2001. Utilization of rare bone materials in medieval Novgorod. In: Choyke, A.M., Bartosiewicz, L. (Eds.), Crafting Bone: Skeletal Technologies through Time and Space. Archaeopress, Oxford, pp. 9–17.

Smirnova, L., 2005. *Comb-making in Medieval Novgorod (950-1450): an industry in transition*. British Archaeological Reports International Series S1369. Hadrian Books, Oxford.

Smith, L., 2016. The tactile account of Anglo-Saxon ivory (550–1066): Image, status, materiality and economics. In: Hulsman, G., Whelan, C. (Eds.), Occupying Space in Medieval and Early Modern Britain and Ireland. Peter Lang, Oxford, pp. 211–233.

Söderberg, A. Personal communication, emails of 31 March 2017 and 12 September 2018.

Søvsø, M. Personal communication, email of 28 August 2017.

Star, B., Barrett, J.H., Gondek, A.T., Boessenkool, S., 2018. Ancient DNA reveals the chronology of walrus ivory trade from Norse Greenland, *Proceedings of the Royal Society B* 285, 20180978.

Steane, K., Darling, M.J., Jones, M.J., Mann, J., Vince, A., Young, J., 2016. *The Archaeology of the Lower City and Adjacent Suburbs*. Lincoln Archaeological Studies No. 4. Oxbow Books, Oxford.

Storå, N., 1987. Russian walrus hunting in Spitsbergen. Études/Inuit/Studies 11, 117–138.

Storli, I., 2018. *Ottars Verden: En reiseberetning fra 800-tallet*. Orkana Akademisk, Stamsund.

Sveinsson, E.Ó., Þórðarson, M., 1935. *Eyrbyggja saga með Brands þáttr orva -- Eiríks saga rauða – Groenlendinga saga – Groenlendinga þáttr*, Hið Íslenzka Fornritafélag, Reykjavík.

Sydvestjyske Museer, n.d. Samlingen OnLine, Sydvestjyske Museer, Ribe. http://sol.-sydvestjyskemuseer.dk/?mode = detail&genstandsnr = 200003570&side = 1&antal = 21 &indexno = 3&search = hvalrostand&sid = d27681d889941266621523b9ef83629&tt = 3. Accessed 25 May 2020.

Tegengren, H., 1962. Valrosstanden i världshandeln, *Nordenskiöld-samfundets Tidskrift* 22, 3–37.

Þorláksson, H., 2010. King and commerce. The foreign trade of Iceland in medieval times and the impact of royal authority. In: Imsen, S. (Ed.), The Norwegian Domination and the Norse World c. 1100-1400. Tapir Academic Press, Trondheim, pp. 149–173.

Ulbricht, I., 1984. *Die Verarbeitung von Knochen, Geweih und Horn im mittelalterlichen Schleswig*, Karl Wachholtz Verlag, Neumünster.

van Bree, P.H.H., 1968. Über "verzierte" Walroßschädel. Zeitschrift für Säugetierkunde 33, 312–315.

van Deinse, A.B., 1944. Over resten van fossiele en recente Pinnipedia, aangetroffen in Zeeland en elders Nederland. De Levende Natuur 48 (10), 119–125.

Vésteinsson, O., 2009. A medieval merchants' church in Gásir, North Iceland. Hikuin 36, 159–170.

Wahløø, C., 1992. Chess piece, fragmentary, in: Roesdahl, E., Wilson, D.W. (Eds.), From Viking to Crusader: Scandinavia and Europe 800-1200, Nordic Council, Uddevalla, Sweden, p. 390.

Wallace, P., 1987. The economy and commerce of Viking Age Dublin, in: Düwel, K., et al. (Eds.), *Untersuchungen zu Handel und Verkehr der vor - und Frühgeschichtlichen Zeit in Mittel- und Nordeuropa. Teil IV. Der Handel der Karolinger- und Wikingerzeit*, Abhandlungen der Akademie der Wissenschaften in Göttingen. Philo.-Hist. Klasse 3, Nr. 156, Göttingen, pp. 200–245.

Wallace, P.F., 2015. *Viking Dublin: The Wood Quay Excavations*. Irish Academic Press, Dublin.

Wikström, A., 2011. *Fem stadsgårdar: arkeologisk undersökning i kv. Trädgårdsmästaren 9 & 10 i Sigtuna 1988-90*. Sigtuna Museum, Sigtuna.

Williamson, P., 2010. *Medieval Ivory Carvings: Early Christian to Romanesque*. V&A Publishing, London.

Williamson, P., Davies, G., 2014. *Medieval Ivory Carvings: 1200-1550*. V&A Publishing, London.

Wilson, D.M., 1960. An Anglo-Saxon ivory comb. British Museum Quarterly 23, 17–19.

Woods, A., 2013. The coinage and economy of Hiberno-Scandinavian Dublin. In: Duffy, S. (Ed.), Medieval Dublin XIII, Four. Courts Press, Dublin, pp. 43–69.

CHAPTER

9

Modern European commercial walrus exploitation, 1700 to 1960 CE

Ian Gjertz

Department for Oceans, Research Council of Norway, Oslo, Norway

Chapter Outline

Introduction

Atlantic walruses (*Odobenus rosmarus rosmarus*) have been hunted throughout their natural range. Hunting by humans can be divided into two main categories. One is a subsistence hunt conducted by peoples inhabiting areas where walruses were found. The other was a hunt motivated by profit, and dominated by both British whalers and Norwegian sealers. This chapter begins by examining the hunting methods, uses and economy of walrus hunting from the 1700s to 1960 CE. As hunting varied considerably across geographic regions, this chapter concludes with a systematic examination of practices, hunting intensity and timing for five key geographic areas.

The Atlantic Walrus. DOI: https://doi.org/10.1016/B978-0-12-817430-2.00003-0

Walrus hunting methods

European commercial walrus hunting was conducted on land, on ice and in the water. Walruses hauled-out in large numbers on land for days at a time were vulnerable. Hunters would attack them from the shoreline to prevent the walruses' escaping back to the safety of the water. The animals closest to shore were killed first using lances, their bodies creating an obstacle for the rest of the herd. Such slaughter could last for many hours. In a well-documented case, 16 men killed 900 walruses in seven hours on a single island, now known as Håøya in south-eastern Svalbard (Lamont, 1861; Wollebæk, 1901). In his review of walrus harvests from the European Arctic, Lønø (1972) provides more examples of large-scale slaughter on land.

The lance was the weapon of choice before the introduction of high-powered rifles. The lance was a large spear-like iron tool that had a broad 40-cm-long, double-edged head fastened to a 2–3 m long wooden shaft. The diameter of the shaft was about 4 cm at the handle and about 6 cm at the front where the shaft was attached to the iron lance head (Lamont, 1861; Wollebæk, 1901; Hoel, 1949).

The tendancy for walruses to haul out on land for extended periods of time led to hunters seeking them out during all seasons of the year, sea-ice permitting. Hunters built cabins and wintered in areas where walruses often came ashore (Scoresby, 1820). Apart from specializing in killing walruses, these trappers would also harvest Arctic foxes (*Alopex lagopus*), reindeer (*Tarandus platyrhynchos*), polar bears (*Ursus maritimus*), as well as the down and eggs of various species of bird. This European method of broad-based winter trapping was first developed by the Russians (Lamont, 1876; Conway, 1906; Storå, 1987), and was practiced from about 1700 to 1850 CE. Norwegians learned the method from the Russians, and started similar winter trapping in 1795 CE, with a peak in the years between the two World Wars (Hoel, 1949; Lønø, 1972).

Unlike at terrestrial haul-outs, walruses hauled out on ice could easily flee into the water when attacked. They would therefore have to be targeted with a rifle or lance while still on the ice. This was difficult, as the rest of the walruses could flee, and an injured walrus might sink if it managed to reach the water. Thus, in most cases, walruses on ice were first harpooned with a rope attached to the harpoon to prevent the walrus from escaping, and then killed with a lance or a rifle (Wollebæk, 1901). Only muzzleloader firearms were available before the 1850's, and these were not commonly used by sealers. Such rifles were slow to load, and the soft lead bullets not well suited for killing walruses (Scoresby, 1820). This changed with the introduction of metallic cartridges and breech-loading rifles. The Remington rolling block rifle became a favourite among sealers from the late 19th century onward (Hoel, 1949). The repeating rifles had greater rates of fire, better accuracy and muzzle velocity, which made them increasingly effective (Stewart et al., 2014). When shooting at walruses on land or ice, it was important to aim just behind the ear, as frontal shots to the head rarely succeeded due to the walrus' thick skull (Lamont, 1876; Wollebæk, 1901).

Swimming walruses were most commonly hunted with harpoons. When the harpoon head penetrated the animal, the harpoon shaft came loose and floated. After a struggle, the walrus was pulled up alongside the boat and killed with a lance or shot in the back of the head (Lamont, 1861; Wollebæk, 1901; Hoel, 1949). One boat could harpoon up to six or seven females at a time, but could only handle a maximum of two large males since these were strong enough to sink a boat (Lamont, 1861; Wollebæk, 1901). The boats were equipped with block and tackle, and the crew would usually haul the dead walrus onto the ice for butchery (Lamont, 1861; Wollebæk, 1901; Hoel, 1949) (Fig. 9.1).

Each harpoon had a one-flue iron head that was usually 8–30 cm long, and was attached to a strong rope of about 20–50 m in length. The other end of the rope was either fastened to the boat or attached to an inflated buoy (Wollebæk, 1901; Hoel, 1949). The wooden harpoon shaft could vary in length, but was usually 3–4.5 m long. The front half of the harpoon shaft was octagonal in shape, with the remainder being rounded. The diameter of the shaft was about 3 cm. A good harpooner could accurately throw a harpoon up to 20 m (Wollebæk, 1901). When the harpoon rope was attached to the boat, there was a danger of the boat being pulled under. Thus, a knife had to be at hand so the rope could be cut if necessary.

FIGURE 9.1

Harpooning walruses in Svalbard waters.

Illustration by William Livesay in Lamont, J., 1861.

Boats used for hunting

Whaling ships were historically quite large and had crews numbering between 40 and 50 people (Scoresby, 1820; Lubbock, 1937). These larger vessels formed floating bases; once in the whaling area, smaller whaling boats carried small crews that conducted the actual hunting. The whaling vessel would have up to eight whale boats, each manned by six people (Scoresby, 1820).

A typical 19th century Norwegian walrus boat would be carvel-built and double-ended. It usually had a crew of between four and five persons (Lamont, 1861). A four-man crew consisted of one harpooner, two rowers and someone who would steer the boat. A rudder was never used when hunting, as it could break when a harpooned walrus came up to the boat (Lamont, 1861, Wollebæk, 1901). Instead, the boat was steered in the aft by a man using two short oars (Fig. 9.2).

A Norwegian walrus sloop in the 19th and early 20th century usually had a 10-man crew. The sloop had two walrus boats, each would be manned by four crew members. The remaining two crew members would stay aboard the sloop, assist the two boats if necessary and could attend to the sloop if weather or ice conditions deteriorated (Nordgaard, 1895). By the 20th century, walrus hunting in eastern Canada and Greenland was often conducted from small motorised Peterhead-like boats. These would often have crews of between two and three persons (Born et al., 1995; Stewart et al., 2014).

FIGURE 9.2

Two harpooned walruses pulling a Norwegian walrus boat. The walruses are about to be killed with a lance.

Illustration by Gerhard von Yhlen in Chydenius, K., 1865.

Products of walrus hunting

Once caught, the tusks were removed from the walrus carcass in one of three ways. The front of the walrus skull (the maxilla bone) was chopped open with an axe and the tusks broken loose. Alternatively, the entire front end of the skull would be chopped off between the frontal and parietal bones. This portion of the skull would then be brought back with the tusks still attached. Finally, one could bring back the whole skull intact with tusks. Male walrus tusks were known to reach up to 80 cm length and weigh 4 kg each, but were usually about 60 cm long and 2 kg a piece. Female tusks rarely exceeded 50 cm in length and would weigh 1.5 kg each (Lamont, 1861; Hoel, 1949).

The tusks of the walrus are composed of hard, white and compact ivory (see Chapter 5, this volume), and were used by dentists to create false teeth (Scoresby, 1820, see Stewart et al., 2014 for details). In addition to being used in dentures, walrus ivory was also used for making smaller objects, such as chess pieces, whistles and umbrella and cane handles (Lamont, 1861; Hoel, 1949). Walrus tusks also have a thick pulp cavity of secondary dentine, considered at the time of little value. Male tusks usually have many cracks, whereas female tusks are smaller, have a smoother structure, thinner cores and are usually free of cracks (Wollebæk, 1901; Hoel, 1949).

The skins of large male walruses could weigh between 200–400 kg, while female skins generally weighed about 100 kg (Wollebæk, 1901; Hoel, 1949). Walrus hides were difficult to handle because of their weight and awkward size. Therefore, large hides were usually cut in two halves dorso-ventrally before the blubber was removed from the skin (Wollebæk, 1901). This was relatively easy on land but difficult in the water. When killed in the water, walruses would be hauled up on sea-ice with block and tackles before the skin was stripped from the carcass. The walrus could also be towed into shallow water at high tide, with the hide removed once the tide fell (Born et al., 1997).

According to Scoresby (1820) walrus skin was used instead of mats to protect the yards and rigging of ships from being chafed by friction against each other. When cut into strips, and plaited into cordage, the hides were well suited for wheel ropes as they were much stronger and more durable than hemp. When tanned, walrus skins could be turned into a soft porous leather; but they was most useful and durable when left untreated. Hoel (1949) describes how the leather was used for machine belts and soles for boots for long periods of time. In the middle of the 20th century the thickest walrus leather was often cut into circular discs and used for polishing metal. Walrus hide was also used to encase rollers in machines used for cleaning cotton fibres, for bicycle seats, carrying bags and lining of automobile tyres (Stewart et al., 2014). No alternative to walrus hide was known at the time for many of its uses. The hides of adult males were worth more than the hides from young animals. When split in two and tanned, hides of young walruses were used for belts, wallets and other such small items (Stewart et al., 2014).

Walruses were also hunted for the oil that could be rendered from their blubber. Large males would yield 300 kg of blubber. Walrus blubber was considered inferior to other seal blubber, and less valuable per kilogram as it is leaner and yields only 90% of the oil output compared with seal blubber. Walrus oil, usually mixed with seal oil, was used in the making of train oil, for human consumption (margarine), in the fat industry, for the production of soap, as well as for tanning and softening of finer leathers (see Hoel, 1949).

Baculums (penis bones) were sought after as curiosities (Stewart et al., 2014). Born et al. (1997 and references therein) mentions penis bones as one of the walrus products that was exported from Greenland. Nordgaard (1895) writes that penis bones were sold for use as chair legs, but such use was presumably not common.

The walrus meat was occasionally used by European trappers for dog food (Born et al., 1997) but had no commercial value, and was often discarded by walrus hunters (Wollebæk, 1901). Stewart et al. (2014) show that large numbers of walruses were taken for meat by explorers and researchers in eastern Canada and western Greenland.

Live walruses, especially calves, were occasionally taken to be sold, but calves proved very difficult to keep alive, since they would often refuse to eat (Nordgaard, 1895; Stewart et al., 2014).

The economy of hunts

The monetary value of walrus tusks, skin and blubber are only occasionally mentioned in the historical literature. When available, values are most often quoted in either British pounds or Norwegian kroner. These values have been converted for this chapter into present-day British currency and referred to as (GBP) using British Government Inflation statistics (Allen, 2012), Statistics Norway Consumer Price Index price calculator (see www.SSB.no), and the present 2019 exchange rate of 1 GBP equals 11 Norwegian kroner.

Walrus tusks were paid for by the kilogram and prices fluctuated according to availability. Prices in the 19th and 20th centuries varied between GBP 13–30/kg (see Southwell, 1897; Wollebæk, 1901; Hoel, 1949; Stewart et al., 2014). At the turn of the 20th century, tourism to the Arctic became more popular, and as a result the value of whole walrus skulls with intact tusks increased. These skulls were sometimes cleaned and then sold to tourists for GBP 60–300 per skull (Wollebæk, 1901).

Prior to the 1870's, prices were fairly consistent for all walrus skins, with payment based on the number of skins rather than the size or weight of individual pieces. However, starting in the 1870's skins were sold according to weight (Wollebæk, 1901). Half-skins cut dorso-ventrally of at least 75 kg, were sold for GBP 7/kg in 1870. Smaller skins earned GBP 1.3–4/kg depending on their size.

In England, the price of walrus leather increased from GBP 3.6/kg in 1863 to GBP 9/kg in 1865, largely because of its use for transmission belts

(Ytreberg, 1946). In the late 1890's, the best hides came from the Davis Strait, with male walruses preferred due to their skin being much thicker than those of females or calves. A male skin would, on average, fetch about GBP 540. Hides from Franz Josef Land, which were mostly from females and young animals, were sold for as little as GBP 2/kg (Southwell, 1897). Prices for the highest quality hides peaked in 1897 at GBP 16/kg to the makers of bicycles. Blubber was sold by weight. Prices fluctuated and were heavily reliant on economic trends and soared during wartime. Walrus oil was sold for GBP 1.9/kg in 1897 (Southwell, 1897). Between 1928 and 1948 blubber prices varied between GBP 0.4 and 0.5/kg, before reaching an all-time high of GBP 3.7/kg just after World War II (Hoel, 1949). The mean value of a whole dead walrus, based on the sum of all the different products varied and was from 1832 to 1910 reported to be between GBP 236 and GBP 575 (Hoel, 1949).

Economics of ship-based walrus hunting

The whaling boats from each whaling ship would often work together to catch a whale, however when harvesting seals and walruses, each boat would most often work alone. The area a boat could cover from the mothership was limited, implying that unless there was an abundance of seals or walruses in the area only a few boats could contribute to the harvest at a time. When hunting large whales this was not a problem, since each whale had a high value and gave work for all. For most of the crew, their pay was based on the proceeds from the catch (Scoresby, 1820), leading to falling incomes as whales became less abundant. The economy of the enterprise therefore became more dependent on income from also harvesting seals and walruses. Walrus herds were never targeted by whalers in the same way that the large seal herds were. Most ships killed a few walruses during the voyage, but this could vary significantly between years (Scoresby, 1820; Lubbock, 1937).

The overall decline in the whaling economy led to the development of more specialized vessels, which were smaller than the large whaling vessels. Despite their small size, these specialized ships would include crews of up to 40–50 people, as locating, killing and processing large groups of seals comprising thousands of animals required a lot of work for all aboard. When walruses started to become scarce in the middle of the 19th century, hunting methods changed again. Ships specialized for walruses would be smaller, often with crews of about 10 people. They could search vast areas for walrus, and if found, all the crew members worked together in the hunt (Lamont, 1861). Crews would also harvest anything else they encountered along the way, such as beluga (*Delphinapterus leucas*), narwhal (*Monodon monoceros*), polar bears, reindeer, muskoxen (*Ovibos moschatus*), seabird eggs, geese, ducks and even Greenland sharks (*Somniosus microcephalus*). These smaller vessels could be also be used for other purposes off-season, such as fishing (Lubbock, 1937; Hoel, 1949).

Historical harvests by geographical area

The following section looks at the historical European walrus harvest in five geographical areas, extending from the Russian European Arctic westwards to the eastern Canadian Arctic. Common to all areas was the impacts of this harvest; a significant reduction in walrus numbers and threatening of walrus populations. Harvest restrictions were eventually introduced in all five areas through national legislation (see Chapter 11, this volume).

Russian areas

Little information is available about sealing and harvesting in the present-day Russian Arctic areas. This is partly due to language barriers but also the fact that most Russian files are kept in closed archives (Gjertz and Wiig, 1994). Most of the available information available is from Norwegian hunters who operated in Russian waters from about 1860 to 1930. Laptev walruses are now considered as the westernmost population of the Pacific walrus (*O.r. divergens*) (Lindqvist et al., 2008) and therefore not included in this review.

A total of 237 small vessels with crews totalling just over two thousand persons were hunting around Svalbard and Novaya Zemlya in 1799 (Iversen, 1927). Some of these were wintering expeditions to Svalbard, which these hunters considered to be Russian territory, but many were probably limited to the summer months (Iversen, 1927). For unknown reasons, there was a dramatic subsequent drop in the number of ships taking part in this harvest; for example, in 1802 only four ships with total crews of 47 persons took part in the walrus hunt (Iversen, 1927).

Historically, walruses were common in the White Sea. Their total extirpation in Mezenskiy Bay was due to hunting and occurred in the second half of the 19th century (Timoshenko, 1986). In the late 19th century, Russian authorities claimed the whole White Sea and surrounding areas as territorial waters and barred foreign vessels from hunting there (Iversen, 1927). Walrus hunting around Novaya Zemlya and in the Kara Sea started in the 17th century and was of little importance before the 1830's (Chapskii, 1939). In this decade, up to 80 Russian vessels hunted seals and walruses annually at Novaya Zemlya (Knipowitsch, 1907; Iversen, 1927). By 1859 and 1860, this fleet was reduced to six vessels, and few Russian ships conducted sealing in these waters hereafter, probably due to the poor economic prospects (Knipowitsch, 1907; Iversen, 1927).

The hunt for walruses in Russian waters appears to have been dominated by Norwegian sealers in the second half of the 19th century, possibly due to less heavy sea-ice and therefore easier access for vessels (Shapiro et al., 2003). Norwegian vessels first went to Novaya Zemlya and Kolguev in 1867 (Iversen, 1927). One year later, skipper Elling Carlsen hunted walruses with great success along the west coast of Novaya Zemlya, and many other sealers were to follow (Ytreberg, 1946).

The Norwegians harvested several hundred walruses each year (see Lønø, 1972 for details), with the peak year being 1887, when more than 1200 walruses were taken. Norwegian hunting at Novaya Zemlya and in the Kara Sea continued until the end of the 1920's (Chapskii, 1939; Lønø, 1972). In the years after World War I, Russians purchased large steel vessels from Newfoundland and resumed their hunting in the White Sea area (Iversen, 1927).

Franz Josef Land was discovered in 1873, commercial hunting in the archipelago began in 1897 and lasted until 1934 (Gjertz et al., 1992; Gjertz et al., 1998). Accessing Franz Josef Land in the earliest years after discovery was very difficult with heavy sea-ice, and almost continuous ice cover even during summer. Smaller vessels therefore had little chance of reaching these islands before the turn of the century, when conditions improved and there was less sea ice late in the season (Shapiro et al., 2003). Three large steam whalers reached Franz Josef Land in 1897 and began hunting walruses (Southwell, 1897; Lubbock, 1937; Gjertz et al., 1998). Catches varied greatly over the years. During the period 1911–21, it is not clear if sealing vessels arrived successfully to the islands each year (Lønø, 1972; Gjertz et al., 1998). An increase in catches occurred after 1923 due to the use of powerful ice-reinforced vessels (Lønø, 1972). Catches in the archipelago were negligible after 1931, partly because Soviet authorities banned Norwegian access after annexing the islands in 1929 (Lønø, 1972), but also because walrus numbers had been severely reduced. In 1934 Russian sealers caught fewer than 10 walruses, despite using a spotter aircraft, leading to the end of the harvest (Chapskii, 1939, Gjertz et al., 1998). A total of about 8500 walruses were killed in Franz Josef Land from 1897 to 1934 (Gjertz et al., 1992).

Since 1921, various steps have been taken to minimise the harvest of Atlantic walruses in Russia (Born et al., 1995; Wiig et al., 2014). In 1935, the harvest from Soviet sealing vessels ceased, and in 1949, the killing of walruses by the fishing and sealing industries was prohibited by the Soviets. The Atlantic walrus in Russian waters was totally protected in 1956 (Beloborodov and Timoshenko, 1974).

Svalbard, Arctic Norway

The Svalbard archipelago was uninhabited when discovered in 1596, and had an abundance of whales and walruses. The earliest walrus hunting in Svalbard started at Bear Island in 1604 by English sealers and whalers (see Lønø, 1972 for details). Large numbers of walruses were killed in Svalbard by European whalers throughout the 17th century, but walrus hunting became less important after the 1650's as the whaling fleet became more pelagic. After this, walrus hunting was dominated by Russian Pomor hunters for more than a century and then eventually by Norwegian sealers.

It is disputed when Russian walrus hunting in Svalbard started, but it was most likely around 1700 (Lønø, 1972; Storå, 1987). The Russian hunters, known as Pomors, came to Svalbard from the White Sea area. The beginning of a Russian walrus hunt in Svalbard may have been triggered new legislation brought in by the

Tsar in 1620 that prevented inhabitants from the areas surrounding the Dvina and Pechora rivers from entering the Kara Sea, and thus hunting in the coastal waters of Northwest Siberia (Storå, 1987). In Svalbard, the Russian Pomors hunted by boat during the summer, and also built a large network of cabins for overwintering and trapping on Svalbard (Lønø, 1972; Storå, 1987). There were usually six-to-eight Russian vessels at Svalbard each year before 1808 (Lønø, 1972). During the Anglo-Russian war from 1808 to 1812, Russian walrus hunting largely ceased. After the war, only one or two Russian vessels hunted walruses at Svalbard each year (Lønø, 1972). Various sources describe how Russian trappers could take more than a thousand walruses over a single winter (Lønø, 1972). Russian hunting in Svalbard ceased altogether in 1853.

In the 17th and 18th centuries, walrus hunting was mostly carried out in the western and southern parts of the Svalbard archipelago, as the northern and eastern parts were less accessible due to the heavy sea ice conditions (Lønø, 1972; Gjertz and Wiig, 1994). Walrus numbers in the areas of Svalbard where the hunt took place showed clear signs of decrease by the middle of the 19th century (Lamont, 1861).

Norwegian walrus hunting in Svalbard started in the late 18th century with the introduction of new legislation in Norway. Southern Norway had a trade monopoly preventing any imports or and exports from the towns in the northernmost provinces of Norway. This monopoly was lifted in 1787, and the towns Hammerfest and Vardø were opened as ports of trade in 1789. Tromsø followed in 1794 (Hoel, 1949). A bounty was also introduced based on the amount of cargo these ports handled. This resulted in the first Norwegian winter hunting foray in Svalbard in 1795. From 1821 onwards, Norwegians sent vessels to Svalbard every year. Walrus harvests could vary, but, in the 1830's, they averaged 1000–2000 walruses each year. After 1850, ice conditions in the Barents Sea improved (Shapiro et al., 2003), which allowed access to new hunting areas and increased annual walrus harvests for some years (Lønø, 1972). As walrus stocks decreased during the 1880's, annual harvests rarely exceeded 1000 animals, and were typically between 300–700 walruses (Collett, 1911-12; Lønø, 1972). Walruses in Svalbard were, however, severely depleted by hunting by the beginning of the 20th century, with only negligible catches thereafter (Lønø, 1972). Walruses were totally protected by legislation in Norwegian waters in 1952 (Born et al., 1995; Wiig et al., 2014).

In Svalbard, a total of about 35,000 walruses were registered as caught during the period 1820–1920 (Lønø, 1972). This must be considered a conservative number since information is lacking for many of the years in this period. Also, these numbers do not include animals that were struck and lost, so assuming an estimated hunting loss of about 20% (Born et al., 1997), at least 50,000 walruses were killed in Svalbard over this hundred-year period. The number of walruses taken before the period is poorly known.

Eastern Greenland

Walruses in eastern Greenland stand out among other Atlantic walrus populations, as they are the smallest and most isolated population. Before Europeans arrived there was little or no harvest as the local Inuit population were last seen by Europeans in 1823 and had disappeared from the area (Mikkelsen, 2008).

Norwegian and Danish commercial walrus hunting in eastern Greenland started relatively late due to difficult sea-ice conditions off the coast (Born et al., 1997). The start of the European hunt coincided with the period of greatest sea-ice retreat during April in the Barents Sea (Shapiro et al., 2003). This retreat also occasionally led to a reduction in sea-ice cover along the coast of eastern Greenland, enabling some vessels to access the region. The first European walrus harvest in Northeast Greenland in 1889 was carried out from the Norwegian sealer '*Hekla*', which had a steam engine and a crew of 48 persons (Mikkelsen, 2008). After the turn of the 20th century, the smaller and more economic ice-going vessels were being equipped with engines (Hoel, 1949). They could now work their way through the ice fields to reach the East Greenland coast in years when the extent of sea-ice permitted (Mikkelsen, 2008).

The walrus harvest in north-eastern Greenland was not purely commercially motivated. There was a political dispute between Denmark and Norway over the sovereignty of the region. It was politically Danish, but the Norwegians claimed parts of the region based on historical activities and claims. Both sides based their claims on having had activity in the area. This included economic activity (such as hunting and fishing), cartography, as well as geological and botanical mapping of the region. Both Norway and Denmark established networks of hunting cabins on shore and had wintering parties there (Hoel, 1949; Mikkelsen, 2008). These networks were supplied by sealing vessels in late summer. The vessels would also hunt while in the area. The dispute was settled in the International Court in The Hague in 1933, when the court ruled in favour of Denmark.

Various protection measures for walruses were introduced by Danish authorities in north-eastern Greenland from the late 1930's, and total protection against commercial harvest introduced in 1956 (Wiig et al., 2014). In 1974 large parts of the walrus' natural range in north-eastern Greenland was established as a national park (see Born et al., 1997; Born et al., 1995). Born et al. (1997) and Witting and Born (2014) described the status of walruses in Northeast Greenland and back-calculated the size of this population prior to the start of the European catch. They found that a total of 1680 walruses had been killed over a 65-year period, starting in 1889. The carrying capacity of this population is estimated to be 1200–1500 animals.

Western Greenland

Except for subsistence hunting by the Inuit and earlier Norse hunting (see Chapter 7, this volume), the historical walrus harvest in western Greenland was dominated by European whalers and Norwegian sealers. Whaling ceased early in the 20th century, and legislation stopped sealing by foreign vessels in the middle of the century. Walruses along the western coast of Greenland can be split into two populations, the Central West Greenland population and the Northwest Greenland population (Witting and Born, 2014). The range of both populations extends between Greenland and Canada, and both form the basis of subsistence hunts by Inuit in these two countries, and have been subject to harvest by foreign whalers and sealers.

Dutch whalers first arrived in the Davis Strait in 1719 (Lubbock, 1937), and were later replaced by the British whalers who also hunted walruses as a supplement to whaling. Until 1820, most whaling was conducted along the Greenland side of the Strait (Stewart et al., 2014), before moving to the western Canadian side. Born et al. (1994) have extracted catch data from 145 whaling logbooks and found that a total of 3734 walruses were taken in the Davis Strait and Baffin Bay regions by British whalers between 1859 and 1910. These logbooks represent only 6% of all the known whaling expeditions in the area, so the actual number of walruses taken by the whalers must have been much higher. British whaling in western Greenland ceased in 1911, though the British were replaced by Norwegian sealers who began frequenting the region. Born et al. (1994) give details of where these sealers hunted walruses, and how many they harvested each year. Catches were most often in the hundreds, but in some years, more than a thousand walruses could be taken. In total, Norwegian sealers harvested at least 3285 walruses in western Greenland between 1911 and 1951. In both 1949 and 1951, Norwegian sealers took large catches of walruses in the Davis Strait and Baffin Bay, which in total numbered about 2000 animals. This led to a Danish diplomatic protest since Greenland was part of the Danish kingdom. As a result, Norway introduced hunting regulations in 1952, banning the harvest of walruses by Norwegians worldwide. Wiig et al. (2014) give a detailed review of walrus hunting regulations in western Greenland.

Eastern Canadian Arctic

Walrus hunting in the eastern Canadian Arctic was a mixture of subsistence hunting by Inuit, Canadian land-based groups, Arctic explorers and commercial hunts by American and European whalers and sealers. It complicates matters that walruses in these waters often are shared populations with Greenland, and that the whaling vessels would move back and forth between Greenlandic and Canadian waters. Stewart et al. (2014) have written a comprehensive review of the catch history for the Atlantic walrus in the eastern Canadian Arctic. Walruses in the region can be divided into three largely-discrete populations (Stewart et al., 2014). A High Arctic population in the northern Baffin Bay area, a Central Arctic population stretching from western Greenland to Baffin Island and in through Hudson Strait, and a Low Arctic population in southern Hudson Bay. The so-called Maritimes walrus in southern Quebec and the Atlantic Provinces were extirpated by commercial hunting from the late 1500's to the late 1700's (Stewart et al., 2014).

European commercial harvest of walruses eastern Canadian Arctic started as a supplement to whaling. Most of the whaling prior to 1820 took place on the Greenland side of the Davis Strait and Baffin Bay, with only low numbers of walruses harvested on the Canadian side. Later, the whalers shifted to the west side of the strait, including Lancaster Sound (see Lubbock, 1937; Stewart et al., 2014). Here, the whales were still abundant and there was no need to catch walruses. After two decades of whaling in these waters, whale numbers had declined, and ice conditions deteriorated. As the catches of whales dwindled, the whalers had to change their hunting methods and also catch walruses. From 1859 onward, the steam engine began entering the Arctic fleet. This improved manoeuvrability and increased access to areas with heavy ice. By the turn of the 20th century catches of walruses had become an important part of the whaling economy.

There was no large commercial hunt of walruses by whalers prior to 1885 (Stewart et al., 2014). Between 200 and 300 walruses were taken annually from the High Arctic population, and in the period 1899–1911, up to 1400 animals were caught each year from the Central Arctic population. The Low Arctic population was difficult to reach due to heavy ice conditions, and therefore largely ignored by commercial hunters (Stewart et al., 2014). From 1910 to 1931, land-based traders and Inuit often cooperated in the walrus harvest; these hunts may therefore be considered both subsistence and commercial harvesting. Regulations banning the commercial harvest of walruses by whalers and traders were introduced in Canada in 1928, and from then, walrus hunting was reserved for Indigenous use (see Stewart et al., 2014; Chapter 6, this volume). According to Stewart et al. (2014), commercial European catches by whalers and other non-Canadian harvests were reported to total about 9500 landed walruses from 1820 to 1928. Of these, 1341 were from the High Arctic population, where catches peaked in the 1880's, and about 8100 were from the Central Arctic population, where harvests peaked in 1910. This number does not include walruses that were struck and lost.

Conclusion

Europeans, especially British whalers and Norwegian sealers, hunted Atlantic walruses across large parts of the Arctic. They were motivated by the possibility of commercial gain through the sale of walrus products such as tusks, hides and oil. Detailed walrus catch statistics are available for some parts of the Arctic, but in most areas catch statistics are either of poor quality or lacking. The uncontrolled harvest over decades, or even centuries, led to severe declines in the various walrus populations. As a result, national legislation was introduced in the 20th century primarily to protect the interests of the local Indigenous populations, but in some cases, also to prevent extirpation of entire walrus populations.

References

Allen, G., 2012. Inflation: the value of the pound 1750-2011. House of Commons Research Paper 12/31, 29 May 2012.

Beloborodov, A.G., Timoshenko I.K., 1974. Protection of the Atlantic Walrus. Priroda 3, 97-99. (in Russian, see Fisheries Research Board of Canada Translation Series 3812, 1976).

Born, E.W., Dietz, R., Heide-Jørgensen, M.P., Knutsen, L.Ø., 1997. Historical and present distribution, abundance and exploitation of Atlantic walruses (*Odobenus rosmarus rosmarus* L.) in eastern Greenland. Meddelelser Grønland. Biosci. 46, 5–73.

Born, E.W., Heide-Jørgensen, M.P., Davis, R., 1994. The Atlantic walrus (*Odobenus rosmarus rosmarus*) in West Greenland. Meddelelser Grønland. Biosci. 40, 3–33.

Born, E.W., Gjertz, I., Reeves, R.R., 1995. Population assessment of Atlantic walrus (*Odobenus rosmarus rosmarus* L.). Norsk Polarinstitutt Meddelelser 138, 1–100.

Chapskii, K.K., 1939. Short historical analysis of contemporary conditions of the stocks of walruses in the Barents and Kara Seas. Probl. Arktiki 3, 62–69 (in Russian).

Chydenius, K., 1865. Svenska Expeditionen Till Spetsbergen år 1861 under ledning av Otto Torell. P. A. Norstedt & Söner, Stockholm (in Swedish).

Collett, R., Norges Hvirveldyr, 1911–1912. Vol. 1. Norges pattedyr. H. Aschehoug & Co (W. Nygaard), Kristiania (in Norwegian).

Conway, M., 1906. No Man's Land. Cambridge University Press, Cambridge.

Gjertz, I., Wiig, Ø., 1994. Past and present distribution of walrus in Svalbard. Arctic 47, 34–42.

Gjertz, I., Hansson, R., Wiig, Ø., 1992. The historical distribution and catch of walrus in Franz Josef land. Norsk Polarinstitutt Meddelelser 120, 67–81.

Gjertz, I., Wiig, Ø., Øritsland, N.A., 1998. Backcalculation of original population size for walruses *Odobenus rosmarus* in Franz Josef Land. Wildlife Biology 4, 241–248.

Hoel, A., 1949. Ishavsfangst – fangstnaering. In: Strøm, J. (Ed.), Norsk fiskeri og fangst håndbok. Alb. Cammermeyers Forlag, Oslo (in Norwegian), pp. 709–861.

Iversen, T., 1927. Drivis og selfangst. Aarsberetning vedk. Nor. Fisk. 1927 (1), 1–84 (in Norwegian).

Knipowitsch, N., 1907. Ueber die Biologie der Seehunde und die Seehundjagd im Europaeischen Eismeer. Rapports et Proces-Verbaux 8, 1907, 83–106 and 120–125 (in German with English summary).

Lamont, J., 1861. Seasons with the Sea-Horses; or sporting adventures in the northern seas. Hurst and Blackett, London.

Lamont, J., 1876. Yachting in the Arctic Seas or notes of five voyages of sport and discovery in the neighbourhood of Spitzbergen and Novaya Zemlya. Chatto and Windus, London.

Lindqvist, C., Bachmann, L., Andersen, L.W., Born, E.W., Arnason, U., Kovacs, K.M., et al., 2008. The Laptev Sea walrus *Odobenus rosmarus* laptevi: an enigma revisited. Zoologica Scripta 38, 113–127.

Lønø, O., 1972. The catch of walrus (*Odobenus rosmarus*) in the areas of Svalbard, Novaja Zemlja and Franz Josef Land. Nor. Polarinstitutt Årbok 1970, 199–212.

Lubbock, B., 1937. The Arctic Whalers. Brown, Son and Ferguson Ltd, Glasgow.

Mikkelsen, P.S., 2008. North-East Greenland 1980-60: The Trapper Era. Scott Polar Research Institute, Cambridge.

Nordgaard, O., 1895. Ishavsfangsten. Naturen 19 (4), 105–113 (in Norwegian).

Scoresby, W., 1820. An Account of the Arctic Regions with a Description of the Northern Whale-Fishery. Archibald Constable & Co., Edinburgh.

Shapiro, I., Colony, R., Vinje, T., 2003. April sea ice extent in the Barents Sea 1850-2001. Polar Research 22 (1), 5–10.

Southwell, T., 1897. Notes on the seal and whale fishery of 1896. Zoologist [4th Ser.] 1, 56–60.

Stewart, D.B., Higdon, J.W., Reeves, R.R., Stewart, R.E.A., 2014. A catch history for Atlantic walruses (*Odobenus rosmarus rosmarus*) in the eastern Canadian Arctic. NAMMCO Scientific Publications Series 9, 219–314.

Storå, N., 1987. Russian walrus hunting in Spitsbergen. Études/Inuit/Studies 11 (2), 117–138.

Timoshenko, Yu.K., 1986. About occurrences of walruses in the White Sea and about walrus–harp seal interactions. Abstracts of the 9th All-Union Conference, Arkhangelsk, pp. 380–381 (in Russian).

Wiig, Ø., Born, E.W., Stewart, R.E.A., 2014. Management of Atlantic walrus (*Odobenus rosmarus rosmarus*) in the arctic Atlantic. NAMMCO Sci. Publ. 9, 315–341.

Witting, L., Born, E.W., 2014. Population dynamics of walruses in Greenland. NAMMCO Scientific Publications Series 9, 191–218.

Wollebæk, A., 1901. Hvalrossen (*Odobaenus rosmarus*), Norsk Fiskeritidende 20, 75–81 (in Norwegian).

Ytreberg, N.A., 1946. Tromsø bys historie. Vol. 1. Tell Forlag, Oslo (in Norwegian).

SECTION

Future directions and innovations in atlantic walrus research

CHAPTER

10

Molecular advances in archaeological and biological research on Atlantic walrus

Liselotte W. Andersen[1], Magnus W. Jacobsen[2] and Paul Szpak[3]

[1]*Department of Bioscience, Kalø, Aarhus University, Rønde, Denmark*

[2]*Section for Marine Living Resources, National Institute of Aquatic Resources, Technical University of Denmark, Silkeborg, Denmark*

[3]*Department of Anthropology, Trent University, Peterborough, Ontario, Canada*

Chapter Outline

Introduction

Over the last 20 years, the application of molecular tools in biological and archaeological research has increased concurrently with technological advances (Star et al., 2018; Keighley et al., 2019a,b; Barrett et al., 2020). This has provided new possibilities for conducting genetic, stable isotope and fatty acid (FA) analyses on modern and histori-

The Atlantic Walrus. DOI: https://doi.org/10.1016/B978-0-12-817430-2.00002-9

cal samples to improve our understanding of Atlantic walrus biology and demography. The use of palaeontological, archaeological and historical remains, moreover, shows great promise for disentangling the impact of anthropogenic effects and environmental change on Atlantic walrus populations. Such analyses have the potential to provide insights on particular aspects of the behavioural ecology, demography and life-history of Atlantic walruses, including changes in the migratory routes, population size, dispersal rates, choice of haul-out sites, foraging/diet and mating behaviour, in response to particular environmental or climatic conditions, as well as past and present human activities such as hunting (Keighley et al., 2019a). From an archaeological and historical point of view, much can also be learned about the role of walruses to past human societies, and even historical ivory trading networks and their links to regional hunting grounds can be reconstructed through time using molecular data from bones and teeth (Star et al., 2018; Keighley et al., 2018, 2019a,b; Barrett et al., 2020). Gaining these past insights will play a crucial role in designing future conservation and management strategies for Atlantic walruses and ensuring sustainable futures for humans who still rely upon them.

In this chapter, we describe the different techniques and approaches used in genetic, stable isotope and FAs research and their suitability in providing information on population structure, population demography, life-history, diet or ecology of Atlantic walruses. It is important to consider whether samples are modern, historic or ancient when assessing which questions to ask and which methods to use. In this chapter, we define these three time categories as follows:

Modern: samples collected from living or recently deceased animals (usually associated with detailed information about the specimen: sex, age, size, precise location and date of sampling).

Historical (<300 years before present (BP)): specimens (bone, tooth, skin, etc.) collected from living animals and archives of museums or other research institutes (usually associated with detailed information about the specimen: sex, age, size, precise location and date of sampling).

Ancient (>300 years BP): bones and teeth from archaeological or palaeontological sites (often limited information available about the specimen, relies on chronometric or radiocarbon dating).

Population genetics

Nuclear and mitochondrial DNA

DNA resides within cells and constitutes the genetic code in all organisms from plants to animals. Walrus DNA, like that of all mammals, is located in two different organelles within the cell; the nucleus containing nuclear DNA (nuDNA) and mitochondria containing mitochondrial DNA (mtDNA) (Fig. 10.1). A copy of an individual's entire nuDNA or mtDNA is referred to as nuclear or mitochondrial genomes respectively. While both nuDNA and mtDNA consist of the same four building blocks (nucleotides, or bases), they differ in size and how they are inherited.

MOLECULAR TOOLS IN ARCHAEOLOGICAL AND BIOLOGICAL RESEARCH ON THE ATLANTIC WALRUS

Biologists and archaeologists are increasingly using genetic, isotopic and lipid analyses to uncover current and past walrus behaviour, life history, diet, migration, evolution and human interactions. For each method, and depending on whether studies are performed on live animals or archaeological material, a range of various target tissues are used (e.g. teeth and blubber).

▲ GENETIC RESEARCH

Both hard and soft tissues contain DNA. Soft tissue is normally the preferred source of DNA when available, while teeth and bone can be used to extract DNA from archaeological remains. DNA is found in two places within **the animal cell**. DNA from **the nucleus** is inherited from both parents while DNA from **the mitochondria** is almost entirely inherited from mother to offspring. Hence, while nuclear DNA reveals the genetic history of the entire population, mitochondrial DNA only reflects maternal history. Mitochondrial DNA however is typically more abundant and much simpler than nuclear DNA. This means it is easier to obtain for historic or ancient samples and more commonly included in current genetic research.

Tooth cementum:

- high abundance of mitochondrial DNA (high cell density),
- preferable material for DNA extraction with reduced DNA contamination and higher abundance,
- contains a similar mineral component to bone,
- thickest towards the root tip,
- formed in GLG (growth-layer groups) representing annual deposition.

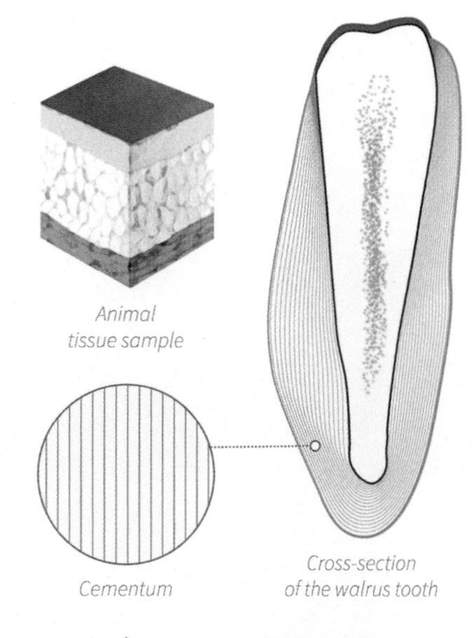

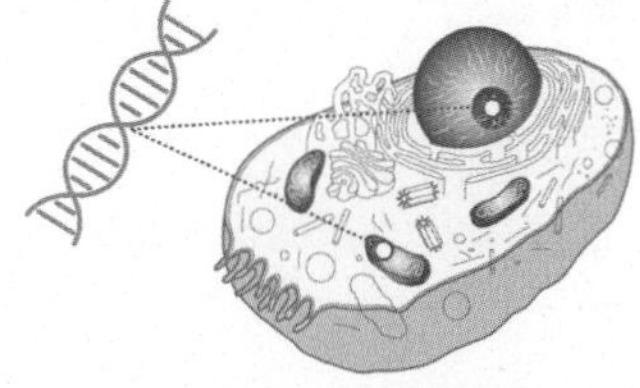

■ ISOTOPIC RESEARCH

Enamel
inorganic fraction

$\delta^{18}O$ $\delta^{208}Pb$

Tooth **enamel** is composed primarily of a calcium phosphate mineral known as bioapatite and can preserve for millions of years. **Oxygen** in tooth enamel is derived from ingested waters and foods. **Lead** is ingested and substitutes for calcium.

Dentin
organic fraction

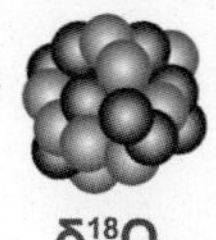
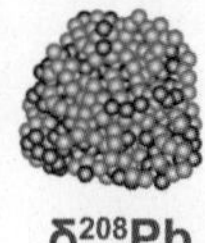

$\delta^{13}C$ $\delta^{15}N$

Dentin has a similar structure to bone and contains 20%–25% organic material by weight, 90% of which is the protein Type I collagen and can preserve for up to 100,000 years. The **carbon** and **nitrogen** in the collagen are derived directly from the diet.

FIGURE 10.1

Description of the molecular tools and their application in archaeological and biological research on the Atlantic walrus (*Odobenus rosmarus rosmarus*). Illustration: Elena Kakoshina.

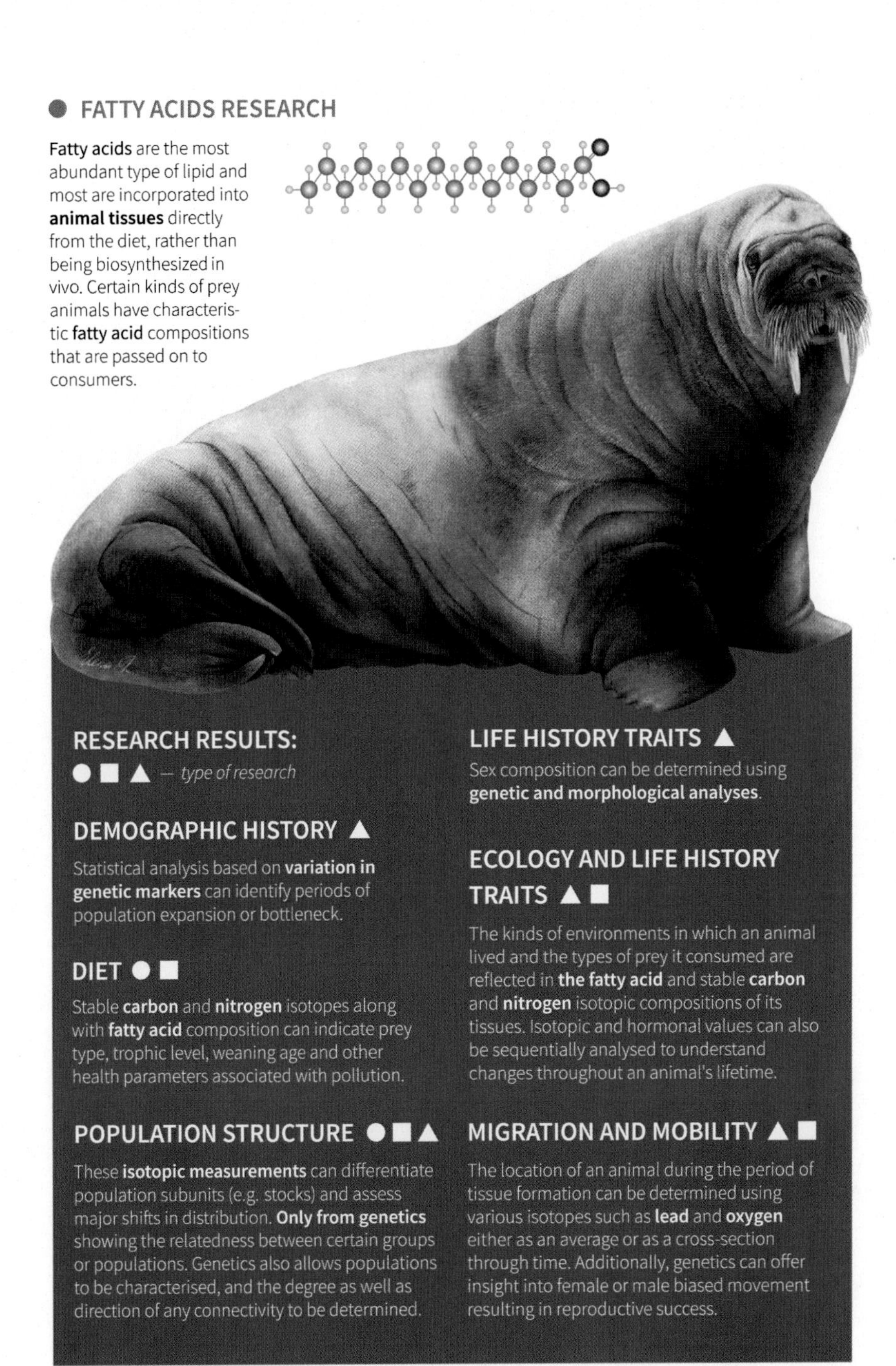

FIGURE 10.1

(Continued).

The nuclear genome is by far the largest, estimated to be approximately ~2400 million bases (megabases, Mb) in walruses (Foote et al., 2015) and is inherited from both parents. The mitochondrial genome (sometimes referred to as the mitogenome) is much smaller, only around 16,000–17,000 bases (base pairs, bp), with 37 genes (13 of which are protein coding) and inherited through the maternal line (Ballard and Whitlock, 2004). Hence, while nuclear DNA reveals the genetic history of the entire population, mitochondrial DNA only reflects maternal history (Ballard and Whitlock, 2004; Galtier et al., 2009). As such, different genetic patterns can be expected in species with sex-biased migration behaviour, as observed for many marine mammals including southern elephant seals (*Mirounga leonina*) and grey seals (*Halichoerus grypus*) (Hoelzel et al., 2001; Klimova et al., 2014). Another important difference between mtDNA and nuDNA is the number of copies of each genome within the cell. While mammals only have a single nucleus in each cell, the mitochondria can number in the thousands (Cole, 2016). The higher copy number of mitochondrial DNA makes it easier to extract sufficiently high DNA yields for downstream applications, even for highly degraded ancient samples (Foote et al., 2012). Additionally, the two sources of DNA differ in the extent of genetic recombination. Unlike the nuclear genome, mitogenomes rarely undergo genetic recombination, meaning that all genes are linked and hence inherited as one block (Galtier et al., 2009). The more conserved gene content and architecture of mitogenomes, their abundance in animal cells and our knowledge about specific regions with relatively slow and fast rates of evolution, has prompted widespread use of mtDNA (Galtier et al., 2009).

In the last 20 years, mtDNA has been the most frequently used marker for making population genetic inferences, and is the preferred marker for so-called genetic barcoding (Hebert et al., 2003). Here a gene, normally Cytochrome c oxidase subunit I, is used for conducting species or population identification in cases where mtDNA is diagnostic (DNA barcoding) (Hebert et al., 2003). However, using a single marker has many limitations as it only provides the evolutionary history of that particular region of the genome, rather than the organism or species being studied, and often cannot resolve detailed genetic patterns on shorter temporal scales (e.g. parentage assignment, individual identification or detailed demographic histories). The greater number of genetic markers available within the nuclear genome provides more information, thus increasing statistical power and genetic resolution. It is also worth noting that mtDNA is assumed to evolve neutrally (although see Galtier et al., 2009). While this is not a problem for most genetic analyses that assume neutrality (such as phylogeography and demographic history inference), it is of little use when investigating natural selection. Consequently, when focusing solely on mtDNA, potential genetic adaptations will go unnoticed (Keighley et al., 2019a). Such adaptations can reflect key differences between populations, which makes them important for informing conservation efforts (Fraser and Bernatchez, 2001; Saánchez-Molano et al., 2013), and understanding a species' or population's response to natural or anthropogenic pressures (Keighley et al., 2019a).

Historical and archaeological samples

DNA degrades over time due to both chemical and biological processes, which results in fragmentation as well as structural and base modifications (Pääbo et al., 2004). Although modern DNA may be more straightforward to analyse as it has not already degraded, postmortem historical or ancient biological material can still contain DNA, even after hundreds of thousands of years (Orlando et al., 2013). The extent of DNA degradation is influenced by factors of the biological material itself as well as the environmental conditions in which it is deposited. For example, while bone is the most common material recovered from walruses at archaeological sites, (Dyke et al., 1999; Gotfredsen et al., 2018) not all skeletal remains protect the DNA equally well. Differences in DNA preservation between different skeletal elements can be related in part to the density, porosity and structure of the bone. Denser, thicker walled and less-porous material is better at protecting DNA from degradation and microbial contamination (Bollongino et al., 2008; Parker et al., 2020; Keighley et al., 2021). Cementum from walrus teeth and tusks has been shown to have a higher percentage of target DNA in walruses from archaeological sites (Keighley et al., 2021). Generally, DNA is expected to degrade faster when exposed to warm and humid conditions, with climatic fluctuating, exposure to direct sunlight and strongly alkaline or acidic soils also negatively impacting on DNA preservation (Sosa et al., 2013; Nielsen-Marsh & Hedges, 2000; Bollongino et al., 2008; Pruvost et al., 2007; Allentoft et al., 2012; Kendall et al., 2017; Lindahl, 1993; Trueman and Tuross, 2002; Foote et al., 2012). Arctic and sub-Arctic environments tend to preserve DNA for longer periods of time, compared to temperate and tropical regions. Samples buried in cultural deposits, natural sediments or caves often exhibit better DNA preservation compared to samples that have been exposed at the surface for centuries or even millennia (Keighley et al., 2021).

To date, mitochondrial DNA has been the focus of the majority of non-human ancient DNA (aDNA) studies on archaeological samples (Foote et al., 2012), including walruses (Lindqvist et al., 2009, 2016; Star et al., 2018; Keighley et al., 2019b). This is largely due to the higher number of mitogenomes per cell compared to nuclear genomes that makes it easier to extract in sufficiently high DNA yield for downstream applications (Foote et al. 2012). However, recent advances in genetic methods have made it increasingly possible to also analyse nuclear DNA (nuDNA) from archaeological samples (see below).

Source of genetic information

As discussed above, DNA in mammals can be found in both the nucleus and mitochondria. Within each of these however, there are different analytical approaches that target particular regions or the entire genome. The choice of which approach depends on the study, that is, the question being asked and the information required to answer it (Table 10.1). Generally, due to the expected

Table 10.1 The different genetic markers used, where they are situated and the questions where they are applicable mentioned in the text.

Molecular marker	Genome	Application
mtRFLP	mtDNA	Population structure/migration
DNA sequences	Nuclear	Population structure/migration, demographic history
DNA sequences	mtDNA	Species identification/DNA barcode
Microsatellites	Nuclear	Individual identification, sex identification, kinship/parentage, population structure/migration, demographic history
SNPs	Nuclear	Individual identification, sex identification, kinship/parentage, population structure/migration, demographic history

mtDNA, *Mitochondrial DNA;* mtRFLP, *mitochondrial restriction enzyme fragment length polymorphism;* SNPs, *single nucleotide polymorphisms.*

genetic diversity (i.e. number of alleles of the different marker or polymorphism connected to the genetic marker) more alleles translates to more detailed information (the higher power) that can be gained from the marker, and more markers provide more information (Dussex et al., 2018).

To date, most genetic studies on walruses have relied on mtDNA restriction enzyme fragment length polymorphism (RFLP), single mtDNA gene sequencing and/or nuclear microsatellite markers (also known as short tandem repeats) (e.g. Andersen et al., 1998, 2009, 2014, 2017). Microsatellites are highly polymorphic, meaning they show large amounts of variation between individuals. This high variability led to a preference for microsatellites over other nuclear markers such as single nucleotide polymorphisms (SNPs). SNPs are mainly biallelic (two alleles) and around 100 SNPs are needed to obtain the same power as 10–20 microsatellites (Kalinowski 2002). However, with recent advances in molecular methods and sequencing technologies, it is increasingly easy and cost-efficient to sequence thousands or even hundreds of thousands of SNPs (Helyar et al., 2011). One of the most important developments for aDNA has been high-throughput DNA sequencing platforms (also referred to as next generation sequencing or NGS), which enable fast and cost-efficient sequencing of millions of short DNA sequence reads in parallel (Goodwin et al., 2016). With the cost of NGS steadily decreasing, it is now feasible to sequence a reduced part of the nuclear genome or even the whole genome for entire populations of nonmodel organisms, thus obtaining thousands of SNP markers for downstream analysis (Puckett, 2016). NGS technologies have furthermore advanced the field of ancient DNA where a tiny part of the extracted DNA may be of endogenous (i.e. target species) content. In these cases, the large sequence output using NGS allows the recovery of millions of target DNA sequences, even in highly degraded samples (Keighley et al., 2018). When endogenous DNA is present in very low amounts, NGS can be used in combination with DNA capture, a molecular method that allows for collection and upconcentration of species-specific target DNA before subsequently sequencing (e.g. Carpenter et al., 2013; Maricic et al., 2010).

Contemporary population structure and migration patterns

Assessment of population structure is one of the cornerstones of modern population genetic research. When a species shows population structure, it means that there has been very little migration (gene flow) between populations due to the occurrence of physical or reproductive barriers that has allowed populations to become increasingly differentiated from each other. When isolated from each other, populations will evolve in different directions because of the cumulative effects of mutation, selection and genetic drift (Mayr, 1942, 1963; Coyne and Orr, 2004). Population structure is not only evidence for a lack of migration, but more importantly, it can lead to the emergence of unique population-specific genetic variants that may be key for the survival of the specific populations (local adaptations), or for the long-term evolutionary potential of the species (Moritz, 1994; Crandall et al., 2000; Fraser and Bernatchez, 2001). For these reasons, the detection of population structure is one of the main criteria for the designation of conservation units (e.g. Moritz, 1994; Crandall et al., 2000; Fraser and Bernatchez, 2001) and several genetic software packages have been developed that allow for testing whether or not population structure exists (e.g. Excoffier and Heckel, 2006).

Multiple studies have analysed the genetic population structure in Atlantic walruses. The first comprehensive study was conducted by Andersen et al. (1998) and investigated population structure between West and East Atlantic walruses based on analysis of 11 nuclear microsatellite markers and RFLPs from the mtDNA. Analysis of the microsatellite markers revealed that Northwest Greenland, East Greenland and Svalbard-Franz Josef Land were separate populations, while the RFLP haplotypes suggested that Northwest Greenland was evolutionary distinct from the East Atlantic walruses. The study further indicated that very little gene flow occurred between West and East Atlantic walruses. After this pioneering study, several followed using the same overall methodology, but adding more populations from a broader geographical area (e.g. Andersen et al., 1998, 2009, 2014, 2017) and sometimes also including SNP markers (Shafer et al., 2014, 2015). Overall, these studies have so far recognised seven contemporary genetically different populations of Atlantic walrus (see Figures 4.2, 4.3 in Chapter 4, Stocks, Distribution and Abundance).

Differences in migration behaviour may exist between male and female walruses, and these patterns can be disentangled by analysing population structure using both nuclear and mitochondrial markers. For example, the study by Andersen and Born (2000) investigated genetic differentiation between walruses from Northwest Greenland (Thule) and off the coast of West Greenland (Attu-Sisimiut). Overall, they found evidence for population structure from both nuclear microsatellites and mtDNA RFLP markers. However, the structure was mainly driven by differences at the mtDNA level (different RFLP haplotypes) suggesting low levels of gene flow between females and high site fidelity (i.e. returning to the same site again and again). Such findings are supported by observed behavioural differences between males and females particularly in areas such as East Greenland, Svalbard-Franz Josef Land, Pechora Sea (Andersen et al., 2017) and are in line with that observed for other species of marine mammals, including southern elephant seals and grey seals (Hoelzel et al., 2001; Klimova et al., 2014). However, walruses sometimes also undertake long-distance migrations or straying behaviour. This is best illustrated by the finding of a single individual with a Pacific walrus mtDNA haplotype in the East Greenland population (Andersen et al., 1998, 2017).

Migration patterns can be analysed directly between populations, which is crucial for understanding potential source sink populations and important knowledge for conservation efforts. Andersen et al. (2009) estimated the strength and direction of historical and contemporary nuclear gene flow between Western Greenland and Hudson Strait subpopulations. The results suggested that walruses in Hudson Strait might be the source for West Greenland walruses, supporting an earlier suggested counter-clockwise movement of walruses in the area (Freuchen, 1921; Vibe, 1950). A more direct method to detect migration between populations is to identify the same individual in several populations using the genetic markers as a genetic tag or DNA profile (Table 10.1, 'Microsatellites', 'SNPs'). If the genetic tag or DNA profile is 'recaptured', it can be possible to estimate census population size through mark-recapture analysis (Kindberg et al., 2011 (e.g. brown bear); Sethi et al., 2016 (e.g. Pacific walrus)). Further, genetic tags can also be used for identifying parentage and kinship, but this has so far not been applied in any genetic studies on Atlantic walrus.

Population structure in historical and archaeological walrus populations

While analysis of contemporary samples allows us to investigate the current population structure and migration behaviour of walruses, it does not give conclusive insight into the overall stability of this structure over time. With the advances in aDNA methods, it is now possible to investigate whether substantial changes have occurred between modern, historical and archaeological walruses in terms of genetic diversity, migration and distribution, which may be associated to particular anthropogenic effects like hunting, or as a response to climatic and environmental shifts (Keighley et al., 2019a). Multiple genetic studies of walruses have focused on historical or archaeological samples. The first genetic study conducted on walrus bone remains aimed to clarify the number of walrus subspecies. At the time there were three recognised subspecies: Atlantic walrus (*Odobenus rosmarus rosmarus*), Pacific walrus (*Odobenus rosmarus divergens*) and Laptev walrus

(*Odobenus rosmarus divergens laptevi*) (Lindqvist et al., 2009). The study extracted and analysed mtDNA sequences from presumed Laptev walrus remains collected from 1832–1936, along with samples from Atlantic and Pacific walruses. Lindqvist et al. (2009) showed that genetic sequences could be obtained from the historical walrus samples and that the presumed Laptev walruses clustered with samples from the North Pacific. This implied that the Laptev walrus was a part of the westernmost population of the Pacific walrus and not a separate subspecies.

Detailed information about population structure can also improve our understanding of past human–walrus interactions. For instance, McLeod et al. (2014) investigated the potential causes behind the disappearance of walruses inhabiting the Canadian Maritimes (waters of the eastern Canadian provinces of Quebec, New Brunswick, Nova Scotia and Prince Edward Island). This area was inhabited by >100,000 walruses in the 17th century, but none exist today (Naughton, 2012). Whether they were extirpated or changed behaviour, abandoning the region due to the heavy hunting pressure was however unknown. Comparing the mtDNA control region sequences from 37 'Maritime' walrus bones to those from 88 contemporary walruses from Baffin Island and Nunavik revealed that none of the haplotypes ($n = 8$) observed in the Maritime walruses were found in the modern walruses. When these genetic results were integrated with results on morphological differences between the contemporary and Maritime walruses, McLeod et al. (2014) concluded that the Maritime walrus represented a unique and possible evolutionary distinct walrus population that went extinct due to anthropogenic impact. Similarly, at Svalbard, walruses were already being hunted by Europeanss from the early 1600s and were exploited on a more industrial scale from around 1820 through to 1952. Comparing genetic sequences from remains collected at historical haul-out sites on Bjørnøya (dated to 1867) and Håøya (dated to 1852) to contemporary samples, Lindqvist et al. (2016) showed that some of the historical haplotypes were not observed in the contemporary walruses, although these clustered around modern haplotypes. Lindqvist et al. (2016) concluded that the intense hunting of Atlantic walruses at haul-out sites on Bjørnøya and Håøya probably resulted in behavioural changes, with a shift in distribution rather than local extinction.

The discovery of population structure within a species also allows for sourcing of archaeological artefacts to distinct populations, thereby offering clues into past hunting grounds and trade routes. For instance, according to historical sources, walrus hunting by the Norse during the Viking Age mainly took place around Disko Bay in West Greenland from at least the 13th century (Enghoff, 2003; Keller, 2010; Frei et al., 2015). However, walruses were also hunted in the eastern Atlantic, as far as the Barents Sea and Russia, as documented in historical records from the 9th century (Roesdahl, 2001; Bately and Englert, 2007). Star et al. (2018) analysed mitogenomes obtained from archaeological walrus remains (around 10th–17th centuries) traded with Western Europe and compared these to published modern and 19th-century samples from the Canadian Arctic, Greenland and the Barents Sea region (Andersen et al., 1998; Lindqvist et al., 2009, 2016;

Andersen et al., 2017). The authors found that out of the seven samples predating 1120 CE, all but one belonged to the eastern clade of Atlantic walruses, and could originate from various places in Greenland, but most likely from the northeastern Atlantic. These results suggested that early phases of the Norse ivory hunt targeted populations in the northeastern Atlantic, whereas the Norse walrus hunt in subsequent centuries increasingly targeted walruses in West Greenland and Canada, probably at a time when the eastern stocks were becoming increasingly depleted (Star et al., 2018). Norse walrus hunting and ivory trade have also been studied in Icelandic walruses by Keighley et al. (2019b) who analysed archaeological samples from a now-extinct Icelandic walrus population. The study found a unique lineage that disappeared shortly after Norse arrival in Iceland, implying that Norse walrus hunting and the international ivory trade played a significant role in this local extinction.

Life-history patterns

Life-history parameters such as age, sex, reproductive status and body condition have important consequences for a species' behaviour, survival and ecology. Such information is not easily obtained using molecular markers. However, sex determination from both soft and hard tissues is possible using genetic approaches that target sex chromosomes within the nuclear genome (Table 10.1). Information about sex is important in population genetic studies and archaeology making it possible to detect any past or present sex-biased dispersal and thus changes in migration or distribution patterns that may differ between male or female walruses. When studying hunted walruses, understanding the sex ratio of the catch offers clues into hunting practices and strategies (Keighley et al., 2019a). Genetic markers have been specifically developed for walruses that allow sex identification by targeting a region of 327 and 288 base pairs from the X and Y chromosome, respectively (Fischbach et al., 2008). Andersen et al. (2017) applied the method in a population genetic study on eastern Atlantic walruses. Andersen et al. (2017)'s finding confirmed observations that walruses along the east coast of Greenland and Svalbard-Franz Josef Land are sexually segregated, with male groups in the southern part and females with calves in the northern part of the distribution area (Wiig et al., 2007). Robertson et al. (2018) developed a simpler sex determination method for pinnipeds using a quantitative polymerase chain reaction assay. However, given the highly degraded state of DNA from historical and ancient samples, genetic markers used on such material need to target short regions (e.g. Foote et al., 2012) or use whole-genome sequencing approaches to compare the obtained sequences (including sex chromosome sequences) to a reference genome (Bro-Jørgensen et al., 2020).

Demographic history and population size change

Changes in genetic diversity reflect a species' evolutionary history, as well as its future evolutionary potential, which is paramount for a species' ability to adapt to environmental change. Generally, large stable populations hold higher genetic diversity, and thus greater adaptive potential, compared to smaller ones (Frankham, 1996). Changes in demographic history of populations are directly linked to genetic diversity. Source populations are expected to show higher genetic diversity compared to newly established populations (Hewitt, 1999). Additionally, population expansion is expected to lead to an increase in genetic diversity, while a severe reduction in effective size (so-called effective population size not equal to the total population size) will lead to rapid loss of genetic variation, fixation of deleterious alleles and inbreeding depression (Frankham, 1996). Changes in population size may be linked to environmental or anthropogenic impacts (Andersen et al., 2009, 2017; Shafer et al., 2015; Keighley et al., 2019b). For instance, population expansion may occur in response to factors such as an increase in available habitat, improved environmental conditions, greater availability of prey items or reduced hunting pressure (Vilá-Cabrera et al., 2019). Conversely, deteriorations in local environmental conditions or increases in human hunting pressures can cause population decline, which in severe cases can create genetic bottlenecks that see the loss of large amounts of genetic diversity.

Currently several genealogy-based (coalescent-based) methods have been developed to estimate effective population size changes for individual populations, or multiple populations simultaneously to account for the effects of potential gene flow (e.g. Cornuet and Luikart, 1996; Piry et al., 1999; Gutenkunst et al., 2009; Excoffier and Foll, 2011; Bouckaert et al., 2019). In addition to the scale and nature of population size change, the timing can be estimated when the genetic mutation rate is known or can be estimated directly from the data. Such methods can be applied to a wide array of genetic markers like microsatellites, sequence data and SNPs and can infer the demographic history of a population based solely on contemporary samples, as well as historical and ancient samples (Drummond et al., 2003; Ramakrishnan et al., 2005). Here, the use of several unlinked genetic markers and/or the use of temporal samples increase the statistical power and genetic resolution.

To date, all contemporary genetically identified populations of Atlantic walrus have been analysed using nuclear microsatellite markers, but no evidence for bottlenecks due to historical hunting pressure have been found (Andersen et al., 2009, 2017). The reasons for this are still unknown. One possibility is simply that populations responded to human hunting by changing migratory behaviour, fleeing to more-remote and less-accessible areas or that the actual reductions were too small to detect given the sample size and genetic markers used (Andersen et al., 2009). Alternatively, any hunting effect may have been hidden due to errors in technical assumption, for example, of mutation rate models (Peery et al., 2012), or in some cases detecting the signals of past demographic events require ancient DNA approaches. The latter is definitely true of extinct populations, such as the Icelandic (Keighley et al., 2019b) and Maritimes walrus (McLeod et al., 2014) that might have left little or no traces of contemporary genetic samples.

Effective population size change in Atlantic walruses has also been analysed across deeper time scales, using different analytical methods and markers. Shafer et al. (2015) analysed 4854 SNPs to reconstruct the demographic history of walruses from the Canadian Arctic by comparing observed genetic variation in walruses with several modelled demographic scenarios. Overall, a model with divergence between northern and southern Canadian Arctic walruses followed by modest and asymmetric gene flow migration from the northern to the southern cluster showed the best fit to the data. A relatively recent bottleneck event was detected in the high Arctic, while an older bottleneck was indicated in the central Arctic. The older bottleneck and population size decrease was potentially caused by the Last Glacial Maximum and subsequently led to the recolonisation of the Northern Canada waters. Contrary to this result, Andersen et al. (2017) detected a constant effective female population size for Atlantic walruses in East Greenland, and an increase in Svalbard-Franz Josef Land and Pechora Sea walruses starting around 40–30 thousand years ago using mtDNA. These differences in demographic history for two Atlantic walrus stocks may be associated with sea ice distribution, which might have caused geographic variability in the availability of food and haul-out sites for walruses (Andersen et al., 2017).

Stable isotopes and fatty acids

Diet and ecology

Stable isotopes

Stable carbon ($\delta^{13}C$), nitrogen ($\delta^{15}N$) and sulphur ($\delta^{34}S$) isotopic compositions of animal tissues reflect the weighted average of the foods consumed during the period of tissue formation. For example, the isotopic composition of liver reflects the average diet over a period of days (Hobson and Clark, 1992; Tieszen et al., 1983), while the isotopic composition of bone collagen reflects the average diet over years or decades (Hedges et al., 2007). Stable isotopes have been used extensively to characterise the diet and ecology of many animal taxa, both extinct and extant. Stable isotope techniques have been especially valuable in characterizing the ecology of marine mammals (Newsome et al., 2010) because of the difficulties associated with observing these animals and the ethical issues surrounding conducting lethal sampling activities (e.g. for analysis of stomach contents) on sufficiently large numbers of individuals.

Stable carbon isotope compositions are generally conserved across trophic levels, and consumer tissue $\delta^{13}C$ values reflect the ultimate source of primary production. In polar seas, pelagic phytoplankton and sea-ice-associated algae have been observed to have $\delta^{13}C$ values that differ from one another by 5%–12% (France et al., 1998; McMahon et al., 2006; Søreide et al., 2006), allowing the contribution of production derived from sea ice algae in higher trophic levels to be quantified (Kohlbach et al., 2016; Søreide et al., 2013; Stasko et al., 2018). Animals that forage in benthic habits tend to have higher tissue $\delta^{13}C$ values than those that forage in pelagic habitats (McConnaughey and McRoy, 1979), a product of differences in discrimination against ^{13}C during uptake and assimilation of dissolved inorganic carbon by benthic versus pelagic algae driven by water turbidity (France, 1995). Because of these benthic–pelagic differences, it is also the case that animals foraging in inshore habitats have higher tissue $\delta^{13}C$ values than animals foraging offshore (Cherel and Hobson, 2007; Hobson, 1993; Hobson et al., 1994). The $\delta^{13}C$ value of CO_2 in the atmosphere and oceans has decreased significantly since the late-19th century due to the industrial and domestic combustion of isotopically light fossil fuels; the rate of this decline has increased over time (Eide et al., 2017; Quay et al., 1992). Accordingly, these changes must be accounted for when $\delta^{13}C$ time series are developed from animal tissues sampled across multiple years, or when comparisons are made between preindustrial and more recent samples (Hilton et al., 2006; Laws et al., 2002).

Stable nitrogen isotopic compositions ($\delta^{15}N$) increase by 3.4% on average with each trophic level (Minagawa and Wada, 1984; Post, 2002). As with $\delta^{13}C$, the $\delta^{15}N$ values at the base of the food web can also vary. Where N_2-fixing cyanobacteria are dominant, the $\delta^{15}N$ values in consumers are relatively low in comparison to areas where phytoplankton predominate (Montoya, 2008). Areas characterised by strong upwelling and anoxic conditions (favouring denitrification) have particularly high $\delta^{15}N$ values throughout the food web (Voss et al., 2001). This variation at the base of the food web is reflected in the tissues of consumers at higher trophic levels, meaning that $\delta^{15}N$ cannot be used to compare the trophic position of a given taxon across space without accounting for variation at the base of the food web (Lorrain et al., 2015; Pethybridge et al., 2018). The $\delta^{15}N$ values at the base of the food web can vary significantly over time (on temporal scales of $10^2–10^6$ years), with variation in sediment core records being interpreted primarily in light of varying rates of denitrification and N_2 fixation (Altabet, 2006). There may be significant age-related variation in mammal tissue $\delta^{15}N$ caused by nursing. Nursing animals possess $\delta^{15}N$ values that are higher than those of the mother (Fogel et al., 1989), making strict interpretations of trophic position less straightforward. This nursing effect has been recorded in both soft tissues with rapid turnover rates and bone collagen for marine mammals (Gorlova et al., 2012; Habran et al., 2010; Newsome et al., 2006).

Relative to other endemic Arctic marine mammals, a relatively small number of light stable isotope compositions have been determined for Atlantic walrus tissues. The bulk of these data have been collected to contextualise other data, such as tissue Hg concentrations (Atwell et al., 1998; Rigét et al., 2007a; Rigét et al., 2007b) and persistent organic pollutants (Muir et al., 1995; Scotter et al., 2019). A few studies have measured the $\delta^{13}C$ and $\delta^{15}N$ values of various marine food web components and include measurements on walrus, although the sample sizes for walruses are always very small (Hobson et al., 2002; Hobson and Welch, 1992; Linnebjerg et al., 2016). When compared to other marine mammals or food web components, walruses occupy the lowest trophic position (on the basis of their $\delta^{15}N$ values) and these studies have concluded that walruses forage primarily on benthic invertebrates, consistent with what is known about their diets based on other lines of evidence (Fay, 1985; Fisher and Stewart, 1997). There are few published $\delta^{34}S$ measurements on Atlantic walrus tissues, but one recent study suggests that these may prove valuable in discerning the importance of production derived from benthic microalgae versus pelagic phytoplankton (Szpak and Buckley, 2020).

As with modern walruses, there have been sparingly few isotopic measurements on archaeological specimens. These data have primarily been generated for the purpose of providing a baseline for reconstructing ancient human diet (Coltrain et al., 2004; Nelson et al., 2012; von Steinsdorff and Grupe, 2006). One exception is the study conducted by Jaouen et al. (2016) for archaeological marine mammals from Little Cornwallis Island dating to the Late Dorset period. Consistent with modern studies, these data demonstrate that walruses occupy the lowest trophic position relative to other endemic Arctic marine mammals, such as polar bear, ringed seal (*Pusa hispida*) and bearded seal (*Erignathus barbatus*). One interesting pattern that has emerged from the few studies that have compared walruses to other taxa is the lack of a distinction in $\delta^{13}C$ between walruses and other pinnipeds. Given that $\delta^{13}C$ values are frequently cited as a distinguisher between benthic and pelagic consumers, it is surprising that walruses possess comparable or even lower $\delta^{13}C$ values than ringed or bearded seals (Fig. 10.2). This may simply be a product of very tight benthic–pelagic coupling and low in situ benthic productivity in the regions that have been studied. Alternatively, because there is a modest increase in consumer $\delta^{13}C$ values with trophic level, the low trophic position of walruses relative to bearded seal and ringed seal may serve to obscure differences in $\delta^{13}C$ related to benthic–pelagic gradients.

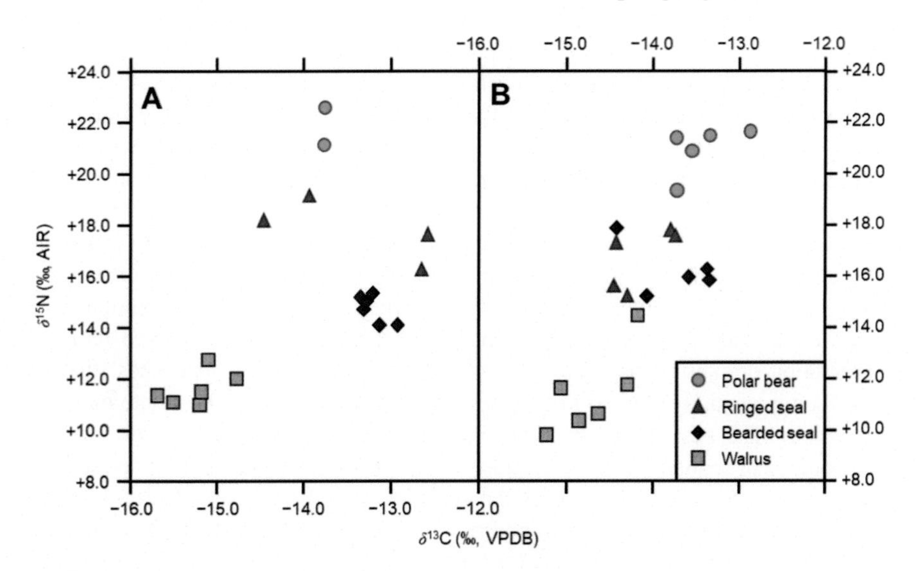

FIGURE 10.2

Stable carbon and nitrogen isotopic compositions for ancient walruses compared to other marine mammals from (A) Little Cornwallis Island in the central Canadian Archipelago (Jaouen et al., 2016) and (B) Skraeling Island, eastern Ellesmere Island, Canada (Szpak Unpublished Data).

A key question that has emerged in relation to walrus diet is the extent to which walruses rely on higher trophic-level prey (pinnipeds and seabirds), a behaviour that has been recorded in both Atlantic and Pacific walruses (Fay, 1960; Fox et al., 2010; Lowry and Fay, 1984; Mallory et al., 2004). This behaviour is noteworthy because it may be linked to climate change, with the occasional or more permanent loss of sea ice habitats potentially increasing the extent to which walruses feed on seals and seabirds (Seymour et al., 2014). These dietary shifts should be easily detectable with stable isotope analyses, although detection of changing feeding strategies in more ancient walrus specimens would require that there had been a significant contribution of higher trophic-level prey over many years due to the slow turnover rate of bone collagen. In the most extensive isotopic study of walruses (Atlantic or Pacific), Clark et al. (2019) examined the $\delta^{13}C$ and $\delta^{15}N$ values of ancient and modern Pacific walruses over the last 4000 years in the Chukchi Sea. They did not find differences in mean isotopic compositions in periods where high or low sea ice productivity was indicated by dinocysts, although there was greater isotopic (and hence dietary) variation when sea ice extent was low. They also observed relatively low $\delta^{15}N$ values in contemporary walruses (especially males) relative to archaeological specimens. Given that these low $\delta^{15}N$ values were not observed in times of low sea ice extent in the past, they suggested that this is indicative of changes in the diet of walruses that may deviate from patterns of long-term stability for the last 4000 years.

Rather than focusing on walrus diet (and changes therein) as reflected in tissue stable isotope compositions, a more productive avenue of research might involve the utilisation of isotopic measurements of ancient walrus tissues to track changes at the base of the food web. Given that their diets are so consistently dominated by bivalves, walrus tissue isotopic compositions are more likely to reflect shifts in nutrient regimes and community composition of primary producers than changes in trophic position (see Newsome et al., 2007; Sherwood et al., 2014). A similar approach has been utilised with ringed seals in both Alaska (Szpak et al., 2018) and the central Canadian Arctic Archipelago (Szpak et al., 2019) to attempt to track changes in the productivity of sea ice algae in the Late Holocene. A common challenge of all such studies utilising archaeological materials is the uneven temporal distribution of sites within a region and the resulting 'floating sequences' that are produced, potentially leading to long periods of time that lack appropriate materials to sample. The best means by which to overcome this challenge is to focus on archaeological sequences where walruses were persistently abundant in archaeofaunal assemblages (Desjardins, 2013; Dyke et al., 2019) or to use palaeontological materials derived from natural strandings (Dyke et al., 1999) that are more likely to produce a more-or-less continuous time series.

Fatty acids

Analysis of the FAs found in the tissues of marine mammals has proven useful for better understanding their diets and food web structure (Budge et al., 2006; Dalsgaard et al., 2003; Iverson et al., 2004; Kelly and Scheibling, 2012). Relative differences in diet composition among taxa, among geographic regions, or over time can be inferred in a qualitative sense using FA signatures (Beck et al., 2007; Falk-Petersen et al., 2004; Iverson et al., 1997). Quantitative FA signature analysis (QFASA) has been proposed as a method that provides more refined estimates of prey composition relative to stable isotope analysis (Iverson et al., 2004) because the isotopic compositions of particular types of prey are rarely unique enough from one another for their proportions in the diet to be accurately estimated, particularly for consumers that eat a large number of different prey (Phillips et al., 2014).

FAs, the primary class of lipids in animal tissues, are liberated from triacyglycerols in the process of digestion but are thereafter generally incorporated with minimal modification (relative to proteins and carbohydrates) into consumer tissues (Iverson et al., 2004). While primary producers can synthesise a wide range of FAs, animals have a very limited capacity to do so (Guschina and Harwood, 2009). Considering marine mammals such as walrus, many FAs are ubiquitous among their prey and are therefore not useful as diet tracers, but in some cases, FAs can trace the relative importance of specific prey items or classes of primary producers at the base of the food web. In practice, a quantitative interpretation of these data is far more complicated and requires both an extensive understanding of the FA composition of all potential prey and taxon-specific 'calibration coefficients' that account for lipid metabolism in the target consumer (Happel et al., 2016; Iverson et al., 2004; Nordstrom et al., 2008).

One example of a specific class of FA that can be traced to specific prey types are the nonmethylene-interrupted (NMI) C20 and C22 FA, which are synthesised by benthic invertebrates such as molluscs (Barnathan, 2009; Joseph, 1982; Zhukova, 1991). These FAs have been used to investigate the extent of dietary niche overlap between Pacific walruses and bearded seal in Alaska (Budge et al., 2007). Both species contained substantial quantities of NMI FA, but the proportions varied between the two taxa (Fig. 10.3). Such elevated levels of NMI FA in marine mammal blubber are rare, suggesting that walruses and bearded seal both consumed benthic invertebrates (or other prey that consumed these invertebrates) to a large extent, but differed in the specific types of prey consumed (Budge et al., 2007). Stable carbon and nitrogen isotope analyses of walruses and bearded seal have also confirmed distinct niches for these two species, with bearded seals consuming a larger quantity of epibenthic fish and hence feeding at a much higher trophic level than walrus (Fig. 10.3) (Finley and Evans, 1983). The lone FA study of Atlantic walruses confirmed NMI FAs in the dermis, inner and outer blubber (Skoglund et al., 2010) but at lower relative abundances than in the study of Pacific walruses from Alaska (Budge et al., 2007). These NMI FAs were also present in all of the potential benthic prey species that were sampled, precluding the possibility of more quantitative assessments of prey composition.

A particularly relevant series of lipid biomarkers (highly branched isoprenoids, HBIs) have been identified in diatoms and subsequently in higher trophic-level consumers such as pinnipeds (Brown et al., 2014; Brown et al., 2013), belugas

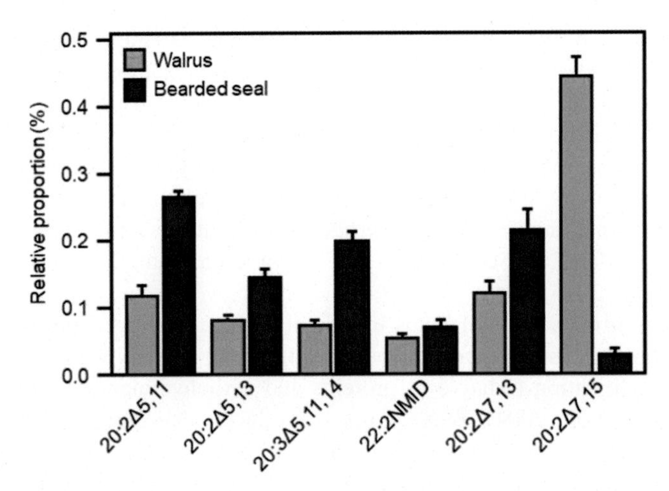

FIGURE 10.3

Proportions of nonmethylene-interrupted fatty acids in Pacific walruses and bearded seal from Little Diomede Island, Alaska, United States.

Data from Budge et al. (2007) Fatty acid biomarkers reveal niche separation in an Arctic benthic food web. Marine Ecology Progress Series, 336:305–309. doi: 10.3354/meps336305.

(Brown et al., 2017) and polar bears (Brown et al., 2018). Several of these HBIs are produced by sea-ice-associated diatoms during the spring bloom, while others are produced by pelagic phytoplankton during the summer bloom (Belt, 2018; Belt and Müller, 2013). An index of these different HBIs has been developed to quantify the relative importance of sympagic and pelagic productivity in Arctic food webs (Brown et al., 2014). These techniques can be used to examine the relative importance of sympagic production that is delivered to the benthos (Oxtoby et al., 2017) and how this might change over time, particularly when combined with stable isotope measurements of these FAs. Such shifts may have significant impacts on walrus populations (Grebmeier et al., 2006).

Compound-specific stable isotope analysis (CSIA) provides an opportunity to examine the isotopic compositions of particular compounds (such as FAs and amino acids) that are isolated via gas chromatography–combustion–isotope ratio mass spectrometry (GC/C-IRMS). A major advantage of these techniques over bulk isotopic measurements is that they reduce equivocality (i.e. ambiguities arising from a single outcome arising from two or more possible scenarios that cannot be differentiated from each other) in the interpretation of the data. For example, $\delta^{13}C$ values of bulk consumer tissues may be influenced by either a change in the diet of a consumer, or a shift in the $\delta^{13}C$ value at the base of the food web. Individual FAs that are unique to particular kinds of producers have $\delta^{13}C$ values that more faithfully reflect the conditions at the base of the food web (Budge et al., 2008). These techniques have been applied to FA derived from Arctic pinnipeds, including Pacific walrus, to quantify the relative importance of pelagic and sympagic production (Oxtoby et al., 2017; Wang et al., 2016). Essential amino acid $\delta^{15}N$ values reflect the $\delta^{15}N$ value at the base of the food web, while nonessential amino acid $\delta^{15}N$ values reflect the trophic position of the consumer (McClelland and Montoya, 2002; Popp et al., 2007). CSIA therefore provides an opportunity to discern shifts in trophic position from shifts in the $\delta^{15}N$ value at the base of the food web. This is a particularly prescient concern for studying ancient ecosystems since we might expect large-scale environmental changes to be accompanied by shifts in the diet of particular consumers.

The analysis of FA from archaeological vertebrate tissue samples remains largely unstudied. It is unclear whether the FAs preserved in ancient bones are comparable to those obtained from blubber biopsy samples, or if FAs trapped within the cortical bone microstructure vary significantly among different taxa. FAs have been isolated from bone samples that are thousands of years old (Kostyukevich et al., 2018) and their carbon isotopic compositions have been measured (Colonese et al., 2015; Colonese et al., 2017). Given the excellent organic preservation of archaeological vertebrate remains at high latitude sites where walruses may occur, the possibility exists that ancient FA composition and isotopic measurements of ancient FAs may be informative with respect to past environmental conditions.

Population structure and migration

Stable isotopes

While carbon and nitrogen isotope compositions of consumer tissues are influenced by the foods that they consume, a range of other isotope systems vary predictably across space, influenced by either the local climate (particularly precipitation regime) or underlying geology (Hobson, 1999). The oxygen ($\delta^{18}O$) and hydrogen ($\delta^{2}H$ or δD) isotopic compositions of meteoric waters vary widely across the globe, in accordance with factors such as temperature, latitude, altitude, distance from the coast and rainfall amount (Gat, 1996). The $\delta^{18}O$ values of ocean waters are relatively homogenous but surface waters do exhibit geographic variation at different latitudes (LeGrande and Schmidt, 2006) and coastal areas, particularly those influenced by large freshwater discharge, may have distinct $\delta^{18}O$ values relative to open ocean (Cooper et al., 2005). The oxygen isotope compositions of many consumer tissues are directly correlated to those of local waters, allowing $\delta^{18}O$ measurements, as well as the related $\delta^{2}H$, to be used for the study of migrations among terrestrial taxa (Bowen et al., 2005).

Strontium and lead substitute for calcium in biological tissues including bones and teeth (Trueman and Tuross, 2002). The isotope ratios of strontium ($^{87}Sr/^{86}Sr$) and lead (e.g. $^{206}Pb/^{207}Pb$, $^{208}Pb/^{207}Pb$, $^{20x}Pb/^{204}Pb$) in consumer tissues reflect the underlying geology and are therefore sensitive tracers of locality (Aggarwal et al., 2008; Bentley, 2006). Sr isotope ratios of human enamel have been used extensively to study migration in ancient populations; however, Sr isotope ratios of bones appear to be less suited as they are often significantly contaminated from the local burial environment (Hoppe et al., 2003) owing to the greater porosity and smaller size of the bioapatite mineral crystallites in bone (Trueman and Tuross, 2002).

Geographic (rather than dietary) isotope systems have been utilised infrequently to study marine mammal ecology and biogeography (relative to $\delta^{13}C$ and $\delta^{15}N$), although a series of case studies involving modern and ancient walruses highlight the potential for future research in this area. For example, it has been shown that oxygen isotope compositions of marine mammal tissues can differentiate animals inhabiting different ocean basins, particularly across a large range of latitudes (Matthews et al., 2016). Walrus tissue $\delta^{18}O$ values are unlikely to be useful discriminators of stock structure in modern populations because the spatial scales over which oceanic $\delta^{18}O$ values vary is quite large (but see Vighi et al., 2016). Sequential analyses of incrementally growing tissues, such as dentine annuli, could provide insight into fine-scale seasonal movements, as has been demonstrated for sperm whales (Borrell et al., 2013). The possibility also exists that oxygen isotope compositions of walrus ivory could aid in eliminating potential source areas for historically traded objects. Positively identifying a particular source area with this technique is, however, unlikely due to the fact that geographically disparate areas may share similar $\delta^{18}O$ values in meteoric waters.

Although Sr isotope ratios have been successfully applied to understand the migrations of fish such as salmon (Brennan et al., 2015; Kennedy et al., 2000; Koch et al., 1992), the single study that has been performed on walruses from the Canadian Arctic revealed no meaningful spatial variation, with all of the walrus $^{87}Sr/^{86}Sr$ ratios falling around the global average for seawater (Davis et al., 1998). On the other hand, significant spatial variation has been found in the lead isotope ratios of modern walrus bulk tooth cementum from the Canadian Arctic and Greenland (Outridge et al., 2003; Outridge and Stewart, 1999). These studies revealed that walrus populations exploited by different communities in the Canadian Arctic tended to have distinct tissue lead isotope ratios (particularly $^{208}Pb/^{204}Pb$ and $^{208}Pb/^{207}Pb$). Walrus tooth lead isotope ratios were related to those of the local geological lead sources; however, lead isotope ratios were more homogenous in the animals than the underlying bedrock. This difference in lead isotope ratios is a common phenomenon and reflects the ability of consumers to effectively average the bioavailable lead or strontium across space (Price et al., 2002). These data suggested that there is a greater number of discrete stocks of Atlantic walruses in the Canadian Arctic than has been previously supposed (Born et al., 1995).

One limitation of using bulk cementum is that multiple years of the animal's life are averaged. Two or more areas with disparate lead isotope ratios in the underlying geological formations may potentially be integrated, resulting in a signature that is unrepresentative of any specific period of the animal's life. These limitations can be mitigated by serial-sampling of the incrementally growing portions of the teeth (Stern et al., 1999), an approach that Stewart et al. (2006) applied to a subset of the individual walruses analysed by Outridge et al. (2003).

Individuals with 'local' bulk cementum lead isotope ratios had little variation in the incrementally sampled tissue, suggesting relatively little long-distance movement throughout their lives. Some walruses that had outlier lead isotope ratios (relative to other 'local' animals) appeared to be either immigrants (those that lived elsewhere and migrated to the area where they were harvested, settling long enough to acquire to local geological signature) or what the authors dubbed 'prodigal sons' (animals that may have been born locally, migrated elsewhere after weaning and reacquired the local signature as mature adults).

The results of the study by Outridge et al. (2003) demonstrated the potential for using lead isotope measurements of walrus ivory to determine its source region and trace the trade in various objects, similar to studies conducted with elephant ivory (van der Merwe et al., 1990; Vogel et al., 1990). This particular application has yet to be fully realised. Frei et al. (2015) conducted a pilot study with modern, historic and ancient walrus bone and ivory to explore the potential of utilising the approach of Outridge et al. (2003). Unlike bone, enamel is highly resistant to diagenesis, and Sr and Pb isotopic compositions are assumed to be relatively stable in the burial environment. Their results demonstrated that it should be feasible to differentiate walrus ivory that was obtained from Greenland, Iceland and the White Sea, opening up the possibility for testing specific hypotheses about the motivation for the Norse colonisation of Iceland and Greenland.

Future perspectives in genetics, stable isotopes and fatty acids research

Given the advances in NGS technologies, it is now possible to conduct genome scale analyses of nonmodel organisms by sequencing either reduced or full genomes of nuclear or mitochondrial DNA (Ellegren, 2014; da Fonseca et al., 2016; Goodwin et al., 2016). Currently, few studies on Atlantic walruses have applied NGS methods (Shafer et al., 2015; Star et al., 2018; Keighley et al., 2019b), but more are underway (e.g. Keighley et al., 2018). The genomic datasets will increase genetic resolution (e.g. Drummond et al., 2003; Foote et al., 2012), which in turn will pave the way for detailed information about walrus biology, demography and human interactions. New insights are likely to include better estimates of divergence timing, demographic event characterisation and the extent of local genetic adaptation within populations (Drummond et al., 2003; Keighley et al., 2019a). As genome-scale studies become increasingly affordable and feasible there is enormous potential to analyse the wealth of hitherto unstudied paleogenetic samples which are now stored in museums worldwide (Keighley et al., 2018). The sequencing of historic and ancient samples will provide novel information on past human–walrus interactions and provide insights into the nature, scale and duration of any impact of anthropogenic practices on Atlantic walruses past and present (Keighley et al., 2019a; Keighley et al., 2019b; Barrett et al., 2020).

Combining genetic and isotopic analyses of skeletal remains can provide highly complementary information. It is often difficult to assign biological sex to vertebrate remains, but molecular sexing techniques can allow for the investigation of sexual segregation in foraging habitats in the ancient past (Szpak et al., 2020; Bro-Jørgensen et al., 2020). In some cases, genetic and isotopic markers can be used in combination to assess the population structure of organisms, both modern (Clegg et al., 2003) and ancient (Eda et al., 2012; Hutchinson et al., 2015). This approach would likely be effective with Atlantic walruses given the genetic and isotopic differences that are known to exist within the range of this species today. Molecular techniques have the potential to identify the presence of populations that are now extinct (Nichols et al., 2007; Keighley et al., 2019b), and isotopic techniques can provide the opportunity to study the ecology of these populations (Moss et al., 2006; Szpak et al., 2012). Combining the two techniques, it will be possible to test hypothesis regarding the possible convergence of changes in feeding ecology and population structure over time, as well as comparing contemporary and archaeological samples of walruses to provide clues as to how climate or humans may have impacted walruses (Alter et al., 2012). Theoretically, this could be modelled and used to predict future consequences of climate change and human activities on walruses across their range.

References

Aggarwal, J., Habicht-Mauche, J., Juarez, C., 2008. Application of heavy stable isotopes in forensic isotope geochemistry: a review. Applied Geochemistry 23, 2658–2666. Available from: https://doi.org/10.1016/j.apgeochem.2008.05.016.

Allentoft, M.E., Collins, M., Harker, D., Haile, J., Oskam, C.L., Hale, M.L., et al., 2012. The half-life of DNA in bone: measuring decay kinetics in 158 dated fossils. Proceedings of the Royal Society B: Biological Sciences 279, 4724–4733. Available from: https://doi.org/10.1098/rspb.2012.1745.

Altabet, M.A., 2006. Isotopic tracers of the marine nitrogen cycle: present and past. In: Volkman, J.K. (Ed.), Marine Organic Matter: Biomarkers, Isotopes and DNA. Springer, Berlin, pp. 251–293.

Alter, S.E., Newsome, S.D., Palumbi, S.R., 2012. Pre-whaling genetic diversity and population ecology in eastern Pacific gray whales: insights from ancient DNA and stable isotopes. PLoS ONE 7, e35039. Available from: https://doi.org/10.1371/journal.pone.0035039.

Andersen, L.W., Born, E.W., 2000. Indications to two genetically different subpopulations of Atlantic walrus (*Odobenus rosmarus rosmarus*) in West and Northwest Greenland. Canadian Journal of Zoology 78, 1999–2009. Available from: https://doi.org/10.1139/z00-118.

Andersen, L.W., Born, E.W., Gjertz, I., Wiig, O., Holm, L.-E., Bendixen, C., 1998. Population structure and gene flow of the Atlantic walrus (*Odobenus rosmarus rosmarus*) in the eastern Atlantic Arctic based on mitochondrial DNA and microsatellite variation. Molecular Ecology 7, 1323–1336. Available from: https://doi.org/10.1046/j.1365-294x.

Andersen, L.W., Born, E.W., Doidge, D.W., Gjertz, I., Wiig, Ø., Waples, R.S., 2009. Genetic signals of historic and recent migration between sub-populations of Atlantic walrus, *Odobenus rosmarus rosmarus* west and east of Greenland. Endangered Species Research 9, 197–211. Available from: https://doi.org/10.3354/esr00242.

Andersen, L.W., Born, E.W., Stewart, R.E.A., Dietz, R., Doidge, D.W., Lanthier, C., 2014. A genetic comparison of West Greenland and Baffin Island (Canada) walruses: management implications. In: Kovacs, K., et al., (Eds.), Studies of the Atlantic Walrus (Odobenus rosmarus rosmarus) in the North Atlantic Arctic, 9. NAMMCO Special Publication, p. 2014.

Andersen, L.W., Jacobsen, M.W., Lydersen, C., Semenova, V., Boltunov, A., Born, E.W., et al., 2017. Walruses (*Odobenus rosmarus rosmarus*) in the Pechora Sea in the context of contemporary population structure of Northeast Atlantic walruses. Biological Journal of the Linnean Society 122, 897–915. Available from: https://doi.org/10.1093/biolinnean/blx093.

Atwell, L., Hobson, K.A., Welch, H.E., 1998. Biomagnification and bioaccumulation of mercury in an arctic marine food web: insights from stable nitrogen isotope analysis. Canadian Journal of Fisheries and Aquatic Sciences 55, 1114–1121. Available from: https://doi.org/10.1139/f98-001.

Ballard, J.W., Whitlock, M.C., 2004. The incomplete natural history of mitochondria. Molecular Ecology 13, 729–744. Available from: https://doi.org/10.1046/j.1365-294X.2003.02063.x.

Barnathan, G., 2009. Non-methylene-interrupted fatty acids from marine invertebrates: occurrence, characterization and biological properties. Biochimie 91, 671–678. Available from: https://doi.org/10.1016/j.biochi.2009.03.020.

Barrett, J.H., Boessenkool, S., Kneale, C.J., O'Connell, T.C., Star, B., 2020. Ecological globalisation, serial depletion and the medieval trade of walrus rostra. Quaternary Science Reviews 229, 106122. Available from: https://doi.org/10.1016/j.quascirev.2019.106122.

Bately, J., Englert, A., 2007. Ohthere's Voyages: A Late 9th-Century Account of Voyages Along the Coasts of Norway and Denmark and Its Cultural Context. Viking Ship Museum, Roskilde, Denmark, ISBN 978–8785180-47–6.

Beck, C.A., Rea, L.D., Iverson, S.J., Kennish, J.M., Pitcher, K.W., Fadely, B.S., 2007. Blubber fatty acid profiles reveal regional, seasonal, age-class and sex differences in the diet of young Steller sea lions in Alaska. Marine Ecology Progress Series 338, 269–280. Available from: https://doi.org/10.3354/meps338269.

Belt, S.T., 2018. What do IP25 and related biomarkers really reveal about sea ice change? Quaternary Science Reviews . Available from: https://doi.org/10.1016/j.quascirev.2018.11.025.

Belt, S.T., Müller, J., 2013. The Arctic sea ice biomarker IP25: a review of current understanding, recommendations for future research and applications in palaeo sea ice reconstructions. Quaternary Science Reviews 79, 9–25. Available from: https://doi.org/10.1016/j.quascirev.2012.12.001.

Bentley, R.A., 2006. Strontium isotopes from the Earth to the archaeological skeleton: a review. Journal of Archaeological Method and Theory 13, 135–187. Available from: https://doi.org/10.1007/s10816-006-9009-x.

Born, E.W., Gjertz, I., Reeves, R.R., 1995. Population Assessment of Atlantic Walrus (Odobenus rosmarus L.), 138. Norsk Polarinstitutt, Meddelelser.

Borrell, A., et al., 2013. Stable isotopes provide insight into population structure and segregation in eastern North Atlantic sperm whales. PLoS ONE 8, e82398. Available from: https://doi.org/10.1371/journal.pone.0082398.

Bollongino, R., Tresset, A., Vigne, J.-D., 2008. Environment and excavation: pre-lab impacts on ancient DNA analyses. Comptes Rendus Palevol 7 (2–3), 91–98. Available from: https://doi.org/10.1016/j.crpv.2008.02.002.

Bouckaert, R., Vaughan, T.G., Barido-Sottani, J., Duchêne, S., Fourment, M., Gavryushkina, A., et al., 2019. BEAST 2.5: an advanced software platform for Bayesian evolutionary analysis. PLoS computational biology 15, e1006650. Available from: https://doi.org/10.1371/journal.pcbi.1006650.

Bowen, G.J., Wassenaar, L.I., Hobson, K.A., 2005. Global application of stable hydrogen and oxygen isotopes to wildlife forensics. Oecologia 143, 337–348. Available from: https://doi.org/10.1007/s00442-004-1813-y.

Brennan, S.R., Zimmerman, C.E., Fernandez, D.P., Cerling, T.E., McPhee, M.V., Wooller, M.J., 2015. Strontium isotopes delineate fine-scale natal origins and migration histories of Pacific salmon. Science Advances 1, e1400124. Available from: https://doi.org/10.1126/sciadv.1400124.

Bro-Jørgensen, M.H., Keighley, X., Ahlgren, H., Scharff-Olsen, C.H., Rosing-Asvid, A., Dietz, R., et al., 2020. Sex identification of ancient pinnipeds using the dog genome. Journal of Archaeological Sciences. Available from: https://doi.org/10.1101/838797.

Brown, T.A., et al., 2013. Identification of the sea ice diatom biomarker IP25 and related lipids in marine mammals: A potential method for investigating regional variations in dietary sources within higher trophic level marine systems. Journal of Experimental Marine Biology and Ecology 441, 99–104. Available from: https://doi.org/10.1016/j.jembe.2013.01.020.

Brown, T.A., et al., 2017. Coupled changes between the H-Print biomarker and $\delta^{15}N$ indicates a variable sea ice carbon contribution to the diet of Cumberland Sound beluga whales. Limnology and Oceanography 62, 1606–1619. Available from: https://doi.org/10.1002/lno.10520.

Brown, T.A., Alexander, C., Yurkowski, D.J., Ferguson, S.H., Belt, S.T., 2014. Identifying variable sea ice carbon contributions to the Arctic ecosystem: a case study using highly branched isoprenoid lipid biomarkers in Cumberland Sound ringed seals. Limnology and Oceanography 59, 1581–1589. Available from: https://doi.org/10.4319/lo.2014.59.5.1581.

Brown, T.A., Galicia, M.P., Thiemann, G.W., Belt, S.T., Yurkowski, D.J., Dyck, M.G., 2018. High contributions of sea ice derived carbon in polar bear (*Ursus maritimus*) tissue. PLoS ONE 13, e0191631. Available from: https://doi.org/10.1371/journal.pone.0191631.

Budge, S.M., Iverson, S.J., Koopman, H.N., 2006. Studying trophic ecology in marine ecosystems using fatty acids: a primer on analysis and interpretation. Marine Mammal Science 22, 759–801. Available from: https://doi.org/10.1111/j.1748-7692.2006.00079.x.

Budge, S.M., Springer, A.M., Iverson, S.J., Sheffield, G., 2007. Fatty acid biomarkers reveal niche separation in an Arctic benthic food web. Marine Ecology Progress Series 336, 305–309. Available from: https://doi.org/10.3354/meps336305.

Budge, S., Wooller, M., Springer, A., Iverson, S., McRoy, C., Divoky, G., 2008. Tracing carbon flow in an arctic marine food web using fatty acid-stable isotope analysis. Oecologia 157, 117–129. Available from: https://doi.org/10.1007/s00442-008-1053-7.

Carpenter, M.L., Buenrostro, J.D., Valdiosera, C., Schroeder, H., Allentoft, M.E., Sikora, M., et al., 2013. Pulling out the 1%: whole-genome capture for the targeted enrichment of ancient DNA sequencing libraries. American Journal Human of Genetics 93, 852–864. Available from: https://doi.org/10.1016/j.ajhg.2013.10.002.

Cherel, Y., Hobson, K.A., 2007. Geographical variation in carbon stable isotope signatures of marine predators: a tool to investigate their foraging areas in the Southern Ocean. Marine Ecology Progress Series 329, 281–287. Available from: https://doi.org/10.3354/meps329281.

Clark, C.T., Horstmann, L., de Vernal, A., Jensen, A.M., Misarti, N., 2019. Pacific walrus diet across 4000 years of changing sea ice conditions. Quaternary Research 1–17. Available from: https://doi.org/10.1017/qua.2018.140.

Clegg, S.M., Kelly, J.F., Kimura, M., Smith, T.B., 2003. Combining genetic markers and stable isotopes to reveal population connectivity and migration patterns in a Neotropical migrant, Wilson's warbler (*Wilsonia pusilla*). Molecular Ecology 12, 819–830. Available from: https://doi.org/10.1046/j.1365-294X.2003.01757.x.

Cole, L.W., 2016. The evolution of per-cell organelle number. Frontiers Cell Developmental Biology 4, 85. Available from: https://doi.org/10.3389/fcell.2016.00085.

Colonese, A.C., et al., 2015. Archaeological bone lipids as palaeodietary markers. Rapid Communications in Mass Spectrometry 29, 611–618. Available from: https://doi.org/10.1002/rcm.7144.

Colonese, A.C., et al., 2017. The identification of poultry processing in archaeological ceramic vessels using in-situ isotope references for organic residue analysis. Journal of Archaeological Science 78, 179–192. Available from: https://doi.org/10.1016/j.jas.2016.12.006.

Coltrain, J.B., Hayes, M.G., O'Rourke, D.H., 2004. Sealing, whaling and caribou: the skeletal isotope chemistry of Eastern Arctic foragers. Journal of Archaeological Science 31, 39–57. Available from: https://doi.org/10.1016/j.jas.2003.06.003.

Cooper, L.W., et al., 2005. Linkages among runoff, dissolved organic carbon, and the stable oxygen isotope composition of seawater and other water mass indicators in the Arctic Ocean. Journal of Geophysical Research: Biogeosciences 110. Available from: https://doi.org/10.1029/2005JG000031.

Cornuet, J.M., Luikart, G., 1996. Description and power analysis of two tests for detecting recent population bottlenecks from allele frequency data. Genetics 144, 2001–2014.

Coyne, J.A., Orr, H.A., 2004. Speciation. Sinauer Associates Inc, Massachusetts, USA, p. 545, ISBN 0878930892.

Crandall, K.A., Bininda-Emonds, O.R.P., Mace, G.M., Wayne, R.K., 2000. Considering evolutionary processes in conservation biology. Trends Ecology and Evolution 15, 290–295. Available from: https://doi.org/10.1016/S0169-5347(00)01876-0.

da Fonseca, R.R., Albrechtsen, Themudo, G., Ramones-Madrigal, J., Sibbesen, J.A., Maretty, L.S., et al., 2016. Next-generation biology: sequencing and data analysis approaches for non-model organisms. Marine Genomics 30, 3–13. Available from: https://doi.org/10.1016/j.margen.2016.04.012.

Dalsgaard, J., John St., M., Kattner, G., Müller-Navarra, D., Hagen, W., 2003. Fatty acid trophic markers in the pelagic marine environment. Advances in Marine Biology 46, 225–340. Available from: https://doi.org/10.1016/S0065-2881(03)46005-7.

Davis, W.J., Outridge, P.M., Stewart, R.E.A., 1998. Determination of strontium isotope ratios in walrus teeth from Hudson Bay and the Arctic Islands. Geological Survey of Canada Current Research 98-F, 77–80.

Desjardins, S.P.A., 2013. Evidence for intensive walrus hunting by Thule Inuit, northwest Foxe Basin, Nunavut, Canada. Anthropozoologica 48, 37–51.

Drummond, A.J., Pybus, O.G., Rambaut, A., Forsberg, R., Rodrigo, A.G., 2003. Measurably evolving populations. TRENDS in Ecology and Evolution 189, 481–488. Available from: https://doi.org/10.1016/S0169-5347(03)00216-7.

Dussex, N., Taylor, H.R., Stovall, W.R., Rutherford, K., Dodds, K.G., Clarke, S.M., et al., 2018. Reduced representation sequencing detects only subtle regional structure in a heavily exploited and rapidly recolonizing marine mammal species. Ecology and Evolution 8, 8736–8749. Available from: https://doi.org/10.1002/ece3.4411.

Dyke, A.S., Hooper, J., Harington, C.R., Savelle, J.M., 1999. The late Wisconsinan and Holocene record of walrus (*Odobenus rosmarus*) from North America: a review with new data from Arctic and Atlantic Canada. Arctic 52, 160–181.

Dyke, A.S., et al., 2019. An assessment of marine reservoir corrections for radiocarbon dates on walrus from the Foxe Basin region of Arctic Canada. Radiocarbon 61, 67–81. Available from: https://doi.org/10.1017/RDC.2018.50.

Eda, M., Koike, H., Kuro-o, M., Mihara, S., Hasegawa, H., Higuchi, H., 2012. Inferring the ancient population structure of the vulnerable albatross *Phoebastria albatrus*, combining ancient DNA, stable isotope, and morphometric analyses of archaeological samples. Conservation Genetics 13, 143–151. Available from: https://doi.org/10.1007/s10592-011-0270-5.

Eide, M., Olsen, A., Ninnemann, United States, Eldevik, T., 2017. A global estimate of the full oceanic ^{13}C Suess effect since the preindustrial. Global Biogeochemical Cycles 31, . Available from: https://doi.org/10.1002/2016GB0054722016GB005472.

Ellegren, H., 2014. Genome sequencing and population genomics in non-model organisms. Trends in Ecology & Evolution 29, 51–63. Available from: https://doi.org/10.1016/j.tree.2013.09.008.

Enghoff, I.B., 2003. Hunting, Fishing and Animal Husbandry at the Farm Beneath the Sand, Western Greenland: An Archaeozoological Analysis of a Norse Farm in the Western Settlement. Danish Polar Centre, Meddelelser om Grønland: Man and Society, Copenhagen, Denmark, p. 28, ISSN 0106–1062.

Excoffier, L., Foll, M., 2011. Fastsimcoal: a continuous-time coalescent simulator of genomic diversity under arbitrarily complex evolutionary scenarios. Bioinformatics 27, 1332–1334. Available from: https://doi.org/10.1093/bioinformatics/btr124.

Excoffier, L., Heckel, G., 2006. Computer programs for population genetics data analysis: a survival guide. Nature Reviews Genetics 7, 745–758. Available from: https://doi.org/10.1038/nrg1904.

Falk-Petersen, S., Haug, T., Nilssen, K.T., Wold, A., Dahl, T.M., 2004. Lipids and trophic linkages in harp seal (*Phoca groenlandica*) from the eastern Barents Sea. Polar Research 23, 43–50. Available from: https://doi.org/10.1111/j.1751-8369.2004.tb00128.x.

Fay, F.H., 1960. Carnivorous walrus and some arctic zoonoses. Arctic 13, 111–122. Available from: https://doi.org/10.2307/40506856.

Fay, F.H., 1985. Odobenus rosmarus. Mammalian Species 1–7. Available from: https://doi.org/10.2307/3503810.

Finley, K.J., Evans, C.R., 1983. Summer diet of the bearded seal (*Erignathus barbatus*) in the Canadian High Arctic. Arctic 36, 82–89.

Fischbach, A.S., Jay, C.V., Jackson, J.V., Andersen, L.W., Sage, G.K., Talbot, S.L., 2008. Molecular method for determining sex of walruses. Journal of Wildlife Management 72, 1808–1812. Available from: https://doi.org/10.2193/2007-413.

Fisher, K.I., Stewart, R.E.A., 1997. Summer foods of Atlantic walrus, *Odobenus rosmarus rosmarus*, in northern Foxe Basin, Northwest Territories. Canadian Journal of Zoology 75, 1166–1175. Available from: https://doi.org/10.1139/z97-139.

Fogel, M.L., Tuross, N., Owsley, D.W., 1989. Nitrogen Isotope Tracers of Human Lactation in Modern and Archaeological Population, 88. Carnegie Institution Washington Yearbook, pp. 111–117.

Foote, A.D., Hofreiter, M., Morin, P.A., 2012. Ancient DNA from marine mammals: studying long-lived species over ecological and evolutionary timescales. Annals of Anatomy 194, 112–120. Available from: https://doi.org/10.1016/j.aanat.2011.04.010.

Foote, A.D., Liu, Y., Thomas, G.W., Vinar, T., Alfoldi, J., Deng, J., et al., 2015. Convergent evolution of the genomes of marine mammals. Nature Genetics 47, 272–275. Available from: https://doi.org/10.1038/ng.3198.

Fox, A.D., Fox, G.F., Liaklev, A., Gerhardsson, N., 2010. Predation of flightless pink-footed geese (*Anser brachyrhynchus*) by Atlantic walruses (*Odobenus rosmarus rosmarus*) in southern Edgeøya, Svalbard. Polar Research 29, 455–457. Available from: https://doi.org/10.3402/polar.v29i3.6089.

France, R.L., 1995. Carbon-13 enrichment in benthic compared to planktonic algae: food-web implications. Marine Ecology Progress Series 124, 307–312. Available from: https://doi.org/10.3354/meps124307.

France, R., Loret, J., Mathews, R., Springer, J., 1998. Longitudinal variation in zooplankton $\delta^{13}C$ through the Northwest Passage: inference for incorporation of sea-ice POM into pelagic foodwebs. Polar Biology 20, 335–341. Available from: https://doi.org/10.1007/s003000050311.

Frankham, R., 1996. Relationship of genetic variation to population size in wildlife. Conservation Biology 10, 1500–1508. Available from: https://doi.org/10.1046/j.1523-1739.1996.10061500.x.

Fraser, D.J., Bernatchez, L., 2001. Adaptive evolutionary conservation: towards a unified concept for defining conservation units. Molecular Ecology 10, 2741–2752. Available from: https://doi.org/10.1046/j.0962-1083.2001.01411.x.

Frei, K.M., Coutu, A.N., Smiarowski, K., Harrison, R., Madsen, C.K., Arneborg, J., et al., 2015. Was it for walrus? Viking Age settlement and medieval walrus ivory trade in Iceland and Greenland. World Archaeology 47 (3), 439–466. Available from: https://doi.org/10.1080/00438243.2015.1025912.

Freuchen, P., 1921. Distribution and migration of walruses along the western coast of Greenland. Fisheries Research Board of Canada (Translation Series No. 2383) 72, 237–249.

Galtier, N., Nabholz, B., Glemin, S., Hurst, G.D., 2009. Mitochondrial DNA as a marker of molecular diversity: a reappraisal. Molecular Ecology 18, 4541–4550. Available from: https://doi.org/10.1111/j.1365-294X.2009.04380.x.

Gat, J.R., 1996. Oxygen and hydrogen isotopes in the hydrologic cycle. Annual Review of Earth and Planetary Sciences 24, 225–262. Available from: https://doi.org/10.1146/annurev.earth.24.1.225.

Goodwin, S., McPherson, J.D., McCombie, W.R., 2016. Coming of age: ten years of next generation sequencing technologies. Nature Reviews, Genetics 17, 333–351. Available from: https://doi.org/10.1038/nrg.2016.49.

Gorlova, E., Krylovich, O., Savinetsky, A., Khasanov, B., 2012. Ecology of the ringed seal (*Pusa hispida*) from the Bering Strait in the late Holocene. Biology Bulletin 39, 464–471. Available from: https://doi.org/10.1134/s1062359012050056.

Gotfredsen, A.B., Appelt, M., Hastrup, K., 2018. Walrus history around the North Water: human-animal relations in a long-term perspective. Ambio 47, 193–212. Available from: https://doi.org/10.1007/s13280-018-1027-x.

Grebmeier, J.M., et al., 2006. A major ecosystem shift in the northern Bering Sea. Science 311, 1461–1464. Available from: https://doi.org/10.1126/science.1121365.

Guschina, I.A., Harwood, J.L., 2009. Algal lipids and effect of the environment on their biochemistry. In: Arts, M.T., Brett, M.T., Kainz, M.J. (Eds.), Lipids in Aquatic Ecosystems. Springer, Dordrecht, pp. 1–24.

Gutenkunst, R.N., Hernandez, R.D., Williamson, S.H., Bustamante, C.D., 2009. Inferring the joint demographic history of multiple populations from multidimensional SNP frequency data. PLoS genetics 5, e1000695. Available from: https://doi.org/10.1371/journal.pgen.1000695.

Habran, S., et al., 2010. Assessment of gestation, lactation and fasting on stable isotope ratios in northern elephant seals (*Mirounga angustirostris*). Marine Mammal Science 26, 880–895. Available from: https://doi.org/10.1111/j.1748-7692.2010.00372.x.

Happel, A., Stratton, L., Kolb, C., Hays, C., Rinchard, J., Czesny, S., 2016. Evaluating quantitative fatty acid signature analysis (QFASA) in fish using controlled feeding experiments. Canadian Journal of Fisheries and Aquatic Sciences 73, 1222–1229. Available from: https://doi.org/10.1139/cjfas-2015-0328.

Hebert, P.D., Ratnasingham, S., deWaard, J.R., 2003. Barcoding animal life: cytochrome c oxidase subunit 1 divergences among closely related species. Proceedings Biological Science 270, S96–S99. Available from: https://doi.org/10.1098/rsbl.2003.0025.

Hedges, R.E.M., Clement, J.G., Thomas, D.L., O'Connell, T.C., 2007. Collagen turnover in the adult femoral mid-shaft: modeled from anthropogenic radiocarbon tracer measurements. American Journal of Physical Anthropology 133, 808–816. Available from: https://doi.org/10.1002/ajpa.20598.

Helyar, S.J., Hemmer-Hansen, J., Bekkevold, D., Taylor, M.I., Ogden, R., Limborg, M.T., et al., 2011. Application of SNPs for population genetics of nonmodel organisms: new opportunities and challenges. Molecular Ecology Resources 11, 123–136. Available from: https://doi.org/10.1111/j.1755-0998.2010.02943.x.

Hewitt, G.M., 1999. Post-glacial re-colonization of European biota. Biological Journal of Linnean Society 68, 87–112. Available from: https://doi.org/10.1111/j.1095-8312.1999.tb01160.x.

Hilton, G.M., Thompson, D.R., Sagar, P.M., Cuthbert, R.J., Cherel, Y., Bury, S.J., 2006. A stable isotopic investigation into the causes of decline in a sub-Antarctic predator, the rockhopper penguin *Eudyptes chrysocome*. Global Change Biology 12, 611–625. Available from: https://doi.org/10.1111/j.1365-2486.2006.01130.x.

Hobson, K.A., 1993. Trophic relationships among high Arctic seabirds: insights from tissue-dependent stable-isotope models. Marine Ecology Progress Series 95, 7–18.

Hobson, K.A., 1999. Tracing origins and migration of wildlife using stable isotopes: a review. Oecologia 120, 314–326. Available from: https://doi.org/10.1007/s004420050865.

Hobson, K.A., Clark, R.G., 1992. Assessing avian diets using stable isotopes I: turnover of ^{13}C in tissues. The Condor 94, 181–188. Available from: https://doi.org/10.2307/1368807.

Hobson, K.A., Welch, H.E., 1992. Determination of trophic relationships within a high Arctic marine food web using $\delta^{13}C$ and $\delta^{15}N$ analysis. Marine Ecology Progress Series 84, 9–18.

Hobson, K.A., Piatt, J.F., Pitocchelli, J., 1994. Using stable isotopes to determine seabird trophic relationships. Journal of Animal Ecology 63, 786–798. Available from: https://doi.org/10.2307/5256.

Hobson, K.A., Fisk, A., Karnovsky, N., Holst, M., Gagnon, J.-M., Fortier, M., 2002. A stable isotope ($\delta^{13}C$, $\delta^{15}N$) model for the North Water food web: implications for evaluating trophodynamics and the flow of energy and contaminants. Deep Sea Research Part II: Topical Studies in Oceanography 49, 5131–5150. Available from: https://doi.org/10.1016/S0967-0645(02)00182-0.

Hoelzel, A.R., Campagna, C., Arnbom, T., 2001. Genetic and morphometric differentiation between island and mainland southern elephant seal populations. Proceedings Biological Science 268, 325–332. Available from: https://doi.org/10.1098/rspb.2000.1375.

Hoppe, K.A., Koch, P.L., Furutani, T.T., 2003. Assessing the preservation of biogenic strontium in fossil bones and tooth enamel. International Journal of Osteoarchaeology 13, 20–28. Available from: https://doi.org/10.1002/oa.663.

Hutchinson, W.F., et al., 2015. The globalization of naval provisioning: ancient DNA and stable isotope analyses of stored cod from the wreck of the Mary Rose, AD 1545. Royal Society Open Science 2. Available from: https://doi.org/10.1098/rsos.150199.

Iverson, S.J., Frost, K.J., Lowry, L.F., 1997. Fatty acid signatures reveal fine scale structure of foraging distribution of harbor seals and their prey in Prince William Sound, Alaska. Marine Ecology Progress Series 151, 255–271. Available from: https://doi.org/10.3354/meps151255.

Iverson, S.J., Field, C., Bowen, W.D., Blanchard, W., 2004. Quantitative fatty acid signature analysis: a new method of estimating predator diets. Ecological Monographs 74, 211–235. Available from: https://doi.org/10.1890/02-4105.

Jaouen, K., Szpak, P., Richards, M.P., 2016. Zinc isotope ratios as indicators of diet and trophic level in Arctic marine mammals. PLoS ONE 11, e0152299. Available from: https://doi.org/10.1371/journal.pone.0152299.

Joseph, J.D., 1982. Lipid composition of marine and estuarine invertebrates. Part II: Mollusca. Progress in Lipid Research 21, 109–153. Available from: https://doi.org/10.1016/0163-7827(82)90002-9.

Kalinowski, S.T., 2002. How many alleles per locus should be used to estimate genetic distances? Heredity 8, 62–65. Available from: https://doi.org/10.1038/sj/hdy/6800009.

Keighley, X., Bro-Jørgensen, M.H., Jordan, P., Tange Olsen, M., 2018. Ancient pinnipeds: what paleogenetics can tell us about past human-marine mammal interactions. SAA Archaeological Record 18, 38–46.

Keighley, X., Tange Olsen, M., Jordan, P., 2019a. Integrating cultural and biological perspectives on long-term human-walrus (*Odobenus rosmarus rosmarus*) interactions across the North Atlantic. Quaternary Research 1–21. Available from: https://doi.org/10.1017/qua.2018.150.

Keighley, X., Pálsson, S., Einarsson, B.F., Petersen, A., Fernáandez-Coll, M., Jordan, P., et al., 2019b. Disappearance of Icelandic walruses coincided with Norse settlement. Molecular Biology and Evolution 36, 2656–2667. Available from: https://doi.org/10.1093/molbev/msz196.

Keighley, X., Bro-Jørgensen, M.H., Ahlgren, H., Szpak, P., Ciucani, M.M., Sánchez Barreiro, F., et al., 2021. Predicting sample success for large-scale ancient DNA studies on marine mammals. Molecular Ecology Resources 21 (4), 1149–1166.

Keller, C., 2010. Furs, fish, and ivory: medieval Norsemen at the Arctic fringe. Journal of the North Atlantic 3, 1–23. Available from: https://doi.org/10.3721/037.003.0105.

Kelly, J.R., Scheibling, R.E., 2012. Fatty acids as dietary tracers in benthic food webs. Marine Ecology Progress Series 446, 1–22. Available from: https://doi.org/10.3354/meps09559.

Kendall, C., Høier Eriksen, A.M., Kontopoulos, I., Collins, M.J., Turner-Walker, G., 2017. Diagenesis of archaeological bone and tooth. Palaeogeography, Palaeoclimatology, Palaeoecology 491, 21–37. Available from: https://doi.org/10.1016/j.palaeo.2017.11.041.

Kennedy, B.P., Blum, J.D., Folt, C.L., Nislow, K.H., 2000. Using natural strontium isotopic signatures as fish markers: methodology and application. Canadian Journal of Fisheries and Aquatic Sciences 57, 2280–2292. Available from: https://doi.org/10.1139/f00-206.

Kindberg, J., Swenson, J.E., Ericsson, G., Bellemain, E., Miquel, C., Taberlet, P., 2011. Estimating population size and trends of the Swedish brown bear *Ursus arctos* population. Wildlife Biology 17, 114–123. Available from: https://doi.org/10.2981/10-100.

Klimova, A., Phillips, C.D., Fietz, K., Olsen, T., Harwood, J., Amos, W., et al., 2014. Global population structure and demographic history of the grey seal. Molecular Ecology 23, 3999–4017. Available from: https://doi.org/10.1111/mec.12850.

Koch, P.L., Halliday, A.N., Walter, L.M., Stearley, R.F., Huston, T.J., Smith, G.R., 1992. Sr isotopic composition of hydroxyapatite from recent and fossil salmon: the record of lifetime migration and diagenesis. Earth and Planetary Science Letters 108, 277–287. Available from: https://doi.org/10.1016/0012-821X(92)90028-T.

Kohlbach, D., Graeve, M., Lange, B.A., David, C., Peeken, I., Flores, H., 2016. The importance of ice algae-produced carbon in the central Arctic Ocean ecosystem: food web relationships revealed by lipid and stable isotope analyses. Limnology and Oceanography 61, 2027–2044. Available from: https://doi.org/10.1002/lno.10351.

Kostyukevich, Y., et al., 2018. Proteomic and lipidomic analysis of mammoth bone by high-resolution tandem mass spectrometry coupled with liquid chromatography. European Journal of Mass Spectrometry 24, 411–419. Available from: https://doi.org/10.1177/1469066718813728.

Laws, E.A., Popp, B.N., Cassar, N., Tanimoto, J., 2002. ^{13}C discrimination patterns in oceanic phytoplankton: likely influence of CO_2 concentrating mechanisms, and implications for palaeoreconstructions. Functional Plant Biology 29, 323–333. Available from: https://doi.org/10.1071/PP01183.

LeGrande, A.N., Schmidt, G.A., 2006. Global gridded data set of the oxygen isotopic composition in seawater. Geophysical Research Letters 33, L12604. Available from: https://doi.org/10.1029/2006GL026011.

Lindahl, T., 2003. Instability and decay of the primary structure of DNA. Nature 362 (6422), 709–715. Available from: https://doi.org/10.1038/362709a0.

Lindqvist, C., Bachmann, L., Andersen, L.W., Born, E.W., Arnason, U., Kovacs, K.M., et al., 2009. The Laptev Sea walrus, *Odobenus rosmarus laptevi*: an enigma revisited. Zoologica Scripta 38, 113–127. Available from: https://doi.org/10.1111/j.1463-6409. 2008.00364.x.

Lindqvist, C., Roy, T., Lydersen, C., Kovacs, K.M., Aars, J., Wiig, Ø., et al., 2016. Genetic diversity of historical Atlantic walruses (*Odobenus rosmarus rosmarus*) from Bjørnøya and Håøya (Tusenøyane), Svalbard, Norway. BMC Research Notes 9, 112. Available from: https://doi.org/10.1186/s13104-016-1907-8.

Linnebjerg, J.F., et al., 2016. Deciphering the structure of the West Greenland marine food web using stable isotopes ($\delta^{13}C$, $\delta^{15}N$). Marine Biology 163, 230. Available from: https://doi.org/10.1007/s00227-016-3001-0.

Lorrain, A., et al., 2015. Nitrogen isotopic baselines and implications for estimating foraging habitat and trophic position of yellowfin tuna in the Indian and Pacific Oceans. Deep Sea Research Part II: Topical Studies in Oceanography 113, 188–198. Available from: https://doi.org/10.1016/j.dsr2.2014.02.003.

Lowry, L.F., Fay, F.H., 1984. Seal eating by walruses in the Bering and Chukchi Seas. Polar Biology 3, 11–18. Available from: https://doi.org/10.1007/BF00265562.

Mallory, M.L., Woo, K., Gaston, A.J., Davies, W.E., Mineau, P., 2004. Walrus (*Odobenus rosmarus*) predation on adult thick-billed murres (*Uria lomvia*) at Coats Island, Nunavut, Canada. Polar Research 23, 111–114. Available from: https://doi.org/10.3402/polar.v23i1.6270.

Maricic, T., Whitten, M., Pääbo, S., 2010. Multiplexed DNA sequence capture of mitochondrial genomes using PCR products. PLoS ONE 5, e14004. Available from: https://doi.org/10.1371/journal.pone.0014004.

Matthews, C.J.D., Longstaffe, F.J., Ferguson, S.H., 2016. Dentine oxygen isotopes ($\delta^{18}O$) as a proxy for odontocete distributions and movements. Ecology and Evolution 6, 4643–4653. Available from: https://doi.org/10.1002/ece3.2238.

Mayr, E., 1942. Systematics and the Origin of Species. Columbia University Press, New York.

Mayr, E., 1963. Animal Species and Evolution. Harvard University Press, Cambridge, Mass.

McClelland, J.W., Montoya, J.P., 2002. Trophic relationships and the nitrogen isotopic composition of amino acids in plankton. Ecology 83, 2173–2180. Available from: https://doi.org/10.1890/0012-9658(2002)083[2173:TRATNI]2.0.CO;2.

McConnaughey, T., McRoy, C.P., 1979. Food-web structure and the fractionation of carbon isotopes in the Bering Sea. Marine Biology 53, 257–262. Available from: https://doi.org/10.1007/BF00952434.

McLeod, B.A., Frasier, T.R., Lucas, Z., 2014. Assessment of the extirpated Maritimes walrus using morphological and ancient DNA analysis. PLoS ONE 9, e99569. Available from: https://doi.org/10.1371/journal.pone.0099569.

McMahon, K.W., et al., 2006. Benthic community response to ice algae and phytoplankton in Ny Ålesund, Svalbard. Marine Ecology Progress Series 310, 1–14. Available from: https://doi.org/10.3354/meps310001.

Minagawa, M., Wada, E., 1984. Stepwise enrichment of 15N along food chains: further evidence and the relation between $\delta^{15}N$ and animal age. Geochimica et Cosmochimica Acta 48, 1135–1140. Available from: https://doi.org/10.1016/0016-7037(84)90204-7.

Montoya, J.P., 2008. Nitrogen stable isotopes in marine environments. In: Capone, D.G., Bronk, D.A., Mulholland, M.R., Carpenter, E.J. (Eds.), Nitrogen in the Marine Environment, 2nd edn Elsevier, Amsterdam, pp. 1277–1302.

Moritz, C., 1994. Defining evolutionarily significant units for conservation. Trends Ecology and Evolution 9, 373–375. Available from: https://doi.org/10.1016/0169-5347(94)90057-4.

Moss, M.L., et al., 2006. Historical ecology and biogeography of north Pacific pinnipeds: isotopes and ancient DNA from three archaeological assemblages. The Journal of Island and Coastal Archaeology 1, 165–190. Available from: https://doi.org/10.1080/15564890600934129.

Muir, D.C.G., Segstro, M.D., Hobson, K.A., Ford, C.A., Stewart, R.E.A., Olpinski, S., 1995. Can seal eating explain elevated levels of PCBs and organochlorine pesticides in walrus blubber from eastern Hudson Bay (Canada)? Environmental Pollution 90, 335–348. Available from: https://doi.org/10.1016/0269-7491(95)00019-N.

Naughton, D., 2012. The Natural History of Canadian Mammals, 824. University of Toronto Press, Toronto, ISBN/ISSN: 978-1-4426-4483-0.

Nelson, D.E., Møhl, J., Heinemeier, J., Arneborg, J., 2012. Stable carbon and nitrogen isotopic measurements of the wild animals hunted by the Norse and the Neo-Eskimo people of Greenland. Journal of the North Atlantic 3, 40–50. Available from: https://doi.org/10.3721/037.004.s304.

Newsome, S.D., Koch, P.L., Etnier, M.A., Aurioles-Gamboa, D., 2006. Using carbon and nitrogen isotope values to investigate maternal strategies in northeast Pacific otariids. Marine Mammal Science 22, 556–572. Available from: https://doi.org/10.1111/j.1748-7692.2006.00043.x.

Newsome, S.D., Etnier, M.A., Kurle, C.M., Waldbauer, J.R., Chamberlain, C.P., Koch, P.L., 2007. Historic decline in primary productivity in western Gulf of Alaska and eastern

Bering Sea: isotopic analysis of northern fur seal teeth. Marine Ecology Progress Series 332, 211–224. Available from: https://doi.org/10.3354/meps332211.

Newsome, S.D., Clementz, M.T., Koch, P.L., 2010. Using stable isotope biogeochemistry to study marine mammal ecology. Marine Mammal Science 26, 509–572. Available from: https://doi.org/10.1111/j.1748-7692.2009.00354.x.

Nichols, C., Herman, J., Gaggiotti Oscar, E., Dobney Keith, M., Parsons, K., Hoelzel, A. R., 2007. Genetic isolation of a now extinct population of bottlenose dolphins (*Tursiops truncatus*). Proceedings of the Royal Society B: Biological Sciences 274, 1611–1616. Available from: https://doi.org/10.1098/rspb.2007.0176.

Nielsen-Marsh, C.M., Hedges, R.E.M., 2000. Patterns of diagenesis in bone I: the effects of site environments. Journal of Archaeological Science 27 (12), 1139–1150. Available from: https://doi.org/10.1006/jasc.1999.0537.

Nordstrom, C.A., Wilson, L.J., Iverson, S.J., Tollit, D.J., 2008. Evaluating quantitative fatty acid signature analysis (QFASA) using harbour seals *Phoca vitulina richardsi* in captive feeding studies. Marine Ecology Progress Series 360, 245–263. Available from: https://doi.org/10.3354/meps07378.

Orlando, L., Ginolhac, A., Zhang, G., et al., 2013. Recalibrating Equus evolution using the genome sequence of an early Middle Pleistocene horse. Nature 499, 74–78. Available from: https://doi.org/10.1038/nature12323. Recalibrating Equus evolution using the genome sequence of an early Middle Pleistocene horse.

Outridge, P.M., Stewart, R.E., 1999. Stock discrimination of Atlantic walrus (*Odobenus rosmarus rosmarus*) in the eastern Canadian Arctic using lead isotope and element signatures in teeth. Canadian Journal of Fisheries and Aquatic Sciences 56, 105–112. Available from: https://doi.org/10.1139/f98-155.

Outridge, P.M., Davis, W.J., Stewart, R.E.A., Born, E.W., 2003. Investigation of the stock structure of Atlantic walrus (*Odobenus rosmarus rosmarus*) in Canada and Greenland using dental Pb isotopes derived from local geochemical environments. Arctic 56, 82–90.

Oxtoby, L.E., et al., 2017. Resource partitioning between Pacific walruses and bearded seals in the Alaska Arctic and sub-Arctic. Oecologia 184, 385–398. Available from: https://doi.org/10.1007/s00442-017-3883-7.

Pääbo, S., Poinar, H., Serre, D., Jaenicke-Despres, V., Hebler, J., Rohland, N., et al., 2004. Genetic analyses from ancient DNA. Annual Reviews Genet 38, 645–679. Available from: https://doi.org/10.1146/annurev.genet.37.110801.143214.

Parker, C., Rohrlach, A.B., Friederich, S., Nagel, S., Meyer, M., Krause, J., et al., 2020. A systematic investigation of human DNA preservation in medieval skeletons. Scientific Reports 10, 18225. Available from: https://doi.org/10.1038/s41598-020-75163-w.

Peery, M.Z., Kirby, R., Reid, B.N., Stoelting, R., Doucet-Bëer, E., Robinson, S., et al., 2012. Reliability of genetic bottleneck tests for detecting recent population declines. Molecular Ecology 21, 3403–3418. Available from: https://doi.org/10.1111/j.1365-294X.2012.05635.x.

Pethybridge, H., et al., 2018. A global *meta*-analysis of marine predator nitrogen stable isotopes: relationships between trophic structure and environmental conditions. Global Ecology and Biogeography 27, 1043–1055. Available from: https://doi.org/10.1111/geb.12763.

Phillips, D.L., et al., 2014. Best practices for use of stable isotope mixing models in food-web studies. Canadian Journal of Zoology 92, 823–835. Available from: https://doi.org/10.1139/cjz-2014-0127.

Piry, S., Luikart, G., Cornuet, J.M., 1999. BOTTLENECK: a computer program for detecting recent reductions in the effective population size using allele frequency data. Journal of Heredity 90, 502–503.

Popp, B.N., et al., 2007. Insight into the trophic ecology of yellowfin tuna, *Thunnus albacares*, from compound-specific nitrogen isotope analysis of proteinaceous amino acids. In: Dawson, T.E., Siegwolf, R.T.W. (Eds.), Stable Isotopes as Indicators of Ecological Change. Academic Press, London, pp. 173–190.

Post, D.M., 2002. Using stable isotopes to estimate trophic position: models, methods, and assumptions. Ecology 83, 703–718. Available from: https://doi.org/10.1890/0012-9658(2002)083[0703:USITET]2.0.CO;2.

Price, T.D., Burton, J.H., Bentley, R.A., 2002. The characterization of biologically available strontium isotope ratios for the study of prehistoric migration. Archaeometry 44, 117–135. Available from: https://doi.org/10.1111/1475-4754.00047.

Pruvost, M., Schwarz, R., Bessa Correia, V., Champlot, S., Braguier, S., Morel, N., et al., 2007. Freshly excavated fossil bones are best for amplification of ancient DNA. Proceedings of the National Academy of Sciences of the United States of America 104 (3), 739–744. Available from: https://doi.org/10.1073/pnas.0610257104.

Puckett, E.E., 2016. Variability in total project and per sample genotyping costs under varying study designs including with microsatellites or SNPs to answer conservation genetic questions. Conservation Genetics Resources 9, 289–304. Available from: https://doi.org/10.1007/s12686-016-0643-7.

Quay, P.D., Tilbrook, B., Wong, C.S., 1992. Oceanic uptake of fossil fuel CO_2: carbon-13 evidence. Science 256, 74–79. Available from: https://doi.org/10.1126/science.256.5053.74.

Ramakrishnan, U., Hadly, E.A., Mountain, J.L., 2005. Detecting past population bottlenecks using temporal genetic data. Molecular Ecology 14, 2915–2922. Available from: https://doi.org/10.1111/j.1365-294X.2005.02586.x.

Rigét, F., Dietz, R., Born, E.W., Sonne, C., Hobson, K.A., 2007a. Temporal trends of mercury in marine biota of west and northwest Greenland. Marine Pollution Bulletin 54, 72–80. Available from: https://doi.org/10.1016/j.marpolbul.2006.08.046.

Rigét, F., et al., 2007b. Transfer of mercury in the marine food web of West Greenland. Journal of Environmental Monitoring 9, 877–883. Available from: https://doi.org/10.1039/B704796G.

Robertson, K.M., Lauf, M.L., Morin, P.A., 2018. Genetic sexing of pinnipeds: a real-time, single step qPCR technique. Conservation Genetic Resources 10, 213–218. Available from: https://doi.org/10.1007/s12686-017-0759-4.

Roesdahl, E., 2001. Walrus ivory in the Viking Age—and Ohthere (Ottar). Offa 58, 33–37. ISSN 0078–371.

Saánchez-Molano, E., Caballero, A., Fernandez, J., 2013. Efficiency of conservation management methods for subdivided populations under local adaptation. Journal of Heredity 104, 554–564. Available from: https://doi.org/10.1093/jhered/est016.

Scotter, S.E., et al., 2019. Contaminants in Atlantic walruses in Svalbard part 1: relationships between exposure, diet and pathogen prevalence. Environmental Pollution 244, 9–18. Available from: https://doi.org/10.1016/j.envpol.2018.10.001.

Sethi, S.A., Linden, D., Wenburg, J., Lewis, C., Lemons, P., Fuller, A., et al., 2016. Accurate recapture identification for genetic mark–recapture studies with error-tolerant likelihood-based match calling and sample clustering. Royal Society Open Science 3, 160457. Available from: https://doi.org/10.1098/rsos.160457.

Seymour, J., Horstmann-Dehn, L., Wooller, M.J., 2014. Proportion of higher trophic-level prey in the diet of Pacific walruses (*Odobenus rosmarus divergens*). Polar Biology 37, 941–952. Available from: https://doi.org/10.1007/s00300-014-1492-z.

Shafer, A.B.A., Davis, C.S., Coltman, D.W., Stewart, R.E.A., 2014. Microsatellite assessment of walrus (*Odobenus rosmarus rosmarus*) stocks in Canada. NAMMCO Scientific Publications 9, 15–32. Available from: https://doi.org/10.7557/3.2607.

Shafer, A.B.A., Gattepaille, L.M., Stewart, R.E.A., Wolf, J.B.W., 2015. Demographic inferences using short-read genomic data in an approximate Bayesian Computation framework: in silico evaluation of power, biases, and proof of concept in Atlantic walrus. Molecular Ecology 2, 328–345. Available from: https://doi.org/10.1111/mec.13034.

Sherwood, O.A., Guilderson, T.P., Batista, F.C., Schiff, J.T., McCarthy, M.D., 2014. Increasing subtropical North Pacific Ocean nitrogen fixation since the Little Ice Age. Nature 505, 78–81. Available from: https://doi.org/10.1038/nature12784.

Skoglund, E.G., Lydersen, C., Grahl-Nielsen, O., Haug, T., Kovacs, K.M., 2010. Fatty acid composition of the blubber and dermis of adult male Atlantic walruses (*Odobenus rosmarus rosmarus*) in Svalbard, and their potential prey. Marine Biology Research 6, 239–250. Available from: https://doi.org/10.1080/17451000903233755.

Søreide, J.E., Hop, H., Carroll, M.L., Falk-Petersen, S., Hegseth, E.N., 2006. Seasonal food web structures and sympagic–pelagic coupling in the European Arctic revealed by stable isotopes and a two-source food web model. Progress in Oceanography 71, 59–87. Available from: https://doi.org/10.1016/j.pocean.2006.06.001.

Søreide, J.E., Carroll, M.L., Hop, H., Ambrose, W.G., Hegseth, E.N., Falk-Petersen, S., 2013. Sympagic-pelagic-benthic coupling in Arctic and Atlantic waters around Svalbard revealed by stable isotopic and fatty acid tracers. Marine Biology Research 9, 831–850. Available from: https://doi.org/10.1080/17451000.2013.775457.

Sosa, C., Vispe, E., Núñez, C., Baeta, M., Casalod, Y., Bolea, M., et al., 2013. Association between ancient bone preservation and DNA yield: a multidisciplinary approach. American Journal of Physical Anthropology 151, 102–109. Available from: https://doi.org/10.1002/ajpa.22262.

Star, B., Barrett, J.H., Gondek, A.T., Boessenkool, S., 2018. Ancient DNA reveals the chronology of walrus ivory trade from Norse Greenland. Proceedings of Biological Science 285. Available from: https://doi.org/10.1098/rspb.2018.0978.

Stasko, A.D., et al., 2018. Benthic-pelagic trophic coupling in an Arctic marine food web along vertical water mass and organic matter gradients. Marine Ecology Progress Series 594, 1–19. Available from: https://doi.org/10.3354/meps12582.

Stern, R.A., Outridge, P.M., Davis, W.J., Stewart, R.E.A., 1999. Reconstructing lead isotope exposure histories preserved in the growth layers of walrus teeth using the SHRIMP II ion microprobe. Environmental Science & Technology 33, 1771–1775. Available from: https://doi.org/10.1021/es980807f.

Stewart, R.E.A., Outridge, P.M., Stern, R.A., 2006. Walrus life-history movements reconstructed from lead isotopes in annual layers of teeth. Marine Mammal Science 19, 806–818. Available from: https://doi.org/10.1111/j.1748-7692.2003.tb01131.x.

Szpak, P., Orchard, T.J., McKechnie, I., Gröcke, D.R., 2012. Historical ecology of late Holocene sea otters (*Enhydra lutris*) from northern British Columbia: isotopic and zooarchaeological perspectives. Journal of Archaeological Science 39, 1553–1571. Available from: https://doi.org/10.1016/j.jas.2011.12.006.

Szpak, P., Buckley, M., Darwent, C.M., Richards, M.P., 2018. Long-term ecological changes in marine mammals driven by recent warming in northwestern Alaska. Global Change Biology 24, 490–503. Available from: https://doi.org/10.1111/gcb.13880.

Szpak, P., Savelle, J.M., Conolly, J., Richards, M.P., 2019. Variation in late holocene marine environments in the Canadian Arctic archipelago: evidence from ringed seal bone collagen stable isotope compositions. Quaternary Science Reviews 211, 136–155. Available from: https://doi.org/10.1016/j.quascirev.2019.03.016.

Szpak, P., Buckley, M., 2020. Sulfur isotopes (δ^{34}S) in Arctic marine mammals: indicators of benthic vs. pelagic foraging. Marine Ecology Progress Series 653, 205–216. Available from: https://doi.org/10.3354/meps13493.

Szpak, P., Julien, M.-H., Royle, T., Savelle, J.M., Yang, D.Y., Richards, M.P., 2020. Sexual differences in the foraging ecology of 19th Century Belugas (*Delphinapterus leucas*) from the Canadian High Arctic. Marine Mammal Science 36, 451–471. Available from: https://doi.org/10.1111/mms.12655.

Tieszen, L.L., Boutton, T.W., Tesdahl, K.G., Slade, N.A., 1983. Fractionation and turnover of stable carbon isotopes in animal tissues: implications for δ^{13}C analysis of diet. Oecologia 57, 32–37. Available from: https://doi.org/10.1007/BF00379558.

Trueman, C.N., Tuross, N., 2002. Trace elements in recent and fossil bone apatite. Reviews in Mineralogy and Geochemistry 48, 489–521. Available from: https://doi.org/10.2138/rmg.2002.48.13.

van der Merwe, N.J., et al., 1990. Source-area determination of elephant ivory by isotopic analysis. Nature 346, 744–746. Available from: https://doi.org/10.1038/346744a0.

Vibe, C., 1950. The marine mammals and the marine fauna in the Thule District (Northwest Greenland) with observations on the ice conditions in 1939–41, 150. Meddelelser Gronland, pp. 1–115.

Vighi, M., Borrell, A., Aguilar, A., 2016. Stable isotope analysis and fin whale subpopulation structure in the eastern North Atlantic. Marine Mammal Science 32, 535–551. Available from: https://doi.org/10.1111/mms.12283.

Vilá-Cabrera, A., Premoli, A.C., Jump, A.S., 2019. Refining predictions of population decline at species' rear edges. Global Change Biology 25 (5), 1549–1560.

Vogel, J.C., Eglington, B., Auret, J.M., 1990. Isotope fingerprints in elephant bone and ivory. Nature 346, 747–749. Available from: https://doi.org/10.1038/346747a0.

von Steinsdorff, K., Grupe, G., 2006. Reconstruction of an aquatic food web: Viking Haithabu versus Medieval Schlewsig. Anthropologischer Anzeiger 64, 283–295.

Voss, M., Dippner, J.W., Montoya, J.P., 2001. Nitrogen isotope patterns in the oxygen-deficient waters of the eastern tropical North Pacific Ocean. Deep Sea Research Part I: Oceanographic Research Papers 48, 1905–1921. Available from: https://doi.org/10.1016/S0967-0637(00)00110-2.

Wang, S.W., Springer, A.M., Budge, S.M., Horstmann-Dehn, L., Quakenbush, L.T., Wooller, M.J., 2016. Carbon sources and trophic relationships of ice seals during recent environmental shifts in the Bering Sea. Ecological Applications 26, 830–845. Available from: https://doi.org/10.1890/14-2421.

Wiig, O., Born, E.W., Gjertz, I., Lydersen, C., Stewart, R.E.A., 2007. Historical sex-specific distribution of Atlantic walrus (*Odobenus rosmarus rosmarus*) in Svalbard assessed by mandible measurements. Polar Biology 31, 69–75. Available from: https://doi.org/10.1007/s00300-007-0334-7.

Zhukova, N.V., 1991. The pathway of the biosynthesis of non-methylene-interrupted dienoic fatty acids in molluscs. Comparative Biochemistry and Physiology Part B: Comparative Biochemistry 100, 801–804. Available from: https://doi.org/10.1016/0305-0491(91)90293-M.

CHAPTER

Atlantic walrus management, regulation and conservation

11

Fern Wickson

North Atlantic Marine Mammal Commission (NAMMCO), Tromsø, Norway

Chapter Outline

Introduction

Walruses are an important food, income and cultural resource for Arctic Indigenous communities, as highlighted by Keighley et al. (2019), Gotfredsen et al. (2018) and various chapters in this book. Although Atlantic walruses have been hunted by such communities for hundreds of years, the urgent need for coordinated management and regulation really arose when heavy exploitation from commercial hunting by European sealers and whalers (19th and early 20th centuries) led to a dramatic drop in population numbers (Gjertz and Wiig, 1995; Wiig et al., 2014; Witting and Born, 2005).

Restrictions on hunting are now applied in countries across the Atlantic walrus' range—that is, Norway, Russia, Canada and Greenland. These restrictions have evolved over time and have also been extended to include the

The Atlantic Walrus. DOI: https://doi.org/10.1016/B978-0-12-817430-2.00005-4

trade in walrus products, as well as the protection of walrus habitats. There is also an increasingly recognised need for management to address a range of other new, and emerging threats facing walruses, including disturbance from noise, shipping and oil and gas exploration, as well as increasing Arctic tourism and climate change (Hauser et al., 2018). The constellation of threats facing Atlantic walruses means that the rules, regulations and regimes for controlling human impacts have also had to become increasingly coordinated at local, regional and international scales.

Since comprehensive reviews of the history of walrus management and the operation of trade regulations are available elsewhere (see Shadbolt et al., 2014; Wiig et al., 2014), and the regulation of all forms of disturbance is extremely wide in scope, this short overview focuses on summarising existing management regimes at different levels, with a particular focus on hunting (Figs. 11.1 and 11.2). It concludes by outlining the key requirements and challenges that will need to be addressed to ensure effective conservation and sustainable future management of remaining walrus populations.

FIGURE 11.1

Historical depiction of a walrus hunt.

From Alfred Frédal (1866) Le monde de la mer. *Paris: Librairie de L. Hachette et Cie.*

FIGURE 11.2

Painting titled "Sea horses".

From James Cook and John Webber (1784) A voyage to the Pacific Ocean.

Range-state summaries

Norway

Concerns regarding severe walrus population declines led to all commercial hunts by Norwegian citizens, inhabitants and companies being banned by Royal Decree in 1952 (Government of Norway, 1952). This effectively outlawed not only walrus hunting in Norwegian territories, but also hunting by Norwegian whalers and sealers in the waters around Greenland and Russia. This ban remains in place today. Several haul-out and feeding sites around the Svalbard archipelago have also been protected by being declared marine reserves. As a result of these management measures, populations of Atlantic walruses in the Svalbard archipelago have been steadily recovering over the last 60 years (Kovacs et al., 2014).

Russia

The Soviet Union began limiting the harvest of Atlantic walruses in specific waters in 1921. This progressively expanded into bans on the state harvesting of walrus in 1935, opportunistic commercial harvesting by the sealing industry in 1949 and prohibitions on hunting of all subspecies of walrus by any Russian citizen in 1956 (with an exception for subsistence hunting by Arctic Indigenous communities) (Bychkov, 1973 cited in Wiig et al., 2014). Eventually, Atlantic walruses received complete protection (including from subsistence hunting) in 1982 when they were listed in the 'Red Data Book' (Danilov-Danilian, 2001; Shadbolt et al., 2014). Their listing as a category 2 species (i.e., a species decreasing in number) protects them from any activity (resulting in direct mortality or habitat disturbance) affecting their conservation (Higdon and Stewart, 2018).

Greenland

Commercial hunting of Atlantic walruses was prohibited in Greenland in 1956 (Higdon and Stewart, 2018) and is now only permitted on a subsistence basis. Furthermore, only full-time hunters (i.e., hunters obtaining over half their income from hunting and fishing) can apply for a licence to hunt walruses; these licences are granted on an individual basis and cannot be transferred or sold (Ugarte, 2015). Although some management measures (e.g., restrictions on hunting season and hunting methods) were in place in the 1900's and early 2000's (Higdon and Stewart, 2018; Wiig et al., 2014), these were significantly expanded in 2006 by the introduction of a new Executive Order (Greenland Home Rule Government, 2006), which implemented the use of stock-specific quotas, protecting all walruses from being hunted while hauled out on land, and adult females and calves from being hunted, except in the Qaanaaq area in North-West Greenland/Baffin Bay). The Executive Order also restricted hunting methods, so that: (1) no motorised vehicles (except marine vessels smaller than 20BRT/15BT) can be used during hunting or transport to and from hunting areas; (2) only rifles with a calibre of 30.06 or larger, only pointed full metal jacket ammunition, and no automatic or semiautomatic weapons are permitted; and (3) before a final death shot is administered, the hunter must harpoon the animal with floats attached, to prevent the animal from being lost. The Executive Order also made the reporting of catches mandatory, including any animals "struck-and-lost", although actual reporting on struck-and-lost individuals is rare (Ugarte, 2015). This Executive Order was undergoing revision at the time of writing, so some changes may be expected in the future.

Canada

Canada first introduced regulations on walrus hunting in 1928 (Canada Privy Council, 1928), restricting permission to subsistence hunting by Canadian Inuit. Further regulation was introduced in 1980 (Canada Privy Council, 1980) limiting the hunt to a maximum of four walruses per 'Indian or Inuk', except in four communities operating with annual community quotas: Coral Harbour (Salliq), Sanikiluaq, Arctic Bay and Clyde River. When this was consolidated in the Marine Mammal Regulations under the Fisheries Act in 1993, restrictions were also placed on the type of firearms permitted (e.g., with a muzzle energy of no less than 1500 foot pounds) (Government of Canada, 1993). Since 1995, a limited amount of sport hunting has been permitted through a licensed permit from the Nunavut Wildlife Management Board (NWMB) (Stewart et al., 2014a). This licensed sport hunt can only occur through reallocation of existing Indigenous quota by an individual or community, and all edible parts of any walruses killed must remain in the village (COSEWIC, 2006). Since 2010, all sport hunters have been required to provide a harvest report (including animals struck-and-lost) to Fisheries and Oceans Canada (DFO) (Frame, 2010). Subsistence hunters are required to keep a record of any walruses taken for up to 2 years (Stewart, 2002) with catches (not including struck-and-lost) reported to local hunters and trappers organisations, who then relay this information to Department of Fisheries and Oceans (DFO). The management of subsistence hunts through collaborative arrangements between national governments and indigenous communities is outlined in Box 11.1.

A new assessment of the high-Arctic and central-low Arctic populations by the Committee on the Status of Endangered Wildlife in Canada was conducted

Box 11.1 Collaborative management of subsistence hunting.

The subsistence hunts of Atlantic walruses in Greenland and Canada are managed through collaborative arrangements between Indigenous community organisations, national bodies and regulatory authorities. This collaboration is essential because effective management requires that all relevant actors and types of knowledge are involved in decision-making and monitoring processes. It is particularly important for walruses given their cultural importance, their distribution over large and remote areas and the limitations associated with currently-available scientific knowledge.

In Canada, while the Federal Government retains ultimate responsibility for wildlife management, under the terms of existing land claims agreements walrus hunts in Nunavut and Nunavik (northern Quebec) must be co-managed together with the NWMB and the Nunavik Marine Region Wildlife Board. Canada's DFO may provide advice, but it is the wildlife boards that manage and administer the hunts. Stock-specific walrus co-management working groups also exist to provide advice and develop IFMPs. These comanagement working groups include members from DFO, Regional Wildlife Organisations, local Hunters and Trappers Organisations, and Inuit elders. Furthermore, the organisations responsible for ensuring that the rights and obligations prescribed in the land use agreements are upheld in each region are also involved, these are Nunavut Tunngavik Incorporated (NTI) and Makivik Corporation.

In Greenland, catch quotas and their regional distribution are formally recommended by Greenland's Department of Fishery, Hunting and Agriculture and decided on by the cabinet of the Home Rule Government. These proposals are, however, informed by scientific advice from North Atlantic Marine Mammal Commission (NAMMCO) (see the section below), the Greenland Institute for Natural Resources, as well as an open public hearing process (Wiig et al., 2014). The set quotas are then administered by regional municipal councils following consultations with the Association of Fishers and Hunters in Greenland (KNAPK) and the national municipalities association (KANUKOKA).

It is interesting to note that for the subsistence hunt of Pacific walruses, interlocal approaches to management (i.e., on a community-to-community level) have developed between Indigenous groups in Russia and the USA (Meek et al., 2008). It has been argued that this approach can enhance the effectiveness of transborder conservation and management as it shortens the distance between the decision-making bodies and the system being managed (Meek et al., 2008). Such arrangements also have the potential to respond more quickly to changing demands than higher-order international negotiations and structures.

in 2017 (COSEWIC, 2017). This reiterated the 2006 assessment that Atlantic walruses should be listed as a species of 'special concern'. This category means that the species is sensitive to human impacts and "likely to become threatened or endangered if conservation measures are not implemented" (COSEWIC, 2019). Total allowable removals were calculated for six of the seven Canadian stocks in 2013 (Stewart and Hamilton, 2013) and the NWMB is currently in the process of establishing total allowable harvest levels, together with basic needs levels (Government of Canada, 2018). An Integrated Fisheries Management Plan (IFMP) for Walrus in Nunavut, which aims to provide direction for future management, became available in 2018 (Government of Canada, 2018) (Fig. 11.2).

Regional management advice

NAMMCO

The North Atlantic Marine Mammal Commission (NAMMCO) was established in 1992 as a regional intergovernmental body to strengthen the conservation, management and study of cetaceans (whales, dolphins and porpoises) and pinnipeds (seals and walruses) in the North Atlantic. The member countries of NAMMCO are the Faroe Islands, Greenland, Iceland and Norway. Canada and Russia have an open invitation to join the organisation and participate in meetings as observers. Both countries also regularly send expert participants to engage in NAMMCO working groups. In recent years, NTI and Makivik Corporation have also participated in NAMMCO meetings. NAMMCO acknowledges the rights of coastal communities to obtain livelihoods from the sustainable use of living marine resources; the organisation provides advice for conservation and management policy based on an approach bringing user knowledge together with the best available scientific evidence.

NAMMCO established its first ad hoc working group on walruses in 1995 to provide information on stock identity, abundance and the long-term effects of removals and environmental change (NAMMCO, 1995). This was followed 10 years later by a request from the NAMMCO Management Committee to receive updated assessments of walruses on matters of stock identity, abundance, sustainable harvest levels and priorities for research. This led to the establishment of a Walrus Working Group (WWG) that has met to develop scientific advice for policy in 2005, 2009, 2013 and 2018 (all reports available at https://nammco.no/topics/wwg_reports/). The work of the WWG led to a special issue of NAMMCO Scientific Publications (https://septentrio.uit.no/index.php/NAMMCOSP) on 'Walrus of the North Atlantic' published in 2014 (Stewart et al., 2014b), and an international symposium on the impacts of human disturbance on Arctic marine mammals (with a focus on belugas, narwhals and walruses) in 2015 (NAMMCO, 2015). There have also been focused workshops on hunting methods for seals and walruses (NAMMCO, 2004), and on the problem of struck-and-lost individuals (NAMMCO, 2006).

The NAMMCO work providing management advice for walruses in the North Atlantic includes scientific assessments from the WWG to estimate abundance, better understand threats from various human activities and recommend sustainable levels for walrus harvesting. However, NAMMCO also provides advice through its Committee on Hunting Methods for the development of responsible hunting methods that maximise animal welfare and reduce struck-and-lost rates. Greenland has clearly stated that advice from NAMMCO on walrus hunting methods significantly influenced the requirements mandated by the 2006 Executive Order (NAMMCO, 2007), and NAMMCO advice on hunting quotas is also regularly used as a basis for decision-making.

NAMMCO advice for walruses is based on the management objective of a 70% probability that the population will increase. The sustainable harvest levels for Greenland stocks (recommended as annual landed catch) are calculated by taking into account the assumed annual catches in Canada for shared stocks, as well as estimated struck-and-lost rates in the different areas; historically, such rates have varied significantly among areas, and there has been disagreement between scientists and subsistence hunting communities on what is accurate (NAMMCO, 2018a; Shadbolt et al., 2014). Establishing reliable rates is made more challenging by Canada not requiring reporting struck-and-lost rates for subsistence hunts, but also from doubts about the sufficiency of self-reporting as a method, as well as a general scepticism among hunters regarding how such information is used (Dale and Armitage, 2011). To handle this, NAMMCO applied a uniform prior from 5% to 25% above landed catches in its most recent assessment (NAMMCO, 2018a). Based on this approach, the 2018 assessment recommended an annual catch (landed animals) of *no more than* 84 walruses in the Baffin Bay area (79 for Greenland), 86 in West Greenland/South-East Baffin Island (74 for Greenland) and 17 in East Greenland (NAMMCO, 2018a). It is also worth noting that in 2013 the NAMMCO WWG saw no biological reason to deny the carry-over of unused quotas over a short-term block (NAMMCO, 2013); this was reiterated as advice in the 2018 WWG report.

In 2018, the NAMMCO Scientific Committee endorsed a recommendation from the WWG that a dialogue between managers and hunters be established to discuss the best method for obtaining reliable data on struck-and-lost individuals. It also recommended that haul-out sites in regular use be protected by an exclusion zone to safeguard them from disturbance (NAMMCO, 2018b). This was viewed as important because the animals can be easily scared into the water, and there is significant uncertainty around the factors influencing whether walruses return to a haul-out site following human disturbance. NAMMCO has also repeatedly requested that Canada provide reliable catch data.

There are a range of international agreements that are of relevance to walrus management and these are outlined in Box 11.2.

Box 11.2 International agreements relevant to walrus management.

International agreements of particular relevance to the management of trade in walrus products:

- The Convention on International Trade in Endangered Species of Wild Fauna and Flora (CITES), under which the Atlantic walrus is listed on Appendix III (for more information see Shadbolt et al., 2014 and https://www.cites.org/)
- The EU Wildlife Trade Regulation, under which the Atlantic walrus is listed on Annex B. In 2006, the Scientific Review Group formed a negative opinion on commercial imports of walrus products from Greenland (for more information see http://ec.europa.eu/environment/cites/legislation_en.htm)

International agreements of particular relevance to the management of walruses and their habitats:

- The Convention on Biological Diversity (for more information see https://www.cbd.int/)
- The Convention on the Conservation of European Wildlife and Natural Habitats (Bern Convention) (for more information see Wiig et al., 2014 and https://www.coe.int/en/web/bern-convention)
- The United Nations Convention on the Law of the Sea (for more information see Jeffries 2016 and https://www.un.org/depts/los/convention_agreements/texts/unclos/UNCLOS-TOC.htm)

Management challenges and ongoing needs

There are a range of challenges facing the effective management and conservation of Atlantic walruses. In the first instance, walruses are particularly sensitive to human disturbance at their haul-out sites (Kovacs, 2016); the rate of their return following a disturbance is unknown (NAMMCO, 2018b). Walruses also currently face multiple stressors, for example, disturbance from noise, shipping, increasing Arctic tourism, oil and gas exploration, habitat destruction and climate change (Laidre et al., 2015; Hauser et al., 2018; see chapter 12, this volume). The cumulative impact from all these different stressors on a sensitive species is difficult to assess. This is compounded by the fact that most formal environmental impact assessments are performed on a project level basis, which does not consider cumulative stress from a range of different projects or activities, either within or between countries. Furthermore, walruses may not necessarily be able to easily adapt to the range of cumulative stressors they now face given their long breeding cycle (i.e., 2–3 years) and occupy a relatively narrow ecological niche, feeding primarily on bivalves, requiring shallow areas that can support these organisms, as well as appropriate haul-out sites nearby to such feeding grounds (see chapter 3, this volume).

From a practical perspective, it is also very difficult to obtain reliable abundance estimates for walruses. This is because they are distributed over large areas in remote locations. Aerial surveys used to gather sightings data are expensive, time-consuming and ultimately incomplete (due to the inherent inability to effectively cover all areas) (see chapter 4, this volume). The challenges in estimating abundance also make monitoring the implementation and effectiveness of management measures particularly difficult. Furthermore, not having reliable information on struck-and-lost rates can make calculating sustainable catch quotas more difficult – especially since it is clear that there will be variation in struck-and-lost rates across regions, seasons and levels of hunter experience. Finally, the lack of a formal joint management scheme between Canada and Greenland for shared stocks creates a challenging situation for effective management.

To be able to continue developing effective management measures, it is important that scientists, hunting communities and policymakers work together to ensure that reliable abundance estimates can be generated and demographic data collected. It is important to ensure that catch reports are validated and that reporting on struck-and-lost individuals is improved so that estimates can be calculated by region and season as precisely as possible. It is also vital that research on the potential impacts from climate change (particularly reductions in the availability of summer sea ice), and the potential for cumulative impacts from multiple stressors, is conducted to inform assessments and management plans.

Conclusion

During the 19th and 20th centuries, walruses were heavily exploited by unregulated commercial hunting; as a result, numerous populations experienced a dramatic decline. Over the last several decades, a broad range of walrus management measures have been introduced across the North Atlantic. Although these have included widespread moratoria on commercial hunting, subsistence hunting has also been increasingly regulated through the introduction of quota systems and restrictions on hunting seasons and methods. Reporting and monitoring requirements have also been implemented and continue to evolve. This has been combined with efforts to protect habitat and regulate trade. As a result, demographic trends in heavily-regulated areas, such as Norway and Greenland, show that populations are stable or slowly increasing. Data deficiencies and threats remain, and it is important that management practices continue to improve; for example, through ongoing efforts to obtain reliable abundance estimates and struck-and-lost rates, as well as the development of formal joint management regimes for shared stocks.

References

Canada Privy Council, 1928. Order in council P.C. 1036 of June 20, 1928. Canada Gazette 61, 4227.

Canada Privy Council, 1980. Order in council P.C. 1980-1216 of May 8, 1980. SOR/80-338. Canada Gazette 114, 1860–1862.

Cook, J., Webber, J., 1784. A voyage to the Pacific Ocean. Champante and Whitrow, London.

COSEWIC, 2006. Assessment and Update Status Report on the Atlantic Walrus (*Odobenus rosmarus rosmarus*) in Canada. Committee on the Status of Endangered Wildlife in Canada, Ottawa, ON.

COSEWIC, 2017. Assessment and Status Report on the Atlantic Walrus (*Odobenus rosmarus rosmarus*), High Arctic Population, Central-Low Arctic Population and Nova Scotia-Newfoundland-Gulf of St. Lawrence Population in CANADA. Committee on the Status of Endangered Wildlife in Canada, Ottawa.

COSEWIC, 2019. Atlantic walrus (*Odobenus rosmarus rosmarus*): consultation, 2019. https://www.canada.ca/en/environment-climate-change/services/species-risk-public-registry/consultation-documents/atlantic-walrus-2019.html. (Accessed 19 July 2019).

Dale, A., Armitage, D., 2011. Marine mammal co-management in Canada's Arctic: knowledge co-production for learning and adaptive capacity. Marine Policy 35, 440–449. Available from: https://doi.org/10.1016/j.marpol.2010.10.019.

Danilov-Danilian, V.I., 2001. Red Data Book of the Russian Federation: Animals. AST and Astrel Publishers, Moscow, Russia.

Frame, S., 2010. Walrus sport hunt reporting for 2010. https://www.nwmb.com/iku/list-all-site-files/nwmb-meetings/regular-meetings/2010/66-december-7-9-2010/72-tab-03-rm-66-dfo-walrus-sport-hunt-reporting-for-2010-eng/file. (Accessed 28 May 2019).

Frédal, A., 1866. Le monde de la mer. Librairie de L. Hachette et Cie, Paris.

Gjertz, I., Wiig, Ø., 1995. The number of walruses (*Odobenus rosmarus*) in Svalbard in summer. Polar Biology 15, 527–530. Available from: https://doi.org/10.1007/BF00237468.

Gotfredsen, A.B., Appelt, M., Hastrup, K., 2018. Walrus history around the North Water: human–animal relations in a long-term perspective. Ambio 47, 193–212. Available from: https://doi.org/10.1007/s13280-018-1027-x.

Government of Canada, 1993. Marine mammal regulations SOR/93-56. https://laws-lois.justice.gc.ca/eng/regulations/SOR-93-56/index.html. (Accessed 31 May 2019).

Government of Canada, 2018. Integrated fisheries management plan – Atlantic walrus in the Nunavut settlement area. http://www.dfo-mpo.gc.ca/fm-gp/peches-fisheries/ifmp-gmp/walrus-atl-morse/walrus-nunavut-morse-eng.htm#toc1. (Accessed 28 May 2019).

Government of Norway, 1952. Fredning av hvalross (Protection of walruses), Kongelig resolusjon.

Greenland Home Rule Government, 2006. Hjemmestyrets bekendtgørelse nr. 20 af 27 Oktober 2006 om beskyttelse og fangst af hvalros (Home Rule Executive Order No. 20 of 27 October 2006 on protection and harvest of walrus).

Hauser, D.D.W., Laidre, K.L., Stern, H.L., 2018. Vulnerability of Arctic marine mammals to vessel traffic in the increasingly ice-free Northwest Passage and Northern Sea Route. Proceedings of the National Academy of Sciences of the United States of America 115 (29), 7617–7622.

Higdon, J.W., Stewart, D.B., 2018. State of Circumpolar Walrus (*Odobenus rosmarus*) Populations. WWF Arctic Programme, Ottawa, ON.

Keighley, X., Olsen, M.T., Jordan, P., 2019. Integrating cultural and biological perspectives on long-term human-walrus (*Odobenus rosmarus rosmarus*) interactions across the North Atlantic. Quaternary Research 1–21. Available from: https://doi.org/10.1017/qua.2018.150.

Kovacs, K.M., 2016. *Odobenus rosmarus ssp.rosmarus*. The IUCN red list of threatened species 2016: e.T15108A66992323. https://doi.org/10.2305/IUCN.UK.2016-1.RLTS.T15108A66992323.en. (Accessed 21 July 2020).

Kovacs, K.M., Aars, J., Lydersen, C., 2014. Walruses recovering after 60 + years of protection in Svalbard, Norway. Polar Research 33, 26034. Available from: https://doi.org/10.3402/polar.v33.26034.

Laidre, K.L., Stern, H., Kovacs, K.M., Lowry, L., Moore, S.E., Regehr, E.V., et al., 2015. Arctic marine mammal population status, sea ice habitat loss, and conservation recommendations for the 21st century. Conservation Biology 29 (3), 724–737.

Meek, C.L., Lovecraft, A.L., Robards, M.D., Kofinas, G.P., 2008. Building resilience through interlocal relations: case studies of polar bear and walrus management in the Bering Strait. Marine Policy 32 (6), 1080–1089.

NAMMCO, 1995. Report of the ad hoc working group on atlantic walrus. https://nammco.no/wp-content/uploads/2019/04/report_wwg_1995.pdf. (Accessed 29 May 2019).

NAMMCO, 2004. Report of the NAMMCO workshop on hunting methods for seals and walrus. http://nammco.wpengine.com/wp-content/uploads/2016/09/Report-from-the-Workshop-on-Hunting-Methods-for-Seals-and-Walrus-2004.pdf.

NAMMCO, 2006. Report of the NAMMCO workshop to address the problems of "struck and lost" in seal, walrus and whale hunting. http://nammco.wpengine.com/wp-content/uploads/2016/09/Report-NAMMCO-Workshop-on-Struck-and-Lost-2006.pdf. (Accessed 31 May 2019).

NAMMCO, 2007. Report of the committee on hunting methods. http://nammco.wpengine.com/wp-content/uploads/2016/09/committee-on-hunting-methods-report-january-2007.pdf. (Accessed 31 May 2019).

NAMMCO, 2013. Stock status of walrus in Greenland. http://nammco.wpengine.com/wp-content/uploads/2016/09/NAMMCO-WWG-2013-Report-FINAL.pdf. (Accessed 29 May 2019).

NAMMCO, 2015. Report from the NAMMCO symposium on the impacts of human disturbance on Arctic marine mammals, with a focus on belugas, narwhal and walrus. http://nammco.wpengine.com/wp-content/uploads/2017/03/report-nammco-disturbance-symposium-2015.pdf. (Accessed 29 May 2019).

NAMMCO, 2018a. Report of the NAMMCO Scientific Committee Working Group on walrus. https://nammco.no/topics/sc-working-group-reports/. (Accessed 29 May 2019).

NAMMCO, 2018b. Report of the 25th scientific committee meeting. https://nammco.no/wp-content/uploads/2017/01/sc-report-2018.final_complete-1.pdf. (Accessed 31 May 2019).

Shadbolt, T., Arnbom, T., Cooper, E.W.T., 2014. Hauling Out: International Trade and Management of Walrus. TRAFFIC and WWF-Canada, Vancouver, BC.

Stewart, R.E.A., 2002. Review of Atlantic Walrus (*Odobenus rosmarus rosmarus*) in Canada. Fisheries and Oceans Canada, Winnipeg.

Stewart, R.E.A., Hamilton, J.W., 2013. Estimating total allowable removals for walrus (*Odobenus rosmarus rosmarus*) in Nunavut using the potential biological removal approach. DFO Can. Sci. Advis. Sec. Res. Doc. 2013/031. iv + 13p.

Stewart, D.B., Higdon, J., Reeves, R.R., Stewart, R.E.A., 2014a. A catch history for Atlantic walruses (*Odobenus rosmarus rosmarus*) in the eastern Canadian Arctic. NAMMCO Scientific Publications 9, 219–314. Available from: https://doi.org/10.7557/3.3065.

Stewart, R.E.A., Kovacs, K.M., Acquarone, M. (Eds.), 2014b. Walrus of the North Atlantic. NAMMCO Scientific Publications.

Ugarte, F., 2015. Third Standing Non-detriment Findings for Exports from Greenland of Products Derived from Atlantic Walrus (Odobenus rosmarus rosmarus) (No. J. nr. 40.00.01.01.45-1/14). Grønlands Naturinstitut, Nuuk.

Wiig, Ø., Born, E.W., Stewart, R.E.A., 2014. Management of Atlantic walrus (*Odobenus rosmarus rosmarus*) in the Arctic Atlantic. NAMMCO Scientific Publications 9, 315–342. Available from: https://doi.org/10.7557/3.2855.

Witting, L., Born, E.W., 2005. An assessment of Greenland walrus populations. ICES Journal of Marine Science 62, 266–284. Available from: https://doi.org/10.1016/j.icesjms.2004.11.001.

CHAPTER

Anthropogenic impacts on the Atlantic walrus

12

Erik W. Born[1], Øystein Wiig[2] and Morten Tange Olsen[3]

[1]Greenland Institute of Natural Resources, Nuuk, Greenland

[2]Natural History Museum, University of Oslo, Oslo, Norway

[3]Globe Institute, University of Copenhagen, Copenhagen, Denmark

Chapter Outline

The Atlantic Walrus. DOI: https://doi.org/10.1016/B978-0-12-817430-2.00013-3

Introduction

The walrus is regarded as a Focal Ecosystem Component (FEC) in the assessments by the Circumpolar Biodiversity Monitoring Program (CAFF, 2017). An FEC is a biological element that is (1) considered central to the functioning of an ecosystem, (2) major importance to Arctic residents and/or (3) likely to be a good proxy of natural and anthropogenic changes in the environment (CAFF, 2017). In this chapter, we evaluate the impacts of various human activities on Atlantic walruses (*Odobenus rosmarus rosmarus*). We summarise the current information on direct and indirect effects of hunting, fishery, shipping activities, exploration and exploitation of nonrenewable resources, pollution and tourism and recreation (the effects on walruses of global warming and decrease in sea ice are discussed in Chapter 13, this volume). Where relevant information is not specifically available for the Atlantic subspecies, we rely on information obtained from studies of the closely related Pacific walrus (*Odobenus rosmarus divergens*).

Prior to our assessment, several studies have reviewed the effects of anthropogenic activities on marine mammals in general (e.g., Wiig et al., 1996; Isaksen et al., 1998; Laidre et al., 2008, 2015; Moore et al., 2012; NAMMCO, 2015; Boertmann et al., 2021; Boertmann and Mosbech, 2021) and walruses in specific (Born et al., 1995; Born, 2005; Stewart, 2002; Boltunov et al., 2010; Garlich-Miller et al., 2011; Kovacs et al., 2015; Kovacs, 2016; COSEWIC, 2017; Higdon and Stewart, 2018; DFO, 2019; NAMMCO, 2020). Our assessment is based on these reviews, a long list of scientific studies on the effects of specific disturbances on walruses and our personal observations.

The walrus' vulnerability to disturbance

Multiple features of walrus behaviour and ecology make them vulnerable to the impacts of human activities (Born et al., 1995; see also Chapter 3, this volume). Walruses have a narrow ecological niche, feeding primarily on sessile benthic invertebrates, and in particular, bivalves that are distributed in shallow coastal waters (100 m or less), implying that any disturbance to their crucial feeding banks will affect the walruses. In winter, when a dense ice cover excludes them from many coastal and inshore feeding areas, walruses depend on the predictable presence of open-water areas. Therefore, walruses typically winter in polynyas (i.e., areas with moving ice and/or open water surrounded by dense ice during winter) with suitable feeding banks, such as Foxe Basin in Canada, and the North Water and Northeast Water in Greenland. Any disturbance to these areas may negatively affect walruses. Finally, walruses are highly gregarious and rely on the presence of terrestrial haul-outs relatively close to their feeding grounds. These haul-out sites (Inuktitut: *uglit*) are used in the summer ice-free 'open-water' period when walruses feed intensively. Walrus gregariousness and dependence on *uglit* made it possible for European and American whalers and sealers to predict seasonal concentrations and exploit walruses intensively (Born et al., 1995) and generally make walruses susceptible to other human activities (e.g., tourism, shipping, exploration) causing disturbance at *uglit*.

The intensity of an animal's reaction to anthropogenic disturbance can be highly variable, and in some cases, walruses may even be attracted by human activities (Fay et al., 1984a,b; Anon., 1993; Born, personal observation). Moreover, whereas some anthropogenic effects on walrus are readily visible, others are subtler and less apparent. For example, hunting and boat collisions have directly visible and typically lethal effect, and when large groups of walruses haul out on land, fleeing from a disturbance such as noise or human presence may take the form of a stampede, during which young animals may be trampled. Such stampedes can be the cause of significant calf and juvenile mortality due to trampling, as well as abortion of foetuses and separation of cow–calf pairs (Born et al., 1995; Udevitz et al., 2013; COSEWIC, 2017; DFO, 2019). However, human activities may also have more subtle, nonvisible, and sublethal effects, altering the behaviour or physiology of walruses and hence acting in the long term to reduce the fitness of individuals and populations (Pirotta et al., 2018). For example, noise may cause social disruptions by masking the walrus' underwater or in-air communication. Moreover, if walruses respond to disturbances by shifting of body position or fleeing, this may result in increased stress and energy expenditure, insufficient nursing of calves and reduced resting periods on land or sea ice, which may disturb natural moulting and wound-healing processes and impair thermoregulation (Born et al., 1995; Wiig et al., 1996; Moore et al., 2012; Stewart et al., 2012). The behavioural response to disturbance often depends on (1) the sex and age of the animal, (2) the size and location of its group (on ice, in water or on land), (3) the distance from the source of disturbance, (4) the type and intensity of the disturbing factor, (5) the animal's previous experience with disturbance, (6) weather conditions and (7) the animal's behavioural state prior to disturbance. For instance, in cold, windy and rainy conditions, walruses are usually more alert than on warm, calm and sunny days (Born, personal observation). Furthermore, on ice or on shore, male walruses tend to be less shy than females, and solitary individuals tend to be less vigilant than groups. Groups consisting of large males appear to be less sensitive to disturbance than groups of females with calves (Born et al., 1995, and references therein; Øren et al., 2018) (Fig. 12.1).

Historical and current hunting

Historical commercial hunting

The most direct and significant anthropogenic impact on Atlantic walruses is hunting. Historically, the Atlantic walrus was subject to intense commercial hunting for centuries: first by the Norse in northern Norway, Iceland and western Greenland (Magnus, 1555; Roesdahl, 2003; Star et al., 2018; Keighley et al., 2019) and later by European and North American sealers and whalers throughout

FIGURE 12.1

In some cases, walruses are attracted to humans. On 27 September 2008 an adult male walrus visited the military detachment Daneborg in north-eastern Greenland. This particular walrus was tracked by use of satellite telemetry for three seasons, and was identified on a haul-out about 5 km from Daneborg over nine summers from 1999–2010 (Born and Acquarone, 2007; E.W. Born, unpubl. data). It had likely habituated to human activity.

Photo: Morten H. Gormsen.

the subspecies' range (Born et al., 1995, and references therein). Exploitation by the Norse led to the extinction of the Icelandic walrus population (Star et al., 2018; Keighley et al., 2019), and a group of walruses in the Canadian Maritimes (the eastern Canadian provinces of New Brunswick, Nova Scotia, and Prince Edward Island), as well as around Newfoundland and the Gulf of St. Lawrence, was hunted to extinction by the mid-1800's (McLeod et al., 2014; COSEWIC, 2017). In the Svalbard archipelago commercial hunting from the early 17th century until protection in 1952 resulted in the local walrus population nearly becoming extinct (Gjertz and Wiig, 1994). In all areas, the number of walruses harvested through commercial operations was significant. For example a total of about 15,000 Atlantic walruses were taken from 1897 to 1956 around Franz Josef Land among a group of animals numbering between around 6000 and 12,500 at the start of the hunting era (Gjertz et al., 1998). Today, most subpopulations of Atlantic

FIGURE 12.2

A walrus 'cemetery' in Svalbard. For centuries, European sealers and whalers hunted walruses on land in the Svalbard and Franz Josef Land archipelagos bringing the subpopulation in these areas almost to extinction before it became completely protected in 1952 (Norway) and 1956 (Russia), respectively (Wiig et al., 2014).

Photo: E.W. Born.

walrus remain below historical levels (see Chapter 3, this volume). The commercial hunt was primarily for products such as ivory, hide and blubber (Born et al., 1995, and references therein; see also Chapter 9, this volume). This hunt ceased in most areas from the end of the 19th to the first half of the 20th century as the international demands for such products dwindled significantly (Reeves, 1978; Born et al., 1995) (Fig. 12.2).

Traditional and contemporary subsistence hunting

Nowadays, there is a limited hunt of Atlantic walrus in Greenland and Canada for subsistence and (in only Canada) sport purposes, whereas they are completely protected in Svalbard and in the Russian Arctic (NAMMCO, 2006, 2018a, 2020; Wiig et al., 2014; COSEWIC, 2017). The walrus has always been important to the Indigenous resource economy; in areas such as northern Baffin Bay and Foxe Basin, the species has been critical for subsistence for millennia (Anderson and Garlich-Miller, 1994; Born et al., 1987; Godtfredsen et al., 2018;

Chapter 5 and 6, this volume). Traditionally, walruses were taken in the open water where kayaks were used, or from ice edges or during 'thin ice hunting' in winter (Born, 1987; Born et al., 1995, 2017). However, from the beginning of the 20th century, walruses have mainly been taken from boats during the open-water season (Orr et al., 1986; Anderson and Garlich-Miller, 1994; Born et al., 1995). Today, modern hunting technology with long-range boats allows hunters access to nearly all the habitats occupied by walruses, and firearms increase the distance at which animals can be killed (Born et al., 1995, 2017; NAMMCO, 2004). Still, the present-day catch of Atlantic walrus for subsistence purpose in Canada and Greenland is far below historical levels (e.g., COSEWIC, 2017; NAMMCO, 2018a,b, 2020).

In addition to the obvious direct effect on walruses, hunting may also in some situations have indirect adverse effects in the form of behavioural reactions. Hunted populations in Northwest Greenland are generally more skittish than populations at Svalbard or in Northeast Greenland not subject to regular hunt (Born et al., 2017; Born, personal observations), and walruses in hunted populations generally behave more calmly during years without a hunt than in years with a hunt (Malme et al., 1989; Wartzok et al., 2003). In some areas where walruses have been hunted while hauling out on land, they have permanently deserted their traditional *uglit* (e.g., Born et al., 1994; Born et al., 1997; COSEWIC, 2017); in other areas, walruses may temporarily leave their summer feeding grounds due to hunting. For example in the Qaanaaq region of northwest Greenland, walruses leave their local winter foraging grounds when the open-water hunt with motorised vessels commences in spring and move westward across Smith Sound to the central east coast of Ellesmere Island where no hunting is carried out. Typically, walruses do not return to the area until fall, when the formation of new sea ice prevents hunting by boat (Born, 1987; Born et al., 1995, 2017) (Fig. 12.3).

Quotas and signs of recovery

Greenland introduced country-wide quotas for the take of walruses in 2007 (Wiig et al., 2014). In Canada, a small number of communities have annual quotas, whereas individual Inuit in all other areas may take up to four walruses annually (NAMMCO, 2018a). In Greenland, the average annual catch was c. 117 walruses for the period 2007–2019, and in Canada, the average annual catch was c. 215 for the period 2007–2015, including sport hunts which constituted an annual average of c. 4% during the same period. Both the Greenlandic and Canadian numbers are markedly below catches in the preceding decades (Witting and Born, 2014; COSEWIC, 2017; NAMMCO, 2018a, 2020).

Under favourable conditions, a walrus population may increase 7%–8% per year (Kovacs et al., 2014; Witting and Born, 2014, and references therein). The current annual subsistence catch of around 330 walruses or 400 if 'not well-documented' losses are included (NAMMCO, 2004; Witting and Born, 2014) amounts

FIGURE 12.3

Hunters in a small skiff approaching a group of walruses in northwest Greenland. Much of the hunting of walrus in Greenland and Canada is carried out from small boats during summer (Orr et al., 1986; Born et al., 2017).

Photo: E.W. Born.

to 1.5%–2% of the total population of Atlantic walrus in Greenland and Canada (20,000–22,000 walruses; COSEWIC, 2017; NAMMCO, 2020). Hence, there are no indications that current catch levels are unsustainable. Furthermore, with protective measures and regulations of the hunt in place and with national and international regulation of trade of walrus products (Wiig et al., 2014), the subspecies does not appear to be directly threatened by the current subsistence hunt.

Recently, there have been several encouraging examples of Atlantic walrus population increases and recolonisation of former habitats. Since their complete protection in 1952, walruses around Svalbard have increased in range and numbers (Gjertz and Wiig, 1994; Kovacs et al., 2014). In the period 2006–2012 the number of walruses in the archipelago increased by about around 8% per year (Kovacs et al., 2014), and in the subsequent period (2012–2018) the overall increase was around 48% (NAMMCO, 2019). Furthermore, while the group of walruses at Svalbard was previously dominated by males (Gjertz and Wiig, 1995), it now includes an increasing proportion of females and calves (Kovacs et al., 2014). Walruses have also gradually re-occupied their former summering habitats in northeast Greenland after legal protections were introduced in the

1950's (Born et al., 1997; Born, 2020). Similarly, during the ice-free season in northwest Greenland, walruses are increasingly being observed in fjords, such as Inglefield Bredning and Wolstenholme Fjord (Born et al., 2017; Andersen et al., 2018; Orbicon, 2020), which had otherwise been devoid of walruses for decades (Born et al., 1995). In recent years, a substantial increase has been observed in number of walruses at haul-outs at the Yamal Peninsula in the Russian Arctic (Gebruk et al., 2020; Vasilyeva, 2020). We suggest that these changes in distribution and numbers may reflect a combination of walrus population size increases and increased use of terrestrial haul-outs due to a reduced hunting pressure and a reduction in sea ice.

Walrus–fishery interactions

Since the mid-20th century, there has been a substantial commercial fishery in several areas occupied by Atlantic walruses, such as along eastern Baffin Island in Canada, along West Greenland, and in the Barents Sea region (Born et al., 1995; Born, 2005; Blomeyer et al., 2015; Haug et al., 2017; NOEP, 2021). Global warming and associated decreases in sea ice will likely result in increased productivity in Arctic seas, establishment of new fishing grounds, and increasing traffic of fishing vessels (Blomeyer et al., 2015; Haug et al., 2017; PAME, 2021), potentially increasing fishery–walrus interactions. Fishing activities may have negative effects on walrus populations, either via direct mortality (e.g., ship strikes and entanglements) or indirectly via competition for benthic food resources, damaging feeding habitat (e.g., by trawling and dredging) and resulting in displacement caused by in-air and underwater noise (see the following section). However, information on the effects of fishing activities on walruses and their habitat is limited geographically and mostly lacks details – in particular, for the Atlantic walrus.

Direct mortality in bycatch

Pacific walruses are sometimes taken accidentally in trawls (Wickens, 1995, and references therein). However, incidental fishery-related mortalities have been as low as 1–3 per year in Alaska since the mid-2000s, and serious injury from fishery–walrus interactions is generally considered insignificant in the Pacific (Wilson and Evans, 2009; US FWS, 2012; FWS, 2014; Higdon and Stewart, 2018). Likewise, Wickens (1995) mentions only a single instance where an Atlantic walrus was caught in an inshore gill net in the Newfoundland–Labrador region in 1990, and we are not aware of other cases for Atlantic walrus more recently. Thus, bycatch in fishing gear are likely very rare events given the walruses' agility in water and their swimming speed of up to 35 km/h (Fay, 1981).

Competition for food–'The Icelandic scallop case'

Walruses feed primarily on benthic bivalves (see Chapter 3, this volume) including to some extent Icelandic scallop, *Chlamys islandica* (Mansfield, 1958; Gjertz and Wiig, 1992; Skoglund et al., 2010; Born et al., 2017). The benthic bivalves that make up the majority of the walrus diet are typically buried in the upper layer of soft-sediment (mud) seafloors with strong tidal currents (Vibe, 1950; Fay, 1982). In contrast, Icelandic scallop is usually found on hard substrate such as shell gravel, gravel, stones, rocks and sand (Garcia, 2006; Garcia et al., 2006; GINR, 2018a, 2020a), where walruses occur more rarely. Still, given the occasional foraging on Icelandic scallop, it has been suspected that commercial dredging for this species competes with walruses for their benthic food items (Born et al., 1995; COSEWIC, 2006, 2017; NAMMCO, 2006; Born et al., 2017).

In Svalbard, the southern Barents Sea and the White Sea, Canada and West Greenland, concentrations of Icelandic scallop are found close to the coast at 20–100 m water depth (Pedersen, 1994; Born et al., 1995; Zolotarev, 2002; GINR, 2018a). Around Svalbard, there was an active fishery for Icelandic scallops from the mid-1980's until the collapse in 1992 and eventual closure in 1995 (Born et al., 1995; Garcia, 2006; Kedra et al., 2013). Exploitation of Icelandic scallops by Russians in the southern Barents Sea and White Sea area began in the late 1980's. The main fishing ground was in the entrance to the White Sea (Zolotarev, 2002; Manushin and Blinova, 2019), an area where Atlantic walruses are known to be frequent (e.g., Higdon and Stewart, 2018). However, overexploitation with up to 14,000 t/year, landed in the 1990's, led to a collapse of the scallop stock around 2009. Consequently, the Russian fishing for scallops in the Barents Sea has been banned since 2018 (Manushin and Blinova, 2019). There has also been some dredging for scallops in Canada, but this activity is widely considered uneconomical and was never initiated on a large scale (Stewart et al., 1993; Lambert and Préfontaine, 1995; COSEWIC (Committee on the Status of Endangered Wildlife in Canada), 2006, 2017). To our knowledge, there are currently no scallop fisheries in areas in Canada where walruses occur regularly.

In West Greenland, there has been dredging for Icelandic scallop during the open-water season since 1983, with catches of around 2000 t/year since 1995 (Pedersen, 1994; Garcia, 2006). The North Atlantic Marine Mammal Commission (NAMMCO) has repeatedly expressed concern that scallop dredging in West Greenland may negatively affect walruses (see NAMMCO, 2006, 2018b). To this end, an interview survey among experienced walrus hunters in West Greenland in 2010 indicated that there could be a direct competition between the walruses and the fishery Icelandic scallop and that walruses may have changed distribution due to impacts from the commercial fishery (Born et al., 2017). However, the relatively small size of the West Greenland scallop fishery indicates that it has had no major effect on the stock of walruses wintering off western Greenland; dredging for scallop is confined to the open-water season when walruses are absent from West Greenland (Born et al., 1994; Dietz et al., 2014), and the scallop beds are mainly located in inshore areas (Garcia, 2006) where walruses rarely occur (Born et al., 2017). Hence, we concur with the conclusions of COSEWIC (2017) and Born et al. (2017) that direct competition for food between walruses and dredging of Icelandic scallop is likely not a substantial concern.

Fishery-induced degradation of walrus foraging habitats

The commercial fishery in the eastern Canadian Arctic, West Greenland and the Svalbard–Barents Sea region for demersal species such as Atlantic cod (*Gadus morhua*), Greenland halibut (*Reinhardtius hippoglossoides*), Northern shrimp (*Pandalus borealis*) and to some extent Icelandic scallop is of particular relevance when considering the putative effects of fishing activities on walrus and their habitats (Born et al., 1995; Blomeyer et al., 2015; Haug et al., 2017; DFO, 2018, 2020). These fisheries involve bottom-trawling (beam and otter trawls), or dredging, which may have a substantial negative impact on the sea bottom and associated fauna by increasing mortality of target and nontarget species, in addition to altering the abiotic environment (Anon, 1990; Messieh et al., 1991; Eleftheriou and Robertson, 1992; Garcia, 2006). Given the walrus' reliance on benthic bivalves for food, any changes to the sea bottom and its fauna may also negatively affect the walrus.

The impacts of bottom-trawling and scallop dredging on benthic habitats and communities depend on the type of fishing gear and seafloor substrate, as well as whether epifauna (living on the surface of the seabed) or infauna (buried in the sediment) and which taxa are considered. However, generally, infauna has a better chance of withstanding trawling and dredging than epifauna (Jones, 1992; Collie et al., 2000; Løkkeborg, 2004; Garcia, 2006; Garcia et al., 2006; ICES, 2009). Benthic macrofauna may suffer substantial mortality or damage, and it may take decades for hard substrate benthic communities to recover after impacts from the use of dredges and bottom-trawls (Kedra et al., 2013; Yesson et al., 2016). The killing or injuring of animals at the surface and uppermost layers of the sediment is the most notable impact in soft bottoms (Tillin et al., 2006; McConnaughey et al., 2000, 2005; Garcia, 2006; Garcia et al., 2006). According to Eleftheriou and Robertson (1992) and Garcia (2006), scallop dredging may have a negative impact on the sea bottom and associated fauna, and according to Anon (1990), the nonselective dredging technique that was used at Svalbard radically altered the seafloor. In a metaanalysis of 39 published studies of the impact of fishing on shelf-sea benthos, Collie et al. (2000) concluded that scallop dredging has the greatest initial effects on benthic biota, while trawling has less effect. Trawling seems to affect the benthic assemblage mainly through resuspension of surface sediment and through relocation of shallow burrowing infauna species to the surface of the seafloor (ICES, 2009). However, shallow-water benthic communities may recover relatively quickly after having been dredged. Twenty years after cessation of heavy fishing for Icelandic scallop, the benthic communities on the Svalbard Bank showed the signs of recovery from the damage they sustained

(Kedra et al., 2013). An experimental study to detect effects of bottom-trawling found that in general, the commercial trawl did not significantly affect the biomass of benthic invertebrates in a sandy seafloor and suggested that storm events may have an overall greater effect on the benthos than do bottom trawls (McConnaughey and Syrjala, 2014).

We find that there is still surprisingly little research-based information about the potential negative effects of commercial fishing activities on the foraging habitats of Atlantic walruses, and the interactions between fisheries and walruses may be more complex than is generally acknowledged. Hence, we caution that no firm conclusions can be drawn on the matter. However, the fact that the current major demersal fisheries (i.e., those of Atlantic cod, Northern shrimp, and Greenland halibut) are situated at greater depths (200–1500 m) than those at which Atlantic walruses usually forage and the fact that the preferred walrus food items are borrowed in sandy substrates lead us to suggest that the indirect effects on Atlantic walrus from degradation of their foraging habitats caused by commercial fishing are likely limited. There might, however, be one exception. On the important walrus wintering grounds in West Greenland, there is direct spatial and temporal overlap between fishing activity and walrus distribution. The West Greenland situation represents one of the few cases – if not, the *only* case – in which intensive commercial fishery for ground fish and shrimp has directly overlapped with important walrus foraging grounds for at least 100 years (Mattox, 1973; Born et al., 1994; Hamilton et al., 2003). During an interview survey in West Greenland, some experienced Inuit subsistence walrus hunters mentioned that walruses may also have changed distribution due to noise from fishery vessels (Born et al., 2017), causing displacement from walrus foraging areas (NAMMCO, 2015). Hence, for several years, the West Greenland walrus–fishery interaction has been a special focus of NAMMCO (e.g., NAMMCO, 1995, 2006, 2018a) and its scientific Working Group on Walrus, which identifies fisheries' interactions among the major stressors on Atlantic walrus populations (NAMMCO, 2018a,b). At the close of this chapter, we have described the situation in West Greenland in greater detail in Box 12.1.

Walruses are sensitive to noise

The ambient noise environment in the Arctic is complex and highly variable due to sounds generated by the seasonally-variable ice cover. However, sounds from vessel traffic and activities related to offshore marine exploration and exploitation also have profound effects on sound levels (see AMSA, 2009; Boertmann and Mosbech, 2021). Walruses produce a variety of sounds, such as in-air grunts, rasps, clicks and underwater songs, which are used for social interactions, including mating and mother–calf contact. These sounds range in frequency from around 0.1 to 4–8 kHz (see Box 12.2) and can overlap with the frequencies of noise from shipping, seismic surveys, drilling and aircraft support (Kastelein et al., 1993a, 1996, 2002; Gordon et al., 2003; Erbe et al., 2016; PAME, 2019; Reichmuth et al., 2020).

This overlap in frequencies is of particular concern if the sound levels perceived by the walrus are above 120 dB re 1 μPa rms (root mean squared) (Lawson and Lesage, 2012; Shell, 2015; Erbe et al., 2019, and references therein). However, sources describing reactions of walruses to ships and boats often do not report details on vessel size, engine power and speed, which makes it difficult to interpret and categorise behavioural responses in relation to shipping activity. Furthermore, to our knowledge, there have been no systematic, quantitative studies of sound levels (or types of sounds) that may elicit disturbance response in walrus in the wild.

Potential negative effects of commercial shipping activities

Shipping activities in the Arctic

The Arctic is experiencing increased vessel traffic in the form of icebreakers, bulk transporters, fishing vessels, cruise ships and supply vessels servicing nonrenewable resource exploration and exploitation (e.g., oil rigs, pipelines and mines). Obviously, such ongoing and projected vessel traffic raise great concerns about potential negative effects – at the population or subpopulation level – on Arctic marine mammals, including the two subspecies of walrus (see AMSA, 2009; Anon., 2012; Melia et al., 2016; Hauser et al., 2018; PAME, 2019, 2020; Wilson et al., 2020; Boertmann and Mosbech, 2021). For decades, there has been intensive shipping activity in areas occupied by walruses in the western Russian Arctic and the Barents Sea region (Timoshenko, 1984; Wiig et al., 1996; Boltunov et al., 2010; Wilson et al., 2020). In Greenland and Canada, ship traffic has generally increased in recent years, and the essential walrus winter foraging banks in the area have been identified as areas where heightened awareness in relation to the impacts of shipping is needed (Yurkowski et al., 2018; Christensen et al., 2018). In Foxe Basin and Hudson Strait, mining-related shipping and development of ports and other major infrastructure are likely to increase, and these have been identified as areas in which walrus habitat and future industrial activity will overlap (COSEWIC, 2017; NAMMCO, 2018a; WWF, 2019; Stewart et al., 2020).

Shipping activities may affect walruses in several ways, including through (1) noise from engines and breaking of ice, which may cause behavioural reactions and masking of social sounds; (2) risk of direct collisions between ships and walruses; (3) introduction of invasive species via ballast water (Christensen et al., 2015); (4) contamination of walrus benthic prey (Garlich-Miller et al., 2011) and (5) emission of CO_2 and 'black carbon' into the atmosphere and consequently on the sea-ice surface, which will increase the absorption of energy from sunlight, thereby enhancing the sea-ice melt (Anon., 2012; Lawson and Lesage, 2012; Garlich-Miller et al., 2011; AMSA, 2009; Boertmann and Mosbech, 2021). The following section highlights the walrus' sensitivity to noise and the response to icebreaking and commercial bulk carriers (Fig. 12.4).

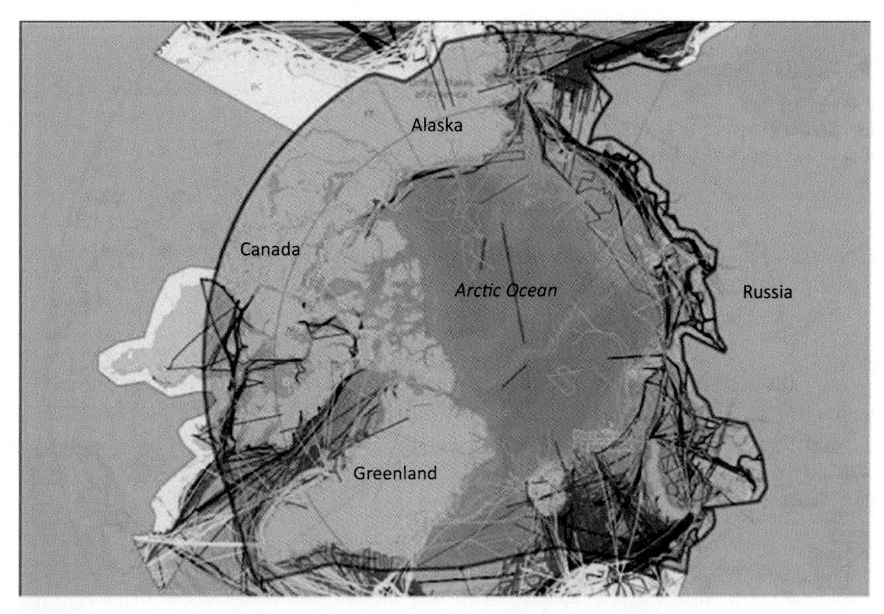

FIGURE 12.4

The intensity of vessel traffic in the Arctic illustrated by ship tracks during September 2019. The thick blue line delineates Arctic areas as defined by the IMO Polar Code to provide for safe ship operation in polar waters (PAME, 2020). IMO, International Maritime Organization.

Disturbance from icebreakers and larger vessels

To force their way through sea ice, large icebreaking vessels often must make accelerations, turns and reversals of direction, which in addition to engine noise, propeller-cavitation sound, generators and deck activities, can cause substantial underwater and/or airborne noise (Garlich-Miller et al., 2011; Lawson and Lesage, 2012, and references therein). Walruses react to icebreaker activities, but responses vary according to the intensity of the ice-breaking (Brueggeman et al., 1990, 1993), and Inuit have confirmed that walruses are frightened by large ships, although their reactions are variable (DFO, 2019). For instance, Pacific walruses hauled out on sea ice have been reported to react to icebreaking activities when only c. 2 km away (Fay et al., 1984a). Females with young usually went into the water when the ship was 500–1000 m away, and when it was 100–300 m away, male walruses went into the water. Apparently, the animals' reactions were triggered both by sound and sight and, according to Fay et al. (1984a), more by the sounds of the ice-breaking than by engine noise. In the Chukchi Sea, Pacific walruses have been reported to move 20–25 km away from the icebreaking operations deeper into the pack ice. Once the ice breaking activities stopped, the walruses began returning to previously-occupied areas. Under these circumstances, walruses displayed some behavioural responses that rapidly decreased at distances beyond 400–500 m from the icebreaker (Brueggeman et al., 1990, 1993). A study on walruses at Round Island in Bristol Bay, Alaska, found that larger ships that remained outside of a c. 4.8 km (3 miles) restriction zone generally did not cause noticeable reactions (DFO, 2019). Another study on the behaviour of Pacific walruses in water and on ice at various distances found that diving and changing swimming course or speed occurred primarily between 300 and 500 m of the vessel (McFarland and Aerts, 2015; McFarland et al., 2017), and in yet another study, about 600–800 m has been proposed as a safe distance to walruses for ice-breaking vessels (Wilson, Goodmann, 2016).

In addition to triggering escape behaviour, shipping noise may affect walrus social interactions, reproduction and mother–calf bonding. Walruses haul out on floes of sea ice and vocalize intensively underwater during the mating season from January–April when there is sea-ice cover. Fay et al. (1984a) assumed that inhibition of communication could take place if the noise level became high enough, thus preventing individuals from locating one another, as well as 'masking' vocalisations of displaying bulls, thereby preventing them from attracting females in oestrous. In addition, walruses are characterised by very strong mother–calf bonds, and the calf usually remains with its mother wherever she goes for up to two years (see Fay, 1982). Calves have been reported to be abandoned by females in several cases during ice-breaking activities (Fay et al., 1984a,b). However, mother–calf recognition is well developed in walruses (Insley et al., 2003; Charrier et al., 2009), and research on captive walruses has indicated that females are able to discriminate the voice of their own calf from that of other calves (Charrier et al., 2009). Hence, we suggest that permanent separation between mother and calf from such activities is uncommon. Wilson et al. (2020) conducted an analysis of the risk of collisions (and mother–calf separation) due to shipping during the spring (March–May) breeding season of Arctic ice-associated pinnipeds, including Atlantic walruses. The authors identified the Pechora Sea as an area of strong overlap between industrial shipping activities and walrus breeding habitat. However, no data were available for the size or speed of vessels; this makes it difficult to quantify levels of impact. More studies are needed on vessel disturbance and collision risks in walrus breeding habitats.

Future impacts of large-scale trans-Arctic shipping routes?

It has been estimated that by the mid-21st century, the duration of navigable periods for Arctic transit shipping will have doubled for standard open-water vessels, with routes across the central Arctic becoming available for moderately ice-strengthened vessels for 10–12 months per year by the late 21st century (Melia et al., 2016). Ongoing or planned Arctic shipping routes all overlap with walrus habitat, including the Northeast Passage (or Northern Sea Route) north of Scandinavia and Russia, the Northwest Passage through the Canadian High Arctic, the Transarctic Sea Route from Alaska to Europe and the Arctic Bridge route from the Barents Sea–Pechora Sea area to Arctic Canada.

Hauser et al. (2018) analysed the relative vulnerability of various marine mammals to potential disturbances associated with shipping via the Northwest Passage and the Northern Sea Route during the ice-free season. Generally, vulnerability was relatively high among the five walrus subpopulations studied (the Pacific walrus in the Chukchi Sea and Laptev Seas, respectively, and three Atlantic walrus subpopulations; Hauser et al., 2018). Among Atlantic walrus subpopulations, the 'W Greenland–SE Baffin Island' and 'Lancaster Sound–Penny Strait' subpopulations will potentially be affected by the Northwest Passage Route, whereas the Northern Sea Route may affect walruses in the 'White Sea–Barents Sea–Novaya Zemlya–Pechora Sea–Kara Sea' subpopulation 'complex'. Hauser et al. (2018) estimated that 42% of the individuals in the walrus subpopulations in their analysis may be exposed to ship traffic. We may add that in terms of the number of individuals *potentially* affected, the effects of shipping on Pacific walruses will likely be far greater than on the three Atlantic walrus subpopulations, which in total number only about 10% of the Pacific walruses (Laidre et al., 2015; NAMMCO, 2020). However, due to their smaller size, any negative impact from shipping may be relatively larger for the subpopulations of Atlantic walrus.

Reaction to aircraft

Walruses are susceptible to disturbance by aircraft – in particular, while hauling out on land (Born et al., 1995, and references therein). The most common reactions of walruses to aircraft noise are head-lift, orientations towards the water and escape into the water of some or all individuals in a group (Salter, 1979). However, stampedes of walruses on land induced by aircraft disturbance resulting in death of several calves have been reported in the eastern Russian Arctic (see Tomilin and Kibal'chich, 1975); in Alaska, walruses stampeded as a result of airliners flying overhead at approximately 30,000 ft (c. 10,000 m) (Garlich-Miller et al., 2011). With decreasing sea ice, Pacific walruses will likely haul out on land in still larger concentrations. Hence, in the future, stampedes caused by disturbance may increasingly be a mortality factor (Udevitz et al., 2013). This may also be the case for Atlantic walruses, large herds of which have been reported on land in recent years (Stewart et al., 2014; Vasilyeva, 2020).

Walrus reactions to aircraft noise depend on age and sex, group size and behaviour, and also vary with aircraft type, range, flight pattern, altitude and environmental factors, such as wind speed and wind direction (Born et al., 1995, and references therein; Garlich-Miller et al., 2011). Walruses are particularly sensitive to the sight and sound of an aircraft moving rapidly overhead and sudden changes in pitch of the engine. Aircraft making turns overhead may be particularly disturbing (see Fay et al., 1984a,b; Garlich-Miller et al., 2011). Moreover, it seems that helicopters generally are more likely to elicit responses than fixed-wing aircraft (Salter, 1979; Garlich-Miller et al., 2011). In a study by Salter (1979), response (e.g., head-lift) was observed when a helicopter (Bell 206) was as far as 8 km away. Orientation towards, and escape into, water occurred at a maximum distance of 1.3 km, and low altitude approaches to within 2.5 km elicited a major response. Responses (mainly head-lift) from various types of fixed-wing propeller aircraft flying at 150–1500 m altitude were elicited at a distance of around 1–5 km, but very few animals escaped into the water. Born and Knutsen (1990) reported the reactions to aircraft (De Havilland Twin-Otter fixed-wing, Hughes 500 and Ecureuil 350 B1 helicopters) sounds in an all-male herd at a haul-out in Northeast Greenland. All aircraft sounds audible to human ears induced reactions in the herd (head-lift, orientation towards the sea). A head lift reaction was induced already at 4–5 km distance both by the helicopter and fixed-wing aircraft. However, only when the Hughes 500 flew directly over or close to the haul-out at ≤200 m altitude did some individuals escape into the water. Return to normal behaviour took between a few minutes to some hours. Interestingly, a reaction (orientation towards water) was also seen when a large flock of cackling barnacle geese (*Branta bernicla*) flew over the herd at c. 75 m altitude. Based on their observations, Born and Knutsen (1990) concluded that air traffic should not go closer than 5 km to haul-out sites. Similarly, in a long-term study of walruses at Round Island in Alaska, animals typically dispersed when aircraft (helicopters and propeller planes) were 165–2500 m above the ground level either flying directly overhead or within several kilometres of haul-out sites. However, there have been several instances of dispersal from aircraft passing at higher altitudes, including a propeller plane at around 6000 m altitude and several commercial jets at around 9000 m above sea level (DFO, 2019, and references therein).

In addition to the immediate response, disturbance from aircraft may cause longer-term effects. For example, in Hudson Bay, Canada, walruses have been observed abandoning haul-out sites for up to three or four days after being disturbed by human activities (Mansfield and Aubin, 1991). Similarly, according to DFO (2019), long-term datasets from Round Island in Alaska suggest walruses have not habituated to disturbance by aircraft over the more than 20-year monitoring period. In contrast, Richardson et al. (1995) noted that some walruses that were exposed to repeated aircraft disturbance at haul-out sites close to airstrips – for example at Cape Lisburne (Alaska) – seemed to become more tolerant of the aircraft noise. To mitigate disturbance to walruses from aircraft (as well as boats), multiple regulations have been introduced in the range states of Atlantic walrus to control the access to walrus haul-outs, the distance of approaches by aircraft (and boats) and the permitted flight altitudes (Born et al., 1995; Wiig et al., 2014; 2017; COSEWIC, 2017; NAMMCO, 2018a; DFO, 2019).

Finally, a recent study tested the behavioural reactions to drones – specifically, a Mavic Phantom 4 – by male walruses on a terrestrial haul out in western Svalbard. The authors found that the reactions were related to the preexperimental agitation level of the group and its composition (i.e., whether or not young individuals present or not). However, generally, walruses reacted little to most approaches at 40–60 m height; however, on three occasions, they fled the shoreline, though stopped short of entering the water (Gonzalez, 2019).

Negative effects of oil, gas and mining activities

Some areas in which Atlantic walrus are distributed potentially still hold large oil and gas reserves (Wikipedia, 2020a). The decrease in sea ice can facilitate the development of oil and gas fields in several parts of the Atlantic Arctic – including West and East Greenland and the southern Barents, Pechora and Kara seas – bringing increased risks to walruses of offshore industrial operations (Isaksen et al., 1998; AMAP, 2007; Lydersen et al., 2012; NAMMCO, 2015; Gulas et al., 2017; Helle et al., 2020; Sutterud and Ulven, 2020; Boertmann and Mosbech, 2021; Boertmann et al., 2021). In the Pechora Sea the oil fields overlap with the summer distribution of walruses; consequently, large-scale petroleum activities pose a potential threat to walruses (Boltunov et al., 2010; Lydersen et al., 2012; Semyonova et al., 2015; Gebruk et al., 2020; Vasilyeva, 2020). In West and East Greenland, oil exploration activities in walrus-distribution areas have been ongoing for decades, with licence areas for oil exploration in West Greenland directly overlapping with the two most important wintering grounds for walruses (AMAP, 2007; Boertmann et al., 2021; Boertmann and Mosbech, 2021). In Canada, there has been oil exploitation and shipping of oil in areas where walruses occur (e.g., Cornwallis, Bathurst and Cameron Islands in the Canadian High Arctic, and Hudson Bay; Wikipedia (2020b)). However, in 2016, Canada introduced a moratorium on new oil and gas leasing in the Canadian Arctic to remain in effect until 2022 (iPolitics, 2019; Vigliotti, 2019).

Noise from offshore seismic surveys and drilling

Drilling operations and seismic surveys for the detection of nonrenewable resources such as oil, gas and minerals produce noise that may disturb marine mammals in the form of habitat displacement and masking of underwater communication (NAMMCO, 2015; Boertmann and Mosbech, 2021). For example, the noise from a drill ship in operation can be substantial and may extend up to c. 40 km from the ship (Kyhn et al., 2011). Seismic surveys with air-gun arrays produce underwater sounds with very high source-level energies; in extreme cases where individual pinnipeds are close to the sound source, they may experience tissue injury and auditory damage from sound waves, including permanent threshold shifts to hearing (IAGC, 2002; Gordon et al., 2003; Garlich-Miller et al., 2011; Boertmann and Mosbech, 2021). Generally, information on the extent to which seismic pulses could damage hearing are difficult to obtain and as a consequence, the impacts on hearing in marine mammals remain poorly known (Gordon et al. 2003). Temporary hearing loss has been observed under experimental conditions (Southall et al. 2007). With reference to studies of other pinnipeds showing avoidance responses to the sound source (Gordon et al. 2003; Boertmann and Mosbech, 2021 and references in these sources) it is likely that walruses will also exhibit initial avoidance and that seismic activity will not result in impairment of hearing unless in individuals remaining very close to the sound source. It is important to note that specific knowledge about noise-induced hearing damage in walrus is very limited.

Few long-term studies exist on the behavioural reactions of walruses to offshore industrial activities involving shipping, flying and drilling. In a 7-year study (2006–2012) including several thousand observations of individual Pacific walruses in the north-eastern Chukchi Sea, 70% of the animals did not exhibit an observable reaction to offshore oil and gas exploration activities, 23% of the animals exhibited an 'attention' response (i.e., looking at a vessel) and the remaining 7% of the walruses reacted by approaching, avoiding or fleeing exploration activities (Shell, 2015: Tables 4–11). Brief, small-scale masking episodes may have few long-term consequences for individuals or groups of walruses. The consequences might be more serious, however, in areas where many acoustic surveys are occurring simultaneously (Garlich-Miller et al., 2011, and references therein). Noise from seismic surveys may also have indirect negative effects on marine mammals by affecting prey availability (Gordon et al., 2003). However, to our knowledge, very little is known about the effects of seismic surveys on bivalves. A study in Australia did not find evidence of seismic-induced mortality in two species of scallops (*Pecten fumatus* and *Mimachlamys asperrima*) but could not exclude sublethal effects (Przeslawski et al., 2018); according to NAMMCO (2015) and Boertmann and Mosbech (2021), the effects of seismic activities on walrus prey species are unknown.

In most areas within the range of Atlantic walrus, regulations exist for seismic operations to mitigate potential impacts on marine mammals (COSEWIC, 2017; DFO, 2019; Boertmann and Mosbech, 2021). Moreover, due to the presence of sea ice and harsh weather during winter and spring, drilling operations and seismic surveys are mostly carried out during the open-water season ('summer') when Atlantic walruses are generally at or close to their coastal or inshore foraging areas. Thus, there is usually no temporal overlap between drilling and seismic surveys and the walruses' winter breeding season when they vocalize intensively (see Chapter 3, this volume). Thus, disturbance from these activities is likely mainly in the form of triggering an escape reaction in walruses during the open-water season. However, global warming may open more Arctic areas to exploration and exploitation of nonrenewable resources in the future, potentially posing a threat to walruses during the winter.

Shipping activities related to exploitation of nonrenewable resources

It is anticipated that disturbances from increased industrial shipping may negatively affect walrus subpopulations in Canada and West Greenland (NAMMCO, 2018a; WWF, 2019). For instance, Baffinland Corporation's Mary River iron ore–mining project on north-central Baffin Island appears to involve icebreakers up to around 85,000 deadweight tonnage (DWT) and bulk carriers up to c. 100,000 DWT with numerous transits during its year-round operation (WWF, 2019) (Fig. 12.5). Similarly, in 2018, the NAMMCO scientific Working Group on

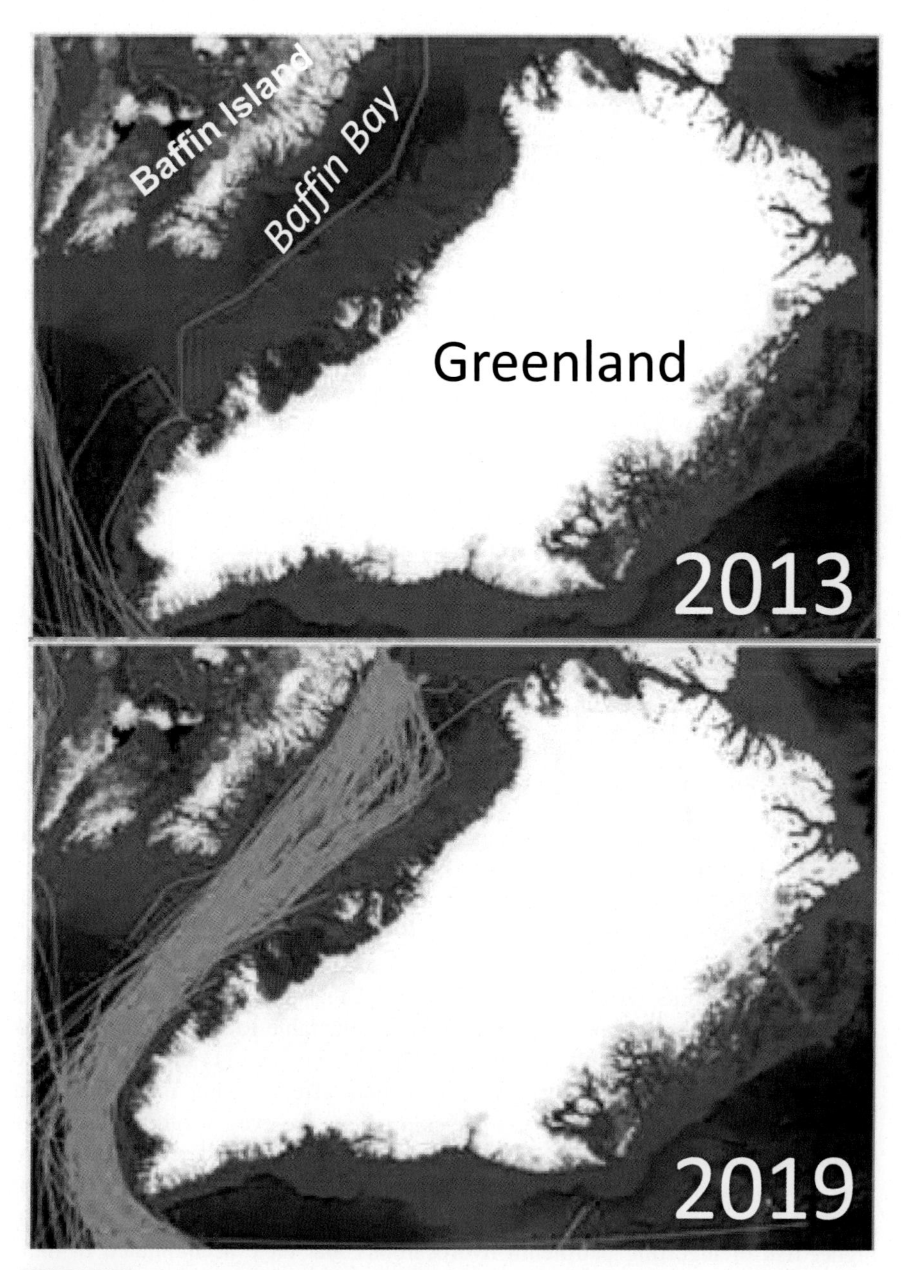

FIGURE 12.5

Bulk carrier traffic to and from the Mary River Mine on northern Baffin Island in 2013 and 2019, respectively (PAME, 2020). The mining involves the seasonal shipping of c. 3.5 million tons of iron ore from Milne Inlet on the north coast of Baffin Island (Baffinland, 2021). The shipping goes through, or close to, important walrus grounds in Baffin Bay (i.e., eastern Baffin Island and West Greenland).

Walrus (NAMMCO, 2018a) expressed concern that a planned mining operation to extract ilmenite ('black sand') from the onshore raised beaches in Northwest Greenland (Bluejay Mining, the Dundas Ilmenite Project; DIP, 2019) may negatively affect walruses from the northern Baffin Bay–North Water walrus subpopulation. In December 2020, the mining company received permission from the Greenland government to extract black sand in a 64-km^2 large area on land on the northern coast of the Wolstenholme Fjord (Sermitsiaq 2020; Jamasmie 2020). According to Orbicon (2019), the plan is to export the ilmenite concentrate from late June to late October using large bulk carriers. In addition, other ships will provide the mine with supplies (including fuel) during the same period of the year (Orbicon, 2019). The Wolstenholme Fjord area is usually devoid of walruses from May–June to October–November when walruses return to the areas from their summering grounds along eastern Ellesmere Island (Born, 1987, 2017). The time when the walruses leave has become earlier, and the time of return later, due to a general decrease in sea ice (Born et al., 2017). Hence, the mining area will mostly be devoid of walrus during its 'open-water' operational season. Furthermore, to our knowledge, the traditional walrus foraging banks in the area – specifically, the eastern Wolstenholme Island–Saunders Island–Dalrymple Rock area (Vibe, 1950; Born et al., 2017) – is situated 25–30 km southwest of the on-land mining site. Thus, in our opinion, any direct or indirect negative effects on walruses from shipping in connection with mining may be mitigated by avoiding shipping routes through or close to these walrus foraging banks.

Oil spills and other marine fouling

In addition to noise, marine oil and gas, and mining exploration, exploitation and transport may pose a threat to Atlantic walrus and/or their benthic food sources through oil spills, discharge of drill cuttings, mud, and pollutants – either from platforms or cargo ships (e.g., Born et al., 1995; Wegeberg et al., 2018). Indeed, it has been suggested that the greatest threat to Arctic ecosystems in relation to shipping is oil spills (AMSA, 2009). We argue that several features of walrus behaviour and ecology make this species more vulnerable to such harmful effects compared to other marine mammals. These features include (1) their high degree of gregariousness, implying that an oil spill may affect several individuals and oil-fouled walruses may rub oil onto the skin or into the eyes of other individuals (Born et al., 1995); (2) their preference for coastal areas and areas of relatively loose pack ice, where spilled oil is likely to accumulate (Griffiths et al., 1987); (3) their (more or less) strict dietary reliance on benthic invertebrates, implying a higher risk of ingesting petroleum hydrocarbons that have accumulated in molluscs, and a destruction of feeding areas (Richardson et al., 1995); and (4) the long nursing period and multiyear calving intervals, implying a relatively low reproductive rate – and hence longer recovery period from oil spills and other disturbances – than other pinnipeds which breed annually. In addition to oil spills, oil exploitation uses water, which after use constitutes the largest by-product of the production process. This 'produced' water contains low concentrations of oil and chemicals (e.g., toxic, radioactive, heavy metals, hormones with disruptive effects or nutrients that influence primary production) that are harmful to the environment. However, due to dilution effects, discharges of water and chemicals into the water column appear to have acute effects on marine organisms only in the immediate vicinity of the extraction activities, and the effects further away are low (Boertmann and Mosbech, 2021, and references therein).

Helle et al. (2020) used a semiquantitative, probabilistic analysis framework to assess the risks an oil spill may pose to walruses, ringed seals (*Pusa hispida*), and polar bears (*Ursus maritimus*) in the Kara Sea. They found that the risk to walruses from an oil spill depends highly on season (i.e., whether there is a sea-ice cover or not, and variability in weather), with the highest risk of impact being during summer and fall. Furthermore, walruses were found to be relatively sensitive to oil spills with heavy oil, which has a higher likelihood of sinking and covering the benthic fauna and also takes longer to break down in the environment than lighter oils. Similar results were found by Isaksen et al. (1998) who made an assessment of the potential impacts on seabirds and marine mammals of petroleum activity in parts of the Northern Barents Sea. The current and planned offshore oil and gas, as well as mining, operations theoretically pose a threat to groups of Atlantic walruses in parts of their range. However, an obvious lack of detailed information from systematic studies dedicated to assessing type and magnitude of negative impacts from industrial activities prevents further insights on this issue.

Pollution from terrestrial industrial sources

Except for local activities related to oil, gas and mining, the Arctic is generally devoid of major industrial complexes. However, various contaminants are transported to the Arctic from far-away industrial sources via ocean currents, rivers and atmospheric circulation. Contaminants such as mercury (Hg) and persistent organic pollutants (POP's) eventually end up in and accumulate in Arctic marine food webs (Braune et al., 2005; Dietz et al., 2013, 2018). In species such as polar bears, which forage at the highest trophic level, the accumulation of such contaminants may have multiple sublethal health effects (Dietz et al., 2019).

Walruses feed primarily on benthic invertebrates low in the food web, and consequently retain generally low levels of POP's and/or Hg, as evident from studies of walruses in Hudson Bay, Foxe Basin, Northwest and Northeast Greenland, Alaska and Russia (Born et al., 1981; Muir et al., 2000; Braune et al., 2005; Kucklick et al., 2006; Riget et al., 2007, 2011; Trukhin and Simokon, 2018). For example, Hg levels in walrus kidneys and livers are well below threshold levels for observed anatomical or physiological effects, and typically much lower than in other Arctic pinnipeds (Dietz et al., 2013). However, in some cases, adult male walruses from eastern Hudson Bay, Northeast Greenland, Svalbard and the Pechora Sea (Russia) were found to have relatively high levels of POP's, approaching those observed in polar bears (Muir et al., 1995; Muir et al., 2000; Wickens, 1995; Wolkers et al., 2006; Boltunov et al., 2019). Some research has suggested (Muir et al., 1995; Muir et al., 2000; Wiig et al., 2000) that the relative high POP levels in these individuals reflected their occasional seal-eating habits. This hypothesis was supported by observations on Svalbard in the late 1990's and early 2000's that walruses with different contaminant loads also had different stable isotope and fatty acid profiles reflecting differences in diet (Wickens, 1995, 2006). Generally, concentrations of POP's in walrus may vary geographically, which has been seen in Greenland with higher levels of POP in West than in East Greenland walrus (Muir et al., 2000). To complicate matters, two recent studies of Svalbard walruses found that all animals foraged at low trophic levels, despite some animals having high blood POP concentrations (Routti et al., 2019; Scotter et al., 2019). The authors suggested that due to their close proximity to the Northeast Atlantic and the Gulf Current, Svalbard walruses experience relatively higher contaminant loads than walruses elsewhere, despite their reliance on molluscs.

In recent decades the concentration of many POP's has declined, whereas concentrations of polychlorinated biphenyls (PCB's) and Hg have remained constant. However, new synthetic chemicals are produced in the thousands, of which some, such as flame retardants, are emerging as a major concern (Rigét et al., 2011; Dietz et al., 2019; Rigét et al., 2019). Moreover, similar to other marine ecosystems, the Arctic is experiencing increasing pollution of nano- and micro-plastics, which in themselves and through interactions with chemical stressors pose new risks to Arctic animals (Rowlands et al., 2021). These risks may be further increased by ongoing and future global warming causing physiological stress and changes in trophic interactions (McKinney et al., 2015). In light of the ongoing large-scale climate changes in the Arctic, we recommend continuous, long-term monitoring of the cumulative effects of contaminants and other stressors on walruses, and their prey.

Tourism and recreational activities

Scenic landscapes, unique cultural traditions and wildlife are increasingly drawing tourists from around the world seeking to explore the Arctic from ships, snowmobiles, aircraft or on foot (Christensen et al., 2016; Chanteloup, 2013; Lasserre and Têtu, 2015; Johnston et al., 2017; Huddart and Stott, 2020; Øian and Kaltenborn, 2020). Tourism began in Svalbard over hundred years ago but intensified beginning in the 1980's, with numbers steadily increasing. In 2018 more than 130,000 tourists visited the archipelago (Øian and Kaltenborn, 2020) (Fig. 12.6).

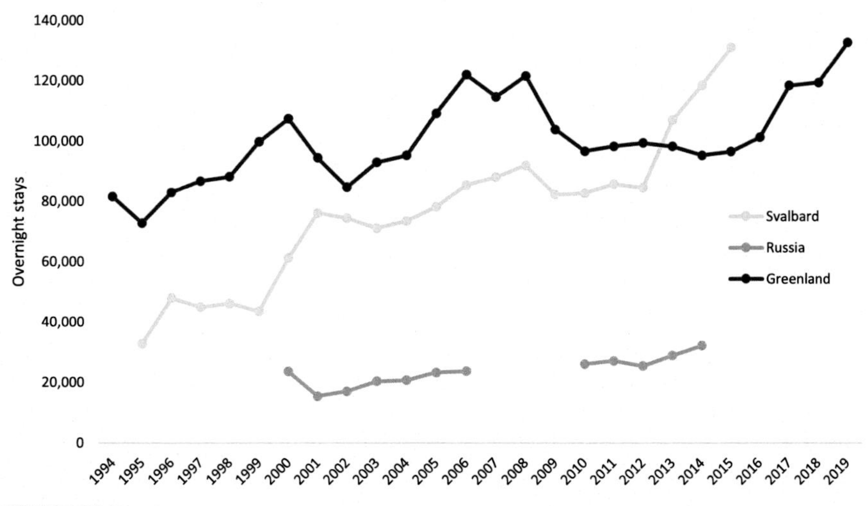

FIGURE 12.6

Tourism in the Arctic is increasing, as illustrated by the number of holiday and business overnight stays by nonresidents in Svalbard (Longyearbyen), Russia (Murmansk and Nenets in the Barents Sea region), and Greenland.

Data from StatBank Greenland (https://bank.stat.gl) and National Ocean Economics Program (https://www.oceaneconomics.org/).

Likewise, tourism in Greenland has been steadily increasing with around 100,000 visitors per year in 2018 and 2019, with an equal proportion of air and cruise ship passengers (Visit Greenland, 2019). In Arctic Canada, the cruise industry has also been growing (Stewart et al., 2011), and areas such as Churchill (Manitoba) and Nunavut are experiencing the same overall increase in tourism as experienced in Greenland, Svalbard and Russia (Lemelin et al., 2010; Stewart et al., 1993; Lasserre and Têtu, 2015; Huddart and Stott, 2020). In 2018, 51,000 people from outside visited Nunavut (Tranter, 2019) where the majority of Canadian Atlantic walruses are distributed. Tourism in the Arctic is mostly limited to the spring, summer and fall due to the remoteness and harsh weather and light regime of the region (Huddart and Stott, 2020; Øian and Kaltenborn, 2020). However, global warming may increase the length of the tourist season, thereby providing further opportunities for tourism-related economic development (Johnston et al., 2017). The increasing interest in 'last-chance tourism' (i.e., the opportunity to experience pristine landscapes, or, in some cases, endangered or threatened wildlife) may also have negative impacts on Arctic nature (Lemelin et al., 2010) through, for instance, long-distance travels to the Arctic contributing to greenhouse gas emissions (Dawson et al., 2010).

From a scientific perspective, surprisingly little is known about the effects of tourism and recreational activities on walruses. Effects of large cruise ships in terms of noise, pollution and fragmentation of sea-ice habitats are expected to be overall similar to those caused by other large vessels. However, cruise ships often carry smaller boats, such as zodiacs and small ferries, for making trips to land or along the coast. Small boats are also increasingly used for short 'animal spotting' trips launched from local towns and settlements by local tour operators. Importantly, in many parts of the Arctic, traffic by small boats constitutes the majority of local traffic and is increasingly used for recreational activities by local communities (Øian and Kaltenborn, 2020). In a study to assess the degrees of disturbance caused by tourists arriving in terrestrial walrus haul-outs in zodiacs, Øren et al. (2018) used monitoring cameras over a five-year period at several haul-outs in Svalbard. The study found that neither tourists nor small boats significantly disturbed or altered the behaviour of the hauled-out walruses, with only one reported incident over the study period of a zodiac rubber dinghy approaching too fast and scaring the walruses into the water. Likewise, low-to-moderate levels of disturbance only caused short-term interruption of resting behaviour in Pacific walruses (Jay et al., 1998). Observations during multiple research projects on a mixed herd of walruses at a haul-out in Northeast Greenland also suggest that walruses may react little to human presence (including biopsy darting and placing electrodes in unrestrained individuals) if certain precautions are taken, such as moving slowly, being as silent as possible, crouching or creeping, and remaining upwind and not silhouetted against the sky (Born et al., 2011; Bertelsen et al., 2006).

Clearly, it is difficult to establish general rules about how walruses react to human activities on land or small boats. In some cases, drowsy walruses hauled out on sea ice or land can be approached slowly – to within about 10 m or less – in small boats. In such cases the final escape reaction may as well be triggered by the sense of smell or the surprise of the sight (Born, personal observations). Walruses are sensitive to smells (DFO, 2019; Box 12.2) and alertness or escape reactions may be triggered by boat exhaust fumes. However, generally, engine power, speed and acceleration, as well as wind direction (whether approaching from down or upwind direction), are the main determinants of type and level of disturbance and therefore also the distance at which walruses may react to small boats (Kruse, 1997; Born, personal observation).

Overall, the above studies and observations suggest that walruses will tolerate low-to-moderate disturbances of short duration. However, given the likely cumulative effects of climate change, tourism, recreational activities and other human activities, there is a need for codes of conducts for tourists when viewing wildlife and enjoying nature (Mason, 1997). To this end, the AECO (2021) has developed guidelines for walrus encounters stipulating that (1) landings must be made a minimum distance of 300 m from the haul-out site; (2) animals should be approached slowly, against the wind and in a single group of people; (3) a minimum distance of 30 and 150 m should be maintained from adult males and mothers with calves, respectively and (4) and approaches to swimming walruses should be avoided (www.aeco.no). According to NAMMCO (2015), regulations in Greenland require that tourists keep a distance of 400 m from walruses on land, and 75 m for walruses in water. In Svalbard, tourist access to the traditional haul-out Moffen is prohibited from May until September, and to reduce disturbance to walruses and other wildlife in the Northeast Svalbard Nature Reserve, access is prohibited year round; this includes a 500-m buffer zone from the coast (Øian and Kaltenborn, 2020). To reduce disturbance at terrestrial haul-outs, NAMMCO (2018b) suggested that all Atlantic walrus haul-out sites should include such a marine-protection buffer zone.

General conclusions and recommendations

Across most of their range, Atlantic walruses have shown encouraging signs of population recovery in recent decades following increased protection from hunting and other disturbances, and a reduction in catches partially reflecting a decrease in demand for walrus products and sound management based on scientific advice. However, shipping, industrial development, pollution, tourism and recreational activities may lead to increased negative anthropogenic impacts in the future, the effects of which we do not yet fully understand. Thus, to mitigate such potential future impacts, we call for the protection of critical habitats, such as terrestrial haul-outs, coastal foraging grounds and restricted wintering areas. Moreover, given the a surprising lack of scientific studies directly qualifying and quantifying the mechanisms behind potential anthropogenic impacts on walruses, we strongly recommend further research in this area. This could include developing a modelling framework linking the type, magnitude and duration of disturbance with sea-ice availability, energy expenditure, body condition and walrus demography (Jay, 2015). These and other insights are crucial for informing management strategies and allowing the full recovery of walruses from the legacies of past industrial-scale commercial harvesting.

Box 12.1 The West Greenland case.

The West Greenland marine ecosystem is one of the most biologically productive areas across the entire Arctic and supports large populations of fish, seabirds and marine mammals including Atlantic walruses (Linnebjerg et al., 2016). Due to its species diversity and productivity, this area has been categorised as particularly important ecologically (Christensen et al., 2012). There is a spatial overlap between fishery activities and distribution of walruses at their important winter foraging banks off central West Greenland (between around 66° 30′ N and c. 70° 30′ N, and between the coast west to c. 56° W). In this area, there has been substantial fishing activity since about 1920 (Hamilton et al., 2003; Born, 2005: 38; Garcia, 2006; Jørgensen and Hammeken, 2013; GINR, 2018b, 2020b; Burmeister and Rigét, 2020). Experienced subsistence walrus hunters living in central West Greenland have suggested that walruses have changed distribution locally due to fishing activities (Born et al., 2017) (Fig. 12.7).

(*Continued*)

Box 12.1 The West Greenland case. (Continued)

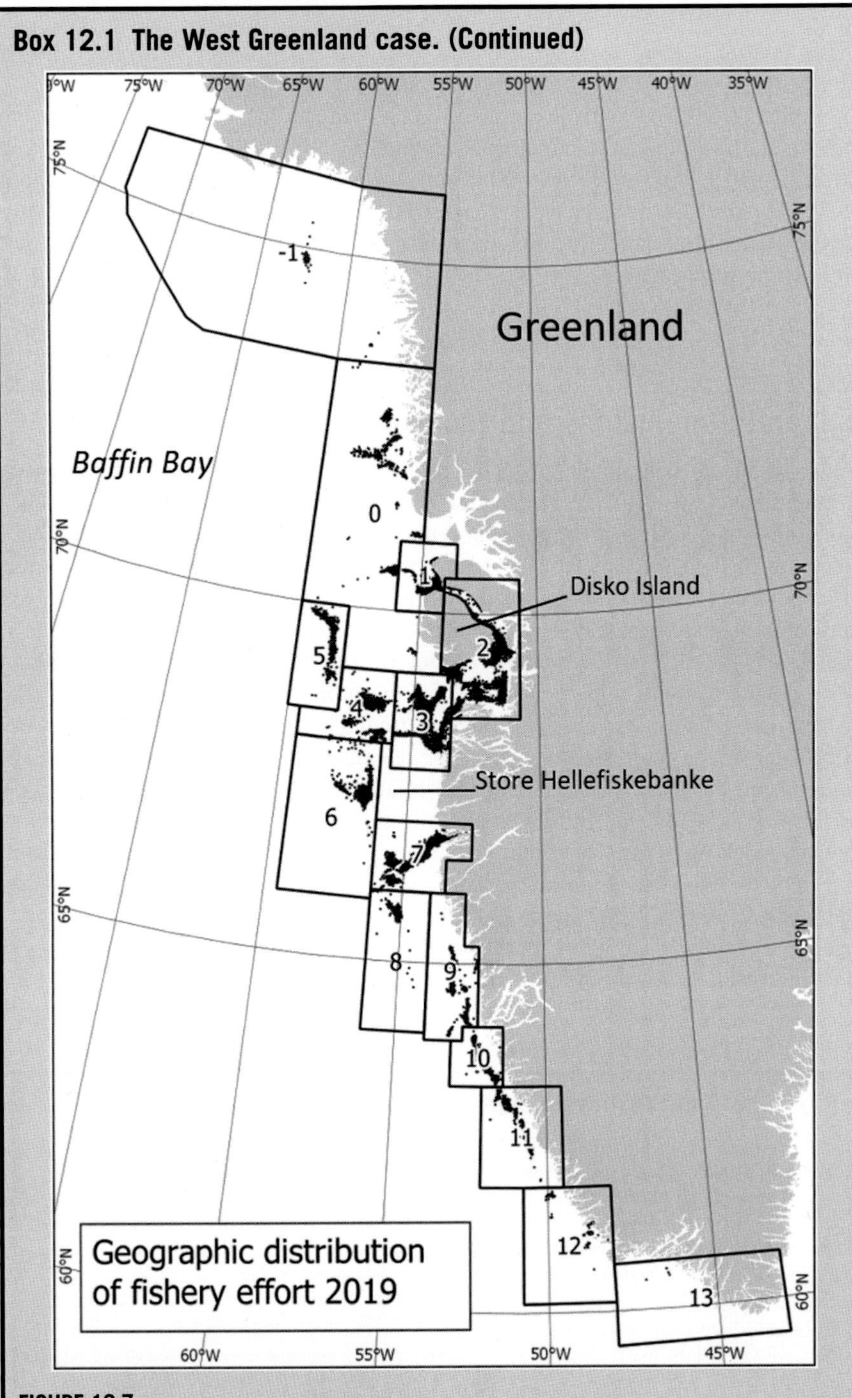

FIGURE 12.7

In central West Greenland there is a substantial fishery in areas where walruses have their foraging grounds. Locations of catch of Northern shrimp in West Greenland during 2019. The boxes indicate the delineations of local fishery management areas (Burmeister and Rigét, 2020).

(Continued)

Box 12.1 The West Greenland case. (Continued)

Distribution of walruses

Walruses from the W Greenland–SE Baffin Island subpopulation concentrate in winter and spring on their West Greenland foraging banks at the eastern edge of the Baffin Bay pack ice ('the West Ice'). During spring (March–May) the walruses emigrate from West Greenland to summer along eastern Baffin Island; they re-appear in West Greenland sometime in late fall (Born et al., 1994, 2017; Dietz et al., 2014). In their wintering grounds, walruses prefer relatively-dense pack ice in waters less than 100 m deep (Born et al., 1994; Dietz et al., 2014; Heide-Jørgensen et al., 2014) where they are known to feed on the bivalves *Mya* sp. and *Serripes* sp., and Northern sand lance/sand eel, *Ammodytes dubius* (Born et al., 1994, 2017). There is a notable spatial segregation of male and female walruses on the banks, with males occurring farther offshore in deeper waters than females and their calves (Born et al., 1994, 2017; Dietz et al., 2014). In West Greenland the walruses show a great affinity to their foraging grounds irrespective of variation among years in the extent and density of sea ice in the area (Dietz et al., 2014; NAMMCO, 2018a). However, according to Inuit subsistence hunters in West Greenland, in recent years, the general decrease in sea ice has pushed walruses to occur farther offshore during winter–spring, allowing (or forcing) them to emigrate west to Baffin Island earlier in the spring (Born et al., 2017).

The fishery

In West Greenland, commercial fishing for Northern shrimp increased in the late 1960's, as those of Atlantic cod declined (Born et al., 1994; Hamilton et al., 2003). At the beginning of the 1970's, trawling for shrimp was initiated in parts of the walruses' main wintering ground in West Greenland (Store Hellefiskebanke in the Sisimiut area). Today, large trawlers operate at the edges of the banks, and some smaller vessels operate in coastal waters in central West Greenland, as sea-ice conditions permit (Garcia et al., 2006; Viðarsson et al., 2015; Burmeister and Rigét, 2020; Wilson et al., 2020). The shrimping operations take place all year round, with some of the catch being taken during the walruses' mating season in February–March (A. Burmeister, Greenland Institute of Natural Resources, *in litt.* December 2020). After the collapse of the stock of Atlantic cod in the mid-1980's (Hamilton et al., 2003), Greenland halibut became the most exploited ground-fish resource in Greenland. Halibut is caught both inshore and offshore. During summer and fall, large trawlers fish halibut (Jørgensen and Hammeken, 2013) at the edges of the walrus wintering grounds. Since the mid-1990's there has been a limited catch of Snow crab (*Chionoecetes opilio*) on the walrus foraging banks (Burmeister and Siegstad, 2008; GINR, 2018b). However, since Snow crabs are taken in crab pots, this fishery is assumed to make only minimal physical impacts on the seafloor (GINR, 2018b).

The benthic habitat

The shallows of the West Greenland banks have not been exposed to trawling activity for many years, and they generally have a high degree of biological benthic diversity (Blicher and Hammeken Arboe, 2017; Blicher et al., 2021). On the important walrus wintering grounds, there is relatively diverse benthic infauna and epifauna, with fairly high biomasses of favoured walrus prey (e.g., bivalves, including *Mya* sp.) (Hansen et al., 2013). Northern sand lance/sand eel, known to be consumed by walruses in central West Greenland (Born et al., 1994, 2017), is abundant on the Store Hellefiskebanke (Hansen and Hjorth, 2013). On the margins of the banks in the 50–100 m depth range where walruses occur winter and spring, the estimated invertebrate faunal biomass (including *Mya*) has been estimated to be c. 700 g ww per m^2 (Hansen and Hjorth, 2013). This infauna biomass is similar to that reported for some other Arctic shelf areas at <100 m depths (Hansen et al., 2013; Kedra et al., 2013). According to Born (2005), there is amble benthic food resources in West Greenland and he suggested that the current number of walruses wintering in the area is likely far below the carrying capacity of the benthos at the foraging banks.

Trends in abundance of walrus

Sighting rates during several systematic aerial surveys conducted since 1981 with the purpose of estimating abundance of walrus on their west Greenland wintering grounds have been used to

(*Continued*)

Box 12.1 The West Greenland case. (Continued)

evaluate trends in abundance over time (NAMMCO, 2018a). From 1981 to 1999 no trends in abundance could be detected for the main wintering ground Store Hellefiskebanke (NAMMCO, 2009, 2013). However, although not statistically significant ($\alpha = 0.05$; $P = .06$), the relative abundance of walrus at Store Hellefiskebanke increased somewhat in abundance from 1981 through 2017 (NAMMCO, 2018a). Modelling of trends in abundance of walruses in West Greenland (1977–2018) indicated that trends varied over time in direct response to annual catches (NAMMCO, 2018a). The current number of walruses wintering in West Greenland is estimated to be between 1100 and 1400 (Heide-Jørgensen et al., 2014).

Overall evaluation

Despite intensive and increased commercial fishing activity in the area since the 1960's, there are no clear indications of a decrease of walruses at their West Greenland wintering grounds. Furthermore, available information indicates that there has not been any noticeable habitat degradation or direct negative impacts on suitable walrus food sources as a result of commercial fishing in central West Greenland despite many decades of trawling in areas where walruses occur. Today, the trawl fishery mainly operates at deeper depths (> 100 m) than the habitat of the main bivalve walrus prey (typically found buried in the sediment below 60–70 m). We, therefore, concur with suggestions in Born et al. (1994) and Witting and Born (2014) that the historical decrease in the numbers of walruses in this area was caused mainly by excessive hunting by Greenlanders for subsistence and commercial purposes. It must, however, be noted that recent trawling activity has been located *around* but not *on* the walrus West Greenland foraging banks, *per se*. This geographical distribution of the fishery does not preclude the possibility that noise from fishery activity may negatively affect walruses by, for example, masking their underwater communication during the mating season.

We recommend that systematic studies be conducted in West Greenland to gain insight into the relationship between the distribution and abundance of walrus and the spatial distribution of sea ice, diversity and density of benthos, and fishing activities.

Box 12.2 Sensory abilities of walruses

Walrus reactions to disturbance may be triggered by sounds, smells, and visual cues – or a combination of these stimuli. In particular, the hearing ability in walruses is good, and their sense of smell is well developed, whereas their vision is believed to be relatively poor (Loughrey, 1959; Born et al., 1995, and references therein). The hearing range of walruses in air is 60 Hz to 23 kHz (Reichmuth et al., 2020), which is similar to that of humans (Purves et al., 2001). A study of the hearing ability of hauled-out Atlantic walruses in the wild indicated they are most sensitive to sounds at between 0.250 and 4 kHz (Kastelein et al., 1993a). Sensitivity tests of an adult Pacific walrus in captivity indicated its best hearing underwater was in the 1–12 kHz range, with sensitivity decreasing at higher frequencies (Kastelein et al., 2002). According to Kastelein (2002), the upper limit of underwater hearing of walruses is 16 kHz. Human speech and industrial noise are in the 0.125–8 kHz range. Walruses are therefore sensitive to most anthropogenic noise in both air and underwater (Kastelein et al., 1996; Kastelein, 2002). Modelling of propagation of underwater sounds has indicated that walruses may be able to detect vessel noise around 40 km away in open water and from at least around 70 km away under the ice (Schack and Haapaniemi, 2017). Vocalisations play an important role in social interactions among walruses (Stirling et al., 1983, 1987; Sjare and Stirling, 1996; Stirling and Thomas, 2003; Sjare et al., 2003). In-air sounds produced by walruses are between around 0.1 and 4 kHz. During the mating season in late winter

(*Continued*)

Box 12.2 Sensory abilities of walruses (Continued)

and early spring, adult male walruses are highly vocal underwater (see Chapter 3, this volume). Underwater vocalisation ranges between around 0.2 and 8 kHz (Kastelein et al., 1996, and references therein; Sjare et al., 2003). The average source level of the male walruses' rutting whistles is 120 dB re 1 pW (Kastelein et al., 1996, and references therein). These sounds may be transmitted underwater up to at least 10 km from the vocalizing walrus (e.g., Stirling et al., 1987; Anon., 1991).

The behaviour of walruses on land and on ice suggests that they rely to a high degree on their sense of smell to obtain information about their surroundings (Loughrey, 1959; Kastelein, 2002; Born, personal observation). Observations in the wild indicate they may retreat if exposed to strong olfactory stimuli, such as exhaust and other waste gases (Loughrey, 1959; Fay et al., 1984a,b; DFO, 2019).

Analyses of walrus anatomy indicate that their eyes seem to be specialized for short-range vision and that they can see in colour (Kastelein et al., 1993b; Kastelein, 2002). Walrus eye anatomy indicates that vision is similarly good in air and water (Mass and Supin, 2019). According to Loughrey (1959), walruses can probably distinguish large moving objects, such as boats, visually at a distance of about 60 m, but are unable to identify a stationary person who is not silhouetted within six m (Fig. 12.8).

FIGURE 12.8

The sense of smell is important in walruses. The dilated nostrils show clearly that this adult Atlantic walrus male is investigating its surroundings via smell.

Photo: E.W. Born.

Acknowledgements

We wish to thank the editors for inviting us to contribute to this volume. We owe the following colleagues our deepest gratitude for helping us with background information and discussions: Helle Siegstad, Lars Witting, Martin Blicher and David M. Boertmann.

References

AECO (Association of Arctic Expedition Cruise Operators), Wildlife guidelines, 2021. https://www.aeco.no/. (Accessed 27 January 2021).

AMAP (Arctic Monitoring and Assessment Programme), 2007. Oil and Gas 2007. AMAP Working Group, Oslo, pp. 1–57. www.amap.no.

AMSA (Arctic Marine Shipping Assessment), 2009. Arctic Marine Shipping Assessment Report. Arctic Council. pp. 1–189.

Anderson, L.E., Garlich-Miller, J., 1994. Economic analysis of the 1992 and 1993 summer walrus hunts in northern Foxe Basin, Northwest Territories. Canadian Technical Report of Fisheries and Aquatic Science 2011, pp. 1–20.

Andersen, A.O., Heide-Jørgensen, M.P., Flora, F., 2018. Is sustainable resource utilisation a relevant concept in Avanersuaq? The walrus case. Ambio 47, S265–S280. Available from: https://doi.org/10.1007/s13280-018-1032-0.

Anon, 1990. Walrus. Hansson, R., Prestrud, P., Øritsland, N.A. (Eds.), Assessment System for the Environment and Industrial Activities in Svalbard. Norwegian Polar Institute, pp. 71–81.

Anon, 1991. Environmental Assessment/Regulatory Impact Review/Initial Regulatory Flexibility Analysis for Amendment 17 to the Fishery Management Plan for the Groundfish Fishery of the Bering Sea and Aleutian Islands Area and Amendment 22 to the Fishery Management Plan for Groundfish of the Gulf of Alaska and for a Regulatory Amendment to Define Groundfish Pots. Prepared by Members of the Plan Teams for Groundfish Fisheries of the Bering Sea and Aleutian Islands Area and the Gulf of Alaska and by staffs of the North Pacific Fishery Management Council, National Marine Fisheries Service, Alaska Department of Fish and Game/U.S. Fish and Wildlife/LGL Research Associates, Anchorage, AK, pp. 1–71.

Anon, 1993. Disturbances. Report of working group 5. Stewart, R.E.A., Richard, P.R., Stewart, B.E. (Eds.), Report of the Second Walrus International Technical and Scientific (WITS) Workshop, 11–15 January 1993, Winnipeg, MB, Canada. Canadian Fisheries and Aquatic Sciences Technical Report 1940, pp. 33–38.

Anon, 2012. Troubled waters: how to protect the Arctic from the growing impact of shipping. Report From Soft Free and European Federation for Transport and Environment (T&E). European Federation for Transport and Environment (T&E) AiSBL, Rue d'Edimbourg, Brussels, pp. 1–23. www.transportenvironment.org

Baffinland, 2021. The Mary River Mine. https://www.baffinland.com/operation/mary-river-mine/mary-river-mine/. (Accessed 31 January 2021).

Bertelsen, M.F., Acquarone, M., Born, E.W., 2006. Resting heart and respiratory rate in wild adult male walruses (*Odobenus rosmarus rosmarus*). Marine Mammal Science 22, 714–718. Available from: https://doi.org/10.1111/j.1748-7692.2006.00055.x.

Blicher, M., Hammeken Arboe, N., 2017. Evaluation of proposed common standards for benthos monitoring in the Arctic-Atlantic – pilot study in Greenland (INAMon). Technical Report Nr. 105. Greenland Institute of Natural Resources, Nuuk. p. 3657, 1–31 + suppl. appendix. ISBN 87-91214-82-3, ISSN 1397. doi: 87-91214-82-3.

Blicher, M.E., Hammeken Arboe, N., Krawzcyk, D., Hansen, J.L.S., 2021. Biological environment. Benthic fauna. Boertmann, D., Mosbech, A. (Eds.), Disko West – An Updated Strategic Environmental Impact Assessment of Oil and Gas Activities. Scientific Report From DCE – Danish Centre for Environment and Energy. http://dce2.au.dk/pub/TRxx.pdf. In press.

Blomeyer, R., Stopperup, K., Erzini, K., Lam, V., Pauly, D., Raakjaaer, J., 2015. Fisheries management and the Arctic in the context of climate change. Structural and Cohesion Policies B. Directorate-General for International Policies, Policy Department, pp. 1–11ISBN 978-92-823-7917-2; doi:10.2861/567433. Available from: http://www.europarl.europa.eu/studies.

Boertmann, D., Mosbech, A., (Eds.), 2021. Disko West – an updated strategic environmental impact assessment of oil and gas activities. Scientific Report From DCE – Danish Centre for Environment and Energy. Aarhus University, DCE – Danish Centre for Environment and Energy. http://www.dmu.dk/Pub/SRxx.pdf.

Boertmann, D., Blockley, D., Mosbech, A., (Eds.), 2021. Greenland Sea – an updated strategic environmental impact assessment of petroleum activities. 2nd revised edition. Scientific Report From DCE – Danish Centre for Environment and Energy No. 375, pp. 1–386. http://dce2.au.dk/pub/SR375.pdf. In press.

Boltunov, A., Semenova, V., Samsonov, D., Boltunov, N., Nikiforov, V., 2019. Persistent organic pollutants in the Pechora Sea walruses. Polar Biol 42 (9), 1775–1785.

Boltunov, A.N., Belikov, S.E., Gorbunov, Y.A., Menis, D.T., Semenova, V.S., 2010. The Atlantic Walrus of the Southeastern Barents Sea and Adjacent Regions: Review of Present-Day Status. WWF—Russia, Marine Mammal Council, Moscow, pp. 1–17.

Born, E.W., 1987. Aspects of present-day maritime subsistence hunting in the Thule area, Northwest Greenland. Hacquebord, L., Vaughan, R. (Eds.), Between Greenland and America: Cross-Cultural Contacts and the Environment in the Baffin Bay Area. Works of the Arctic Centre No. 10. Arctic Centre, University of Groningen, Groningen, pp. 109–132.

Born, E.W., 2005. An Assessment of the Effects of Hunting and Climate on Walruses in Greenland. Greenland Institute of Natural Resources and University of Oslo, Nuuk, pp. 1–346, ISBN 82-7970-006-4.

Born, E.W., Acquarone, M., 2004. An estimation of walrus (*Odobenus rosmarus*) predation on bivalves in the Young Sound area (NE Greenland). Rysgaard, S., Glud, R.N., (Eds.), Carbon Cycling in Arctic Marine Ecosystems: Case Study – Young Sound. Meddelelser Grønland, Bioscience (Monographs on Greenland) 58, 176–191.

Born, E.W., Gjertz, I., Reeves, R.R., 1995. Population assessment of Atlantic walrus. Norsk Polarinstitutt Meddelelser 138, 1–100.

Born, E.W., Knutsen, L.Ø., 1990. Satellite tracking and behavioural observations of Atlantic walrus (Odobenus rosmarus rosmarus) in NE Greenland in 1989. Grønlands Hjemmestyres Miljø – og Naturforvaltning. Technical Report No. 20, pp. 1–68. Available from: www.natur.gl.

Born, E.W., Kraul, I., Kristensen, T., 1981. Mercury, DDT and PCB in the Atlantic walrus (*Odobenus rosmarus rosmarus*) from the Thule district, North Greenland. Arctic 34, 255–260.

Born, E.W., Heide-Jørgensen, M.P., Davis, R.A., 1994. The Atlantic walrus (*Odobenus rosmarus rosmarus*) in West Greenland. Meddelelser Grønland, Bioscience (Monographs on Greenland) 40, 1–33.

Born, E.W., Dietz, R., Heide-Jørgensen, M.P., Knutsen, L.Ø., 1997. Historical and present status of the Atlantic walrus (*Odobenus rosmarus rosmarus*) in eastern Greenland. Monographs on Greenland 46, 1–73.

Born, E.W., Andersen, S.B., Joensen, S., Knutsen, L.Ø., Coryell-Martin, M., Andersen, L. W., 2011. Walrus studies with a note on birds on Sandøen. In: Jensen, L.M., Rasch, M. (Eds.), Zackenberg Ecological Research Operations, 16th Annual Report, 2010. Aarhus University, DCE – Danish Centre for Environment and Energy, pp. 94–97.

Born, E.W., Heilmann, A., Kielsen Holm, L., Laidre, K.L., Iversen, M., 2017. Walruses in West and Northwest Greenland. An interview survey about the catch and the climate, Monographs on Greenland, Vol. 355. Museum Tusculanum Press, Copenhagen, pp. 1–256, Man and Society Vol. 44.

Braune, B.M., Outridge, P.M., Fisk, A.T., Muir, D.C.G., Helm, P.A., Hobbs, K., et al., 2005. Persistent organic pollutants and mercury in marine biota of the Canadian Arctic: an overview of spatial and temporal trends. Science of the Total Environment 351, 4–56.

Brueggeman, J.J., Malme, C.I., Grotefendt, R.A., Volsen, D.P., Burns, J.J., Chapman, D.G., et al., Shell Western E & P Inc., 1990. 1989 Walrus Monitoring Program: The Klondike, Burger, and Popcorn Prospects in the Chukchi Sea. Report Prepared by Ebasco Environmental for Shell Western E & P Inc., Houston, TX, pp. 1–180.

Brueggeman, J., Green, G.A., Grotefendt, R.A., 1993. Walrus response to offshore drilling operations. Abstracts of Tenth Biennial Conference on the Biology of Marine Mammals, Galveston, TX, USA, 11–15 November 1993, p. 33.

Burmeister, A., Rigét, F., 2020. The fishery for Northern shrimp (*Pandalus borealis*) off West Greenland, 1970–2020. Working Paper. NAFO Scientific Council Research Document 020/054. Northwest Atlantic Fisheries Organization. Serial No. 7128. NAFO/ICES *Pandalus* Assessment Group, pp. 1–39.

Burmeister, A., Siegstad, H., 2008. Assessment of snow crab in West Greenland 2008. Greenland Institute of Natural Resources Technical Report No. 68, pp. 1–47. https://natur.gl/wp-content/uploads/2019/07/68-SSR_West_Greenland_2008_tek_rap_nr_68.pdf.

CAFF (Convention of Arctic Flora and Fauna), 2017. State of the Arctic Marine Biodiversity Report. Conservation of Arctic Flora and Fauna International Secretariat, Akureyri, pp. 1–197.

Chanteloup, L., 2013. Wildlife as a tourism resource in Nunavut. Polar Record 49 (3), 240–248. Available from: https://doi.org/10.1017/S0032247412000617.

Charrier, I., Aubin, T., Mathevon, N., 2009. Mother–calf vocal communication in Atlantic walrus: a first field experimental study. Animal Cognition 13, 471–482. Available from: https://doi.org/10.1007/s10071-009-0298-9. Epub 2009 Dec 4.

Christensen, T., Falk, K., Boye, T., Ugarte, F., Boertmann, D., Mosbech, A., 2012. Identifikation af sårbare marine områder i den grønlandske/danske del af Arktis (Identification of Vulnerable Marine Areas in the Greenland/Danish Part of the Arctic).

Aarhus Universitet, DCE – Nationalt Center for Miljø og Energi, pp. 1–72 (in Danish with an English abstract).

Christensen, T., Mosbech, A., Geertz-Hansen, O., Johansen, K.L., Wegeberg, S., Boertmann, D., et al., 2015. Analyse af mulig økosystembaseret tilgang til forvaltning af skibstrafik i Disko Bugt og Store Hellefiskebanke (Analysis of the potential of an ecosystem-based approach to the management of ship traffic in the Disko Bay and Store Hellefiskebanke areas). Teknisk rapport fra DCE – Nationalt Center for Miljø og Energi nr. 6. Aarhus Universitet, DCE – Nationalt Center for Miljø og Energi, pp. 1–102. http://dce2.au.dk/pub/TR61.pdf. (in Danish with an English abstract).

Christensen, T., Aastrup, P., Boye, T., Boertmann, D., Hedeholm, R., Johansen, K.L., et al., 2016. Biologiske interesseområder i Vest – og Sydøstgrønland. Kortlægning af vigtige biologiske områder (Areas of biological interest in West and Southwest Greenland. Mapping of biologically important areas). Teknisk rapport fra DCE – Nationalt Center for Miljø og Energi nr. 89. Aarhus Universitet, DCE – Nationalt Center for Miljø og Energi, pp. 1–210. http://dce2.au.dk/pub/TR89.pdf. (in Danish with an English abstract).

Christensen, T., Lasserre, F., Dawson, J., Guy, E., Pelletier, J.-F., 2018. Chapter 9. Shipping. Adaptation Actions for a Changing Arctic: Perspectives From the Baffin Bay/Davis Strait Region. Arctic Monitoring and Assessment Programme (AMAP), Oslo, pp. 243–260, ISBN 978-82-7971-105-6.

Collie, J.S., Hall, S.J., Kaiser, M.J., Poiner, I.R., 2000. A quantitative analysis of fishing impacts on shelf-sea benthos. Journal of Animal Ecology 69, 785–798. Available from: https://doi.org/10.1046/j.1365-2656.2000.00434.x.

COSEWIC Assessment and Update Status Report on the Atlantic Walrus *Odobenus rosmarus rosmarus* in Canada. Committee on the Status of Endangered Wildlife in Canada, Ottawa, pp. 1–65.

COSEWIC (Committee on the Status of Endangered Wildlife in Canada), 2017. COSEWIC Assessment and Status Report on the Atlantic Walrus *Odobenus rosmarus rosmarus*, High Arctic Population, Central-Low Arctic Population and Nova Scotia-Newfoundland-Gulf of St. Lawrence Population in Canada. Committee on the Status of Endangered Wildlife in Canada, Ottawa, pp. 1–89. http://www.registrelep-sararegistry.gc.ca/default.asp?lang = enandn = 24F7211B-1.

Dawson, J., Stewart, E.J., Lemelin, H., Scott, D., 2010. The carbon cost of polar bear viewing tourism in Churchill, Canada. Journal of Sustainable Tourism 18 (3), 319–336. Available from: https://doi.org/10.1080/09669580903215147.

DFO (Department of Fisheries and Oceans), 2018. Northern shrimp and striped shrimp – shrimp fishing areas 0, 1, 4-7, the Eastern and Western Assessment Zones and North Atlantic Fisheries Organization (NAFO) Division 3M. https://www.dfo-mpo.gc.ca/fisheries-peches/ifmp-gmp/shrimp-crevette/shrimp-crevette-2018-002-eng.html. (Accessed 10 November 2020).

DFO (Department of Fisheries and Oceans), 2019. Mitigation buffer zones for Atlantic Walrus (*Odobenus rosmarus rosmarus*) in the Nunavut settlement area. Canadian Science Advisory Secretariat Research Document 2018/055, pp. 1–27. www.dfo-mpo.gc.ca/csas-sccs/. (Accessed 10 November 2020).

DFO (Department of Fisheries and Oceans), 2020. Greenland halibut – Northwest Atlantic Fisheries Organization Subarea 0. https://www.dfo-mpo.gc.ca/fisheries-peches/ifmp-gmp/groundfish-poisson-fond/2019/halibut-fletan-eng.htm#toc1. (Accessed 10 November 2020).

Dietz, D., Sonne, C., Basu, N., Braune, B., O'Hara, T., Letcher, R.J., et al., 2013. What are the toxicological effects of mercury in Arctic biota? Science of the Total Environment 443, 775–790. Available from: https://doi.org/10.1016/j.scitotenv.2012.11.046.

Dietz, R., Born, E.W., Stewart, R.E., Heide-Jørgensen, M.P., Stern, H., Rigét, F., et al., 2014. Movements of walruses (Odobenus rosmarus) between Central West Greenland and Southeast Baffin Island, 2005-2008. NAMMCO Scientific Publications 9, pp. 53–74. https://doi.org/10.7557/3.2605.

Dietz, R., Letcher, R.J., Desforges, J.P., Eulaers, I., Sonne, C., Wilson, S., et al., 2019. Current state of knowledge on biological effects from contaminants on arctic wildlife and fish. Science of the Total Environment 696, 133792. Available from: https://doi.org/10.1016/j.scitotenv.2019.133792.

Dietz, R., Mosbech, A., Flora, J., Eulaers, I., 2018. Interactions of climate, socio-economics, and global mercury pollution in the North Water. Ambio 47 (Suppl. 2), 281–295. Available from: https://doi.org/10.1007/s13280-018-1033-z.

DIP (Dundas Ilmenite Project), 2019. Introduction to the Project, pp. 1–8. https://bluejay-mining.com/wp-content/uploads/2019/08/Project-description_UK2.pdf.

Eleftheriou, A., Robertson, M.R., 1992. The effects of experimental scallop dredging on the fauna and physical environment of a shallow sandy community. Netherlands Journal of Sea Research 30, 289–299. Available from: https://doi.org/10.1016/0077-7579(92)90067-O.

Erbe, C., Reichmuth, C., Cunningham, K., Lucke, K., Dooling, R., 2016. Communication masking in marine mammals: a review and research strategy. Marine Pollution Bulletin 103, 15–38. Available from: https://doi.org/10.1016/j.marpolbul.2015.12.007.

Erbe, C., Marley, S.A., Smith, R.P., Trigg, J.N., Trigg, L.E., Embling, C.B., 2019. The effects of ship noise on marine mammals – a review. Frontiers in Marine Science 6, 1–21. Available from: https://doi.org/10.3389/fmars.2019.00606.

Fay, F.H., 1981. Walrus *Odobenus rosmarus rosmarus* (Linnaeus 1758). Rigway, S.H., Harrison, R.J. (Eds.), Handbook of Marine Mammals. Vol. I. The Walrus, Sea Lions, Fur Seals and Sea Otter. Academic Press, London, pp. 1–23.

Fay, F.H., 1982. Ecology and biology of the Pacific walrus, *Odobenus rosmarus divergens*, Illiger, North American Fauna, vol. 74. United States Department of the Interior, Fish and Wildlife Service, pp. 1–279.

Fay, F.H., Ray, G.C., Kibal'chich, A.A., 1984a. Time and location of mating and associated behavior of the Pacific walrus, *Odobenus rosmarus divergens* Illiger. Fay, F.H., Fedoseev, G.A. (Eds.), Soviet-American Cooperative Research on Marine Mammals. Pinnipeds. National Oceanographic Atmospheric Administration Technical Report NMFS 1, pp. 81–88.

Fay, F.H., Kelly, B.P., Gehnrigh, P.H., Sease, J.L., Hoover, A.A., 1984b. Modern populations, migrations, demography, tropics, and historical status of the Pacific walrus. Environmental Assessment of the Alaskan Continental Shelf: Final Reports of Principal Investigators, vol. 37. U.S. National Oceanic and Atmospheric Administration, Anchorage, AK, pp. 231–376.

FWS (Fish and Wildlife Service), 2014. Pacific Walrus (*Odobenus rosmarus divergens*): Alaska Stock, pp. 1–30. https://www.fws.gov/r7/fisheries/mmm/stock/Revised_April_2014_Pacific_Walrus_SAR.pdf.

Garcia, E., 2006. The fishery for Iceland scallop (*Chlamys islandica*) in the Northeast Atlantic. Advances in Marine Biology 51, 1–55. Available from: https://doi.org/10.1016/S0065-2881(06)51001-6.

Garcia, E., Ragnarsson, E., Steingrímsson, S.A., Nævestad, D., Haraldsson, H.Þ., Fosså, J. H., et al., 2006. Bottom trawling and scallop dredging in the Arctic. Impacts of fishing on non-target species, vulnerable habitats and cultural heritage. Nordic Council of Ministers, Copenhagen. Report TemaNord No. 529, pp. 1–375. ISBN 92-893-1332-3

Garlich-Miller, J., MacCracken, J.G., Snyder, J., Meehan, R., Myers, M., Wilder, J.M., et al., 2011. Status Review of the Pacific Walrus (*Odobenus rosmarus divergens*). US Fish and Wildlife Service, Marine Mammals Management, Anchorage, AK, pp. 1–155.

Gebruk, A., Mikhaylyukova, P., Mardashova, M., Semenova, V., Henry, L.-A., Shabalin, N., et al., 2020. Integrated study of benthic foraging resources for Atlantic walrus (*Odobenus rosmarus rosmarus*) in the Pechora Sea, south-eastern Barents Sea. Aquatic Conservation: Marine and Freshwater Ecosystems 1–14. Available from: https://doi.org/10.1002/aqc.3418.

GINR (Greenland Institute of Natural Resources), 2018a. Svar af 25.07.2018 fra Grønlands Naturinstitut til Departementet for Fiskeri, Fangst og Landbrug vedrørende høring om forsøgsfiskeri efter søpølser og kammuslinger (Answer of 25 July 2018 From the Greenland Institute of Natural Resources to the Department of Fishery, Hunting and Agriculture Regarding Experimental Fishery for Sea Cucumber and Icelandic Scallop). ID. No. 8388284, Version 2 (Nanoq – ID No.8092730), pp. 1–10 (in Danish). Available from: www.natur.gl.

GINR (Greenland Institute of Natural Resources), 2018b. Rådgivning om krabbefiskeriet for 2017-2018 samt status for krabbebestanden (Advice on the Snow Crab Fishery and Status of the Stock), pp. 1–17 (in Danish). https://natur.gl/wp-content/uploads/2019/05/2017-18-DK_Raadgivning_om_krabbefiskeriet_2017_18.pdf.

GINR (Greenland Institute of Natural Resources), Kammuslinger (Icelandic scallop), 2020a. https://natur.gl/arter/kammuslinger/. (in Danish) (Accessed 2 November 2020).

GINR (Greenland Institute of Natural Resources), 2020b. Sammendrag af rådgivning for 2020 om fiskeri på rejebestandene ved Vest – og Østgrønland (Summary of the Advice for 2020 on the Fishery of the Northern Shrimp Stocks in West and East Greenland). November 2019 J.nr. 20.00-11, pp. 1–7 (in Danish). https://natur.gl/wp-content/uploads/2020/10/Raadgivning_rejer_2020_DK.pdf.

Gjertz, I., Wiig, Ø., 1992. Feeding of walrus *Odobenus rosmarus* in Svalbard. Polar Record 28, 57–59.

Gjertz, I., Wiig, Ø., 1994. Past and present distribution of walruses in Svalbard. Arctic 47, 34–42.

Gjertz, I., Wiig, Ø., 1995. The number of walruses (*Odobenus rosmarus*) in Svalbard in summer. Polar Biology 15, 527–530.

Gjertz, I., Wiig, Ø., Øritsland, N.A., 1998. Backcalculation of original population size for walruses *Odobenus rosmarus* in Franz Josef Land. Wildlife Biology 4, 223–230.

Godtfredsen, A.B., Appelt, M., Hastrup, K.B., 2018. Walrus history around the North Water: Human-animal relations in a long-term perspective. Ambio 47 (Suppl. 2), 193–212. Available from: https://doi.org/10.1007/s13280-018-1027-x.

Gonzalez, A.P., 2019. Drones and marine mammals in Svalbard. BIO-3950 Master's Thesis in Biology – Marine Ecology and Resource Biology, pp. 1–45 plus

Appendixes A–H. Faculty of Biosciences, Fisheries and Economics, Department of Arctic and Marine Biology, The Arctic University of Norway.

Gordon, J., Gillespie, D., Potter, J., Frantzis, A., Simmonds, M.P., Swift, R., et al., 2003. A Review of the effects of seismic surveys on marine mammals. Marine Technology Society Journal 37, 16–34.

Griffiths, D.J., Øritsland, N.A., Øritsland, T., 1987. Marine mammals and petroleum activities in Norwegian waters. Fisken og Havet Serie B, No. 1, pp. 1–179.

Gulas, S., Downton, M., D'Souza, K., Hayden, K., Walker, T.R., 2017. Declining Arctic Ocean oil and gas developments: opportunities to improve governance and environmental pollution control. Marine Policy 75, 53–61. Available from: https://doi.org/10.1016/j.marpol.2016.10.014.

Hamilton, L.C., Brown, B.C., Rasmussen, R.O., 2003. West Greenland's cod-to-shrimp transition: local dimensions of climatic change. Arctic 56, 271–282.

Hansen, J.L.S., Hjorth, M., 2013. Box 4. Abundance of sand eel (*Ammodytes dubius*) in the Store Hellefiskebanke area. Boertmann, D., Mosbech, A., Schiedek, D., Dünweber, M. (Eds.), 2013, Disko West. A Strategic Environmental Impact Assessment of Hydrocarbon Activities. Aarhus University, DCE – Danish Centre for Environment and Energy, pp. 94–95. Scientific Report from DCE – Danish Centre for Environment and Energy No. 71.

Hansen, J.L.S., Sejr, M., Josefson, A.B., Batty, P., Hjorth, M., Rysgaard, S., 2013. Benthic invertebrate fauna in the Disko West area with focus on Store Hellefiskebanke. Boertmann, D., Mosbech, A., Schiedek, D., Dünweber, M. (Eds.), Disko West. A Strategic Environmental Impact Assessment of Hydrocarbon Activities. Aarhus University, DCE – Danish Centre for Environment and Energy, pp. 80–86. Scientific Report from DCE – Danish Centre for Environment and Energy No. 71.

Haug, T., Bogstad, B., Chierici, M., Gjøsæter, H., Hallfredsson, E.H., Høines, Å.S., et al., 2017. Future harvest of living resources in the Arctic Ocean north of the Nordic and Barents Seas: a review of possibilities and constraints. Fisheries Research 188, 38–57. Available from: https://doi.org/10.1016/j.fishres.2016.12.002.

Hauser, D.D.W., Laidre, K.L., Stern, H.L., 2018. Vulnerability of Arctic marine mammals to vessel traffic in the increasingly ice-free Northwest Passage and Northern Sea Route. Proceedings of the National Academy of Sciences of the United States of America 115, 7617–7622. Available from: https://doi.org/10.1073/pnas.1803543115.

Heide-Jørgensen, M.P., Laidre, K.L., Fossette, S., Rasmussen, M., Nielsen, N.H., Hansen, R.G., 2014. Abundance of walruses in Eastern Baffin Bay and Davis Strait. Vol. 9. NAMMCO Scientific Publications, Available from: https://dx.doi.org/10.7557.3.2606.

Helle, I., Mäkinen, J., Nevalainen, M., Afenyo, M., Vanhatalo, J., 2020. Impacts of oil spills on Arctic marine ecosystems: a quantitative and probabilistic risk assessment perspective. Environmental Science & Technology 54, 2112–2121. Available from: https://pubs.acs.org/doi/abs/10.1021/acs.est.9b07086.

Higdon, J.W., Stewart, D.B., 2018. State of Circumpolar Walrus (Odobenus rosmarus) Populations. Prepared by Higdon Wildlife Consulting and Arctic Biological Consultants, Winnipeg, MB for WWF Arctic Programme, Ottawa, ON, pp. 1–100.

Huddart, D., Stott, T., 2020. Adventure Tourism – Environmental Impacts and Management. Palgrave Macmillan, London, p. 475. Available from: https://doi.org/10.1007/978-3-030-18623-4.

IAGC (International Association of Geophysical Contractors), 2002. Airgun Arrays and Marine Mammals, pp. 1–16. www.iagc.org.

ICES (International Council for the Exploration of the Sea), 2009. The Barents Sea and the Norwegian Sea. Report of the ICES Advisory Committee, 2009. ICES Advice, 2009. Book 3, pp. 1–81.

Insley, S.J., Phillips, A.V., Charrier, I., 2003. A review of social recognition in pinnipeds. Aquatic Mammals 29.2, 181–201.

iPolitics, 2019. Trudeau Government Expands Moratorium on Oil and Gas Work in Arctic Waters.

Isaksen, K., Bakken, V., Wiig, Ø., 1998. Potential effects on seabirds and marine mammals of petroleum activity in the northern Barents Sea. Norsk Polarinstitutt Meddelelser 154, pp. 1–66.

Jamasmie, C., 2020. 'Greenland Grants Bluejay Exploitation License'. *Mining.com Newsletter*. 14 December. mining.com. (Accessed 24 January 2021).

Jay, C.V., 2015. Pacific walrus population response to reduced sea ice and human-caused disturbance. Symposium on the Impacts of Human Disturbance on Arctic Marine Mammals, With a Focus on Belugas, Narwhals and Walrus. NAMMCO (North Atlantic Marine Mammal Commission), 13–15 October 2015. University of Copenhagen, Copenhagen, Denmark, p. 19. www.nammco.no.

Jay, C.V., Olson, T.L., Garner, G.W., Ballachey, B.E., 1998. Response of Pacific walrus to disturbances from capture and handling activities at a haul-out in Bristol Bay, Alaska. Marine Mammal Science 14, 819–828. Available from: https://doi.org/10.1111/j.1748-7692.1998.tb00765.x.

Johnston, M.E., Dawson, J., Maher, P.T., 2017. Strategic development challenges in marine tourism in Nunavut. Resources 6 (3), 25. Available from: https://doi.org/10.3390/resources6030025.

Jones, J.B., 1992. Environmental impact of trawling on the sea bed: a review. New Zealand Journal of Marine and Freshwater Research 26, 59–67. Available from: https://www.tandfonline.com/doi/pdf/10.1080/00288330.1992.9516500.

Jørgensen, O.A., Hammeken, N., 2013. Distribution of the commercial fishery for Greenland halibut and Northern shrimp in Baffin Bay. Technical Report no. 91. Greenland Institute of Natural resources. Available from: https://natur.gl/forskning/rapporter/.

Kastelein, R.A., 2002. Walrus (*Odobenus rosmarus*). Perrin, W.F., Wursig, B., Thewissen, J.M.G. (Eds.), Encyclopedia of Marine Mammals, second ed. Academic Press, San Diego, CA, pp. 1212–1217.

Kastelein, R.A., Van Ligtenberg, C.L., Gjertz, I., Verboom, W.C., 1993a. Free field hearing tests on wild Atlantic walruses (*Odobenus rosmarus rosmarus*) in air. Aquatic Mammals 19, 143–148.

Kastelein, R.A., Zweypfenning, R.C.V.J., Spekreijse, H., Dubbeldam, J.L., Born, E.W., 1993b. The anatomy of the walrus head (*Odobenus rosmarus*). Part 3: The eyes and their function in walrus ecology. Aquatic Mammals 19, 61–92.

Kastelein, R.A., Mosterd, P., van Ligtenberg, C.L., Verboom, W.C., 1996. Aerial hearing sensitivity tests with a male Pacific walrus (*Odobenus rosmarus divergens*), in the free field and with headphones. Aquatic Mammals 22, 81–93.

Kastelein, R.A., Mosterd, P., van Santen, B., Hagedoorn, M., 2002. Underwater audiogram of a Pacific walrus (*Odobenus rosmarus divergens*) measured with narrow-band frequency-modulated signals. Journal of the Acoustical Society of America 112, 2173–2182. Available from: https://doi.org/10.1121/1.1508783.

Kedra, M., Renaud, P.E., Andrade, H., Goszczko, I., Ambrose Jr., W.G., 2013. Benthic community structure, diversity, and productivity in the shallow Barents Sea bank (Svalbard Bank). Marine Biology 160, 805–819. Available from: https://doi.org/10.1007/s00227-012-2135-y.

Keighley, X., Palsson, S., Einarsson, B.F., Petersen, A., Fernández-Coll, M., Jordan, P., et al., 2019. Disappearance of Icelandic walruses coincided with Norse settlement. Molecular Biology Evolution 36, 2656–2667. Available from: https://doi.org/10.1093/molbev/msz196.

Kovacs, K.M., 2016. *Odobenus rosmarus* ssp. *rosmarus*. IUCN Red List Threatened Species e.T15108A66992323. Available from: https://doi.org/10.2305/IUCN.UK.2016-1.RLTS.T15108A66992323.en.

Kovacs, K.M., Lemons, P., MacCracken, J.G., Lydersen, C., 2015. Walruses in a time of climate change. Arctic Report Card: update for 2015. Tracking recent environmental changes. Available from: http://www.arctic.noaa.gov/reportcard/walruses.html.

Kovacs, K.M., Aars, J., Lydersen, C., 2014. Walruses recovering after 60 + years of protection in Svalbard, Norway. Polar Research 33, 26034. Available from: https://doi.org/10.3402/polar.v33.26034.

Kruse, S., 1997. Behavioral changes of Pacific walrus (*Odobenus rosmarus divergens*) in response to human activities. US FWS Technical Report MMM 97-4, pp. 1–30.

Kucklick, J.R., Krahn, M.M., Becker, P.R., Porter, B.J., Schantz, M.M., York, G.S., et al., 2006. Persistent organic pollutants in Alaskan ringed seal (Phoca hispida) and walrus (Odobenus rosmarus) blubber. Journal of Environmental Monitoring 8 (8), 848–854.

Kyhn, L.A., Tougaard, J., Sveegaard, S., 2011. Underwater Noise From the Drillship Stena Forth in Disko West, Baffin Bay, Greenland. National Environmental Research Institute, Aarhus University, Aarhus, pp. 1–30NERI Technical Report No. 838. Available from: http://www.dmu.dk/Pub/FR838.pdf.

Laidre, K.L., Stirling, I., Lowry, L.F., Wiig, O., Heide-Jorgensen, M.P., Ferguson, S.H., 2008. Quantifying the sensitivity of Arctic marine mammals to climate-induced habitat change. Ecological Applications 18, S97–S125.

Laidre, K.L., Stern, H., Kovacs, K.M., Lowry, L., Moore, Regehr, E.V., et al., 2015. Arctic marine mammal population status, sea ice habitat loss, and conservation recommendations for the 21st century. Conservation Biology 29, 724–737. Available from: https://doi.org/10.1111/cobi.12474.

Lambert, J., Préfontaine, G., 1995. Le pétoncle d'Islande (Chlamys islandica) au Nunavik. Rapport Technique Canadien des Sciences Halieutiques et Aquatiques 2071, pp. 1–41.

Lasserre, F., Têtu, P., 2015. The cruise tourism industry in the Canadian Arctic: analysis of activities and perceptions of cruise ship operators. Polar Record 51 (1), 24–38. Available from: https://doi.org/10.1017/S0032247413000508.

Lawson, J.W., Lesage, V., 2012. A draft framework to quantify and cumulate risks of impacts from large development projects for marine mammal populations: a case study using shipping associated with the Mary River iron mine project. Canadian Science Advisory Secretariat Research Document 2012/154, pp. 1–22.

Lemelin, H., Dawson, J., Stewart, E.J., Maher, P., Lueck, M., 2010. Last-chance tourism: the boom, doom, and gloom of visiting vanishing destinations. Current Issues in Tourism 13 (5), 477–493. Available from: https://doi.org/10.1080/13683500903406367.

Linnebjerg, J.F., Hobson, K.A., Fort, J., Nielsen, T.G., Møller, P., Wieland, K., et al., 2016. Deciphering the structure of the West Greenland marine food web using stable isotopes ($\delta^{13}C$, $\delta^{15}N$). Marine Biology 163, 230. Available from: https://doi.org/10.1007/s00227-016-3001-0.

Loughrey, A.G., 1959. Preliminary investigation of the Atlantic walrus *Odobenus rosmarus rosmarus* (Linnaeus). Canadian Wildlife Service. Wildlife Management Bulletin Series 1, 14, 1–123.

Løkkeborg, S., 2004. Impacts of trawling and scallop dredging on benthic habitats and communities. FAO Technical Paper No. 472. Food and Agricultural Organization of the United Nations (FAO), Rome, pp. 1–58. http://www.fao.org/3/y7135e/y7135e00.htm.

Lydersen, C., Chernook, V.I., Glazov, D.M., Trukhanova, I.S., Kovacs, K.M., 2012. Aerial survey of Atlantic walruses (*Odobenus rosmarus rosmarus*) in the Pechora Sea, August 2011. Polar Biology 35, 1555–1562.

Magnus, O., 1555. Historia de gentibus septentrionalibus. Rome. Available at Free Google Books.

Malme, C.I., Miles, P.R., Miller, G.W., Richardson, W.J., Roseneau, D.G., Thompson, D. H., et al., 1989. Analysis and ranking of the acoustic disturbance potential of petroleum industry activities and other sources of noise in the environment of marine mammals in Alaska. BBN Report 6945, OCS Study MMS 89-0006. Report From BBN Systems and Technological Corporation, Cambridge, MA for U.S. Minerals Management Service, NTIS PB90-188673. Bolt, Beranek, and Newman, Anchorage, AK.

Mansfield, A.W., 1958. The biology of the Atlantic walrus, Odobenus rosmarus rosmarus (Linnaeus) in eastern Canadian arctic. Fisheries Research Board of Canada Manuscript Report Series (Biology) No. 653, pp. 1–146.

Mansfield, A., Aubin St., D.J., 1991. Distribution and abundance of the Atlantic walrus, *Odobenus rosmarus rosmarus*, in the Southampton Island – Coats Island region of northern Hudson Bay. Canadian Field-Naturalist 105, 95–100.

Manushin, I., Blinova, D.Y., 2019. Russian scallop fishery: happiness or experience? Presentation at Shellfish Symposium, and in Book of Abstracts. Institute of Marine Research, 5–7 November 2019, Tromsø, Norway, p. 37.

Mason, P., 1997. Tourism codes of conduct in the Arctic and sub-Arctic region. Journal of Sustainable Tourism 5 (2), 151–165. Available from: https://doi.org/10.1080/09669589708667282.

Mass, A.M., Supin, A.Y., 2019. Eye optics in semiaquatic mammals for aerial and aquatic vision. Brain Behavior and Evolution 92, 117–124. Available from: https://doi.org/10.1159/000496326.

Mattox, W., 1973. Fishing in West Greenland 1910-1966. The development of a new native industry. Meddelelser Grønland (Monographs on Greenland) 197 (1), 1–344. plus appendix, 1–125.

McConnaughey, R.A., Mier, K.L., Dew, C.B., 2000. An examination of chronic trawling effects on soft-bottom benthos of the eastern Bering Sea. ICES Journal of Marine Science 57, 1377–1388.

McConnaughey, R.A., Syrjala, S.E., 2014. Short-term effects of bottom trawling and a storm event on soft-bottom benthos in the eastern Bering Sea. ICES Journal of Marine Science 71, 2469–2483. Available from: https://doi.org/10.1093/icesjms/fsu054.

McConnaughey, R.A., Syrjala, S.E., Dew, C.B., 2005. Effect of chronic bottom trawling on the size structure of sift-bottom benthic invertebrates. Fisheries Society Symposium 41, pp. 425–437.

McFarland, S.E., Aerts, L.A.M., 2015. Assessing disturbance responses of Pacific Walrus (*Odobenus rosmarus divergens*) to vessel presence in the Chukchi Sea (Abstract). Chukchi Sea Environmental Studies Program (CSESP), Olgoonik – Fairweather, Fairweather Science, Anchorage, AK. https://www.chukchiscience.com/Portals/0/Public/Science/MarineMammals/2015_CSESP_MarineMammals_SMM_Abstract_WalrusResponseVessels.pdf.

McFarland, S.E., Wisdom, S., Hetrick, W., 2017. Assessing disturbance responses of marine mammals to vessel presence in the Bering, Chukchi, and Beaufort Seas. Poster presentation. Alaska Marine Science Symposium (AMSS), 23–27 January 2017, Abstract Book, Anchorage, AK, p. 359 (*fide* Wilson et al., 2020).

McKinney, M.A., Pedro, S., Dietz, R., Sonne, C., Fisk, A.T., Roy, D., et al., 2015. A review of ecological impacts of global climate change on persistent organic pollutant and mercury pathways and exposures in arctic marine ecosystems. Current Zoology 61 (4), 617–628.

McLeod, B.A., Frasier, T.R., Lucas, Z., 2014. Assessment of the extirpated Maritimes Walrus using morphological and ancient DNA analysis. PLoS One 9 (6), e99569. Available from: https://doi.org/10.1371/journal.pone.0099569.

Melia, N., Haines, K., Hawkins, E., 2016. Sea ice decline and 21st century trans-Arctic shipping routes. Geophysical Research Letters 43, 9720–9728. Available from: https://doi.org/10.1002/2016GL069315.

Messieh, S.N., Rowell, T.W., Peer, D.L., Cranford, P.J., 1991. The effects of trawling, dredging and ocean dumping on the eastern Canadian continental shelf seabed. Continental Shelf Research 11, 1237–1263.

Moore, S.E., Reeves, R.R., Southall, B.L., Ragen, T.J., Suydam, R.S., Clark, C.W., 2012. A new framework for assessing the effects of anthropogenic sound on marine mammals in a rapidly changing Arctic. BioScience 62, 289–295.

Muir, D., Born, E.W., Koczansky, K., Stern, G., 2000. Temporal and spatial trends of persistent organochlorines in Greenland walrus (*Odobenus rosmarus rosmarus*). Science of the Total Environment 245, 73–86.

Muir, D.C.G., Segstro, M.D., Hobson, K.A., Ford, C.A., Stewart, R.E.A., Olpinski, S., 1995. Can seal eating explain elevated levels of PCBs and organochlorine pesticides in walrus blubber from eastern Hudson Bay (Canada)? Environmental Pollution 90 (3), 335–348.

NAMMCO (North Atlantic Marine Mammal Commission), 1995. Annex 2. Report of the *ad hoc* Working Group on Atlantic walrus. NAMMCO Annual Report, 1995. Tromsø, Norway, pp. 111–131. www.nammco.no.

NAMMCO (North Atlantic Marine Mammal Commission), 2004. Report of the NAMMCO Workshop on Hunting Methods for Seals and Walrus. North Atlantic House, Copenhagen, pp. 1–60. www.nammco.no.

NAMMCO (North Atlantic Marine Mammal Commission), 2006. Annex 3. NAMMCO Scientific Committee Working Group on the stock status of walruses in the North Atlantic and adjacent seas. Annual Report 2005. North Atlantic Marine Mammal Commission, Tromsø, pp. 279–308. https://nammco.no/topics/annual-reports/.

NAMMCO (North Atlantic Marine Mammal Commission), Annual Report, 2009. https://nammco.no/topics/annual-reports/.

NAMMCO (North Atlantic Marine Mammal Commission), Annual Report, 2013. https://nammco.no/topics/annual-reports/.

NAMMCO (North Atlantic Marine Mammal Commission), 2015. Symposium on the Impacts of Human Disturbance on Arctic Marine Mammals, With a Focus on Belugas, Narwhals & Walrus. 13–15 October 2015. University of Copenhagen, Copenhagen, pp. 1–30. https://nammco.no/wp-content/uploads/2017/01/report-nammco-disturbance-symposium-2015.pdf.

NAMMCO (North Atlantic Marine Mammal Commission), 2018a. Report of the NAMMCO (North Atlantic Marine Mammal Commission) Scientific Working Group on Walrus. https://nammco.no/topics/sc-working-group-reports/.

NAMMCO (North Atlantic Marine Mammal Commission), 2018b. Report of the 25th Scientific Committee Meeting, 13–16 November, Norway. Revised 2019, pp. 70–76. https://nammco.no/topics/annual-reports/.

NAMMCO (North Atlantic Marine Mammal Commission), 2019. Report of the Scientific Committee 26th Meeting, October 29–November 1, Tórshavn Faroe Islands. https://nammco.no/topics/scientific-committee-reports/

NAMMCO (North Atlantic Marine Mammal Commission), 2020. North Atlantic Marine Mammal Commission website. Atlantic walrus. https://nammco.no/topics/atlantic-walrus/#1475844711542-eedf1c7b-5dde. (Accessed 20 October 2020).

NOEP (National Ocean Economics Program), 2021. Arctic fisheries. https://www.oceaneconomics.org/arctic/fisheries/. (Accessed 28 January 2021).

Øian, H., Kaltenborn, B., 2020. Turisme på Svalbard og i Arktis. Effekter på naturmiljø, kulturminner og samfunn med hovedvekt på cruiseturisme [Tourism in Svalbard and the Arctic. Effects on the environment, cultural heritage sites and community with special emphasis on cruise ship tourism]. NINA Rapport 1745. Norsk Institutt for Naturforskning, pp. 1–52. ISSN 1504-3312, ISBN 978-82-426-4500-5 (in Norwegian with English Abstract).

Orbicon, 2019. Potential effects of underwater noise on marine mammals from shipping. Report From Orbicon Consultants, Taastrup, Denmark to Dundas Titanium A/S, pp. 1–22. https://www.orbicon.dk/expertises/milj%C3%B8.

Orbicon, 2020. Results of aerial surveys of marine mammals at Moriusaq, NW Greenland. Report From Orbicon Consultants, Taastrup, Denmark to Dundas Titanium A/S, pp. 1–26. https://www.orbicon.dk/expertises/milj%C3%B8.

Øren, K., Kovacs, K.M., Yoccoz, N.G., Lydersen, C., 2018. Assessing site-use and sources of disturbance at walrus haul-outs using monitoring cameras. Polar Biology 41, 1737–1750. Available from: https://doi.org/10.1007/s00300-018-2313-6.

Orr, J.R., Renooy, B., Dahlke, L., 1986. Information from hunts and surveys of walrus (Odobenus rosmarus) in northern Foxe Basin, Northwest Territories, 1982-1984. Polar Biology Canadian Manuscript Reports of Fisheries and Aquatic Sciences No. 1899, pp. 1–24.

PAME (Protection of the Arctic Marine Environment), 2019. Underwater Noise in the Arctic: A State of Knowledge Report. Rovaniemi. PAME Secretariat, Akureyri, pp. 1–59.

PAME (Protection of the Arctic Marine Environment), 2020. The increase in Arctic shipping 2013-2019. Arctic Shipping Status Report (ASSR), pp. 1–25. https://www.pame.is/projects/arctic-marine-shipping/arctic-shipping-status-reports/723-arctic-shipping-report-1-the-increase-in-arctic-shipping-2013-2019-pdf-version/file (Downloaded 30 January 2021).

Pedersen, S., 1994. Population parameters of the Iceland scallop (*Chlamys islandica* (Müller)) from west Greenland. Journal of Northwest Atlantic Fishery Science 16, 75–87.

Pirotta, E., Booth, C.G., Costa, D.P., Fleishman, E., Kraus, S.D., Lusseau, D., et al., 2018. Understanding the population consequences of disturbance. Ecology and Evolution 8 (19), 2045–7758. Available from: https://doi.org/10.1002/ece3.4458.

Przeslawski, R., Huang, Z., Anderson, J., Carroll, A.G., Edmunds, M., Hurt, L., et al., 2018. Multiple field-based methods to assess the potential impacts of seismic surveys on scallops. Marine Pollution Bulletin 129, 750–761. Available from: https://doi.org/10.1016/j.marpolbul.2017.10.066.

Purves, D., Augustine, G.J., Fitzpatrick, D., Katz, L.C., LaMantia, A.S., McNamara, J.O., et al., (Eds.), 2001. Neuroscience. second ed. Sinauer Associates, Sunderland, MAISBN-10: 0-87893-742-0. Available from: https://www.ncbi.nlm.nih.gov/books/NBK10799/.

Reeves, R.R., 1978. Atlantic walrus (*Odobenus rosmarus rosmarus*): a literature survey and status report. Wildlife Research Report 10. United States Department of the Interior, Fish and Wildlife Service Washington, DC, pp. 1–41.

Reichmuth, C., Sills, J.M., Brewer, A., Triggs, L., Ferguson, R., Ashe, E., et al., 2020. Behavioral assessment of in-air hearing range for the Pacific walrus (*Odobenus rosmarus divergens*). Polar Biology 43, 767–772. Available from: https://doi.org/10.1007/s00300-020-02667-6.

Richardson, W.J., Greene, J.C.R., Malme, C.I., Thomson, D.H., 1995. Marine Mammals and Noise. Academic Press, San Diego, CA.

Rigét, F., Bignert, A., Braune, B., Dam, M., Dietz, R., Evans, M., et al., 2019. Temporal trends of persistent organic pollutants in Arctic marine and freshwater biota. Science of the Total Environment 649, 99–110.

Rigét, F., Braune, B., Bignert, A., Wilson, S., Aars, J., Born, E., et al., 2011. Temporal trends of Hg in Arctic biota, an update. Science of the Total Environment 409, 3520–3526. Available from: https://doi.org/10.1016/j.scitotenv.2011.05.002.

Riget, F., Dietz, R., Born, E.W., Sonne, C., Hobson, K.A., 2007. Temporal trends of mercury in marine biota of west and northwest Greenland. Marine Pollution Bulletin 54, 72–80.

Roesdahl, E., 2003. Walrus ivory in the Viking Age – and Ohthere (Ottar). Offa-Jahrbuch 58, 33–37.

Routti, H., Diot, B., Panti, C., Duale, N., Fossi, M.C., Harju, M., et al., 2019. Contaminants in Atlantic walruses in Svalbard Part 2: Relationships with endocrine and immune systems. Environmental Pollution 246, 658–667.

Rowlands, E., Galloway, T., Manno, C., 2021. A polar outlook: potential interactions of micro- and nano-plastic with other anthropogenic stressors. Science of the Total Environment 754, 142379. Available from: https://doi.org/10.1016/j.scitotenv.2020.142379.

Salter, R.E., 1979. Site utilisation, activity budgets, and disturbance responses of Atlantic walruses during terrestrial haul-out. Canadian Journal of Zoology 57, 1169–1180.

Schack, H., Haapaniemi, J., 2017. Potential impact of noise from shipping on key species of marine mammals in waters off Western Greenland-Case Baffinland. Report Prepared for WWF Denmark, Report Number 30-06-2017. https://arcticwwf.org/site/assets/files/

1280/final_report_potential_impact_of_noise_from_shipping_on_key_species_of_marine_mammals_i_1.pdf.

Scotter, S.E., Tryland, M., Nymo, I.H, Hanssen, L., Lydersen, C., Kovacs, K.M., et al., 2019. Contaminants in Atlantic walruses in Svalbard part 1: relationships between exposure, diet and pathogen prevalence. Environmental Pollution 244, 9–18. Available from: https://doi.org/10.1016/j.envpol.2018.10.001.

Sermitsiaq, 2020. 'Dundas Titanium får udnyttelsestilladelse ved Moriusaq (The Dundas Titanium Company Will Get Permission for Exploitation at Moriussaq)'. *Newspaper Sermitsiaq, Greenland*, 11 March. https://sermitsiaq.ag/dundas-titanium-faar-udnyttelsestilladelse-ved-moriusaq (in Danish).

Semyonova, V.S., Boltunov, A.N., Nikiforov, V.V., 2015. Studying and Preserving the Atlantic Walrus in the South-East Barents Sea and Adjacent Areas of the Kara Sea. 2011-2014 Study Results. World Wildlife Fund (WWF), Murmansk, pp. 1–82.

Shell, 2015. Revised Outer Continental Shelf Lease Exploration Plan Chukchi Sea, Alaska. Burger Prospect: Posey Area Blocks 6714, 6762, 6764, 6812, 6912, 6915 Revision 2 (March 2015). Environmental Assessment. OCS EIS/EA BOEM 2015-020, pp. 1–212 + Appendix A, B, C.

Sjare, B., Stirling, I., 1996. The breeding behavior of Atlantic walruses, *Odobenus rosmarus rosmarus*, in the Canadian High Arctic. Canadian Journal of Zoology 74, 897–911.

Sjare, B., Stirling, I., Spencer, C., 2003. Structural variation in the songs of Atlantic walruses breeding in the Canadian high Arctic. Aquatic Mammals 29, 297–318.

Skoglund, E.G., Lydersen, C., Grahl-Nielsen, O., Haug, T., Kovacs, K.M., 2010. Fatty acid composition of the blubber and dermis of adult male Atlantic walruses (*Odobenus rosmarus rosmarus*) in Svalbard, and their potential prey. Marine Biology Research 6, 239–250.

Southall, B.L., Bowles, A.E., Ellison, W.T., Finneran, J.J., Gentry, R.L., Greene, C.R., et al., 2007. Marine mammal noise exposure criteria: initial scientific recommendations. Aquatic Mammals 33, 411–521.

Star, B., Barrett, J.H., Gondek, A.T., Boessenkool, S., 2018. Ancient DNA reveals the chronology of walrus ivory trade from Norse Greenland. Proceedings of the Royal Society B 285, 20180978. Available from: https://doi.org/10.1098/rspb.2018.0978.

Stewart, R.E.A., 2002. Review of Atlantic walrus (Odobenus rosmarus rosmarus) in Canada. Canadian Science Advisory Secretariat Research Document 2002/09, pp. 1-20.

Stewart, R.E.A., Richard, P.R., Stewart, B.E., (Eds.), 1993. Report of the 2nd Walrus International Technical and Scientific (WITS) Workshop, 11–15 January 1993, Winnipeg, Manitoba, Canada. Canadian Technical Report of Fisheries and Aquatic Science 194, pp. 1–91.

Stewart, E.J., Dawson, J., Draper, D., 2011. Cruise tourism and residents in Arctic Canada: development of a resident attitude typology. Journal of Hospitality and Tourism Management 18 (1), 95–106. Available from: https://doi.org/10.1375/jhtm.18.1.95.

Stewart, R.E.A., Lesage, V., Lawson, J.W., Cleator, H., Martin. K.A., 2012. Science technical review of the draft Environmental Impact Statement (EIS) for Baffinland's Mary River project. DFO Canadian Science Advisory Secretariat Research Document 2011/086, pp. 1–6.

Stewart, R.E.A., Born, E.W., Dietz, R., Ryan, A.K., 2014. Estimates of Minimum Population Size for Walrus Around Southeast Baffin Island, 9. NAMMCO Scientific Publications, Nunavut, pp. 141–157. Available from: http://doi.org/10.7557/3.2615.

Stewart, D.B., Higdon, J.W., Stewart, R.E.A., 2020. Threats and Effects Pathways of Shipping Related to Non-Renewable Resource Developments on Atlantic Walruses (*Odobenus rosmarus rosmarus*) in Hudson Strait and Foxe Basin, 3283. Canadian Technical Report of Fisheries and Aquatic Science, Nunavut, pp. 1–59.

Stirling, I., Thomas, J.A., 2003. Relationships between underwater vocalizations and mating systems in phocid seals. Aquatic Mammals 29, 227–246.

Stirling, I., Calvert, W., Cleator, H., 1983. Underwater vocalizations as a tool for studying the distribution and relative abundance of wintering pinnipeds in the High Arctic. Arctic 36, 262–274.

Stirling, I., Calvert, W., Spencer, C., 1987. Evidence of stereotyped underwater vocalizations of male Atlantic walruses (*Odobenus rosmarus rosmarus*). Canadian Journal of Zoology 65, 2311–2321.

Sutterud, T., Ulven, E. (2020). 'Norway Plans to Drill for oil in Untouched Arctic Areas'. *The Guardian*. Wed 26 Aug 2020 13.42 BST. Last modified on Thu 27 Aug 2020. https://www.theguardian.com/environment/2020/aug/26/norway-plans-to-drill-for-oil-in-untouched-arctic-areas-svalbard. (Accessed 29 November 2020).

Tillin, H.M., Hiddink, J.G., Jennings, S., Kaiser, M.J., 2006. Chronic bottom trawling alters the functional composition of benthic invertebrate communities on a sea-basin scale. Marine Ecology Progress Series 318, 31–45.

Timoshenko, I.K., 1984. Concerning the protection and restoration of the western Atlantic population of the walrus. Yablokov, A.V. (Ed.), Marine Mammals (F.H. Fay, Trans., 1–4). Nauka, pp. 100–103 (in Russian).

Tomilin, A.G., Kibal'chich, A.A., 1975. The walruses in the vicinity of Wrangel Island. Zoologicheskij Zhurnal 54, 266–272 (Canadian Fisheries and Marine Service Translation 3721, 1976. 1-1).

Tranter, E. (2019). 'Over 34,000 More People Visited Nunavut in 2018 Than in 2015', *Nunatsiaq News*, 12 November. https://nunatsiaq.com/stories/article/over-34000-more-people-visited-nunavut-in-2018-than-in-2015.

Trukhin, A.M., Simokon, M.V., 2018. Mercury in organs of Pacific walruses (Odobenus rosmarus divergens) from the Bering Sea. Environmental Science and Pollution Research 25, 3360–3367.

Udevitz, M.S., Taylor, R.L., Garlich-Miller, J.L., Quakenbush, L.T., Snyder, J.A., 2013. Potential population-level effects of increased haulout-related mortality of Pacific walrus calves. Polar Biology 36, 291–298. Available from: https://doi.org/10.1007/s00300-012-1259-3.

US FWS (United States Fish and Wildlife Service), 2012. Pacific Walrus (*Odobenus rosmarus divergens*): Alaska Stock. United States Fish and Wildlife Service, Marine Mammals Management, Anchorage, AK. http://alaska.fws.gov/fisheries/mmm/walrus/reports.htm#stock.

Vasilyeva, M. (2020). 'Russian Scientists Discover Huge Walrus Haul-Out in Arctic Circle'. *Reuter. Science News*. 6 November. https://www.reuters.com/article/uk-russia-animals-walruses/russian-scientists-discover-huge-walrus-haulout-in-arctic-circle-idUKKBN27M1SF.

Vibe, C., 1950. The marine mammals and the marine fauna in the Thule District (Northwest Greenland) with observations on the ice conditions in 1939-1941. Meddelelser Grønland (Monographs on Greenland) 150, 1–115.

Vigliotti, M., Trudeau government expands moratorium on oil and gas work in arctic waters. 2019. https://ipolitics.ca/2019/08/08/. (Accessed 28 November 2020).

Viðarsson, J.R., Þórðarson, G., Henriksen, E., Iversen, A., Djurhuus, D., Berthelsen, T., et al., 2015. Coastal fisheries in the North Atlantic. Resources & products. Skýrsla Matís 1–56. ISSN 1670-7192.

Visit Greenland, 2019. Tourism Statistics Report Greenland, pp. 1–22. http://www.tourismstat.gl/resources/reports/en/r31/Tourism%20Statistics%20Report%20Greenland%202019.pdf.

Wartzok, D., Popper, A.N., Gordon, J., Merrill, J., 2003. Factors affecting the responses of marine mammals to acoustic disturbance. Marine Technology Society Journal 37, 6–15.

Wegeberg, S., Johnsen, A., Aamand, J., Lassen, P., Gosewinkel, U., Fritt-Rasmussen, J., et al., 2018. Arctic marine potential of microbial oil degradation. Scientific Report from DCE – Danish Centre for Environment and Energy No. 271. Aarhus University, DCE – Danish Centre for Environment and Energy, pp. 1–54. http://dce2.au.dk/pub/SR271.pdf.

Wickens, P.A., 1995. A review of operational interactions between pinnipeds and fisheries. FAO Fisheries Technical Paper 346. FAO Library A: 361047, pp. 1–86. ISSN 0429-9345.

Wiig, Ø., Belikov, S.E., Boltunov, A.N., Garner, G.W., 1996. Selection of marine mammal valued ecosystem components and description of impact hypotheses in the Northern Sea Route Area. INSROP Working Paper No. 40 – 1996, II.4.3, pp. 1–70. ISBN 82-7613-141-7, ISSN 0805-2522.

Wiig, Ø., Berg, V., Gjertz, I., Seagars, D.J., Skaare, J.U., 2000. Use of skin biopsies for monitoring levels of organochlorines in walruses (*Odobenus rosmarus*). Polar Biology 23, 272–278.

Wiig, Ø., Born, E.W., Stewart, R.E.A., 2014. Management of Atlantic Walrus (*Odobenus rosmarus rosmarus*) in the Arctic Atlantic, vol. 9. NAMMCO Scientific Publications, pp. 315–341. Available from: http://doi.org/10.7557/3.2855.

Wikipedia, 2020a. Petroleum exploration in the Arctic. https://en.wikipedia.org/wiki/Petroleum_exploration_in_the_Arctic#Overview. (Accessed 28 November 2020).

Wikipedia, 2020b. History of the petroleum industry in Canada (frontier exploration and development). https://en.wikipedia.org/wiki/History_of_the_petroleum_industry_in_Canada_(frontier_exploration_and_development)#Arctic_frontiers. (Accessed 24 January 2021).

Wilson, B., Evans, D., 2009. Groundfish Trawl Fishery, Pacific Walrus, and Local Fishery Interactions in Northern Bristol Bay – An Updated Discussion Paper. North Pacific Management Council, p. 36. Available from: https://www.npfmc.org/wp-content/PDFdocuments/catch_shares/Trawl_walrus1209.pdf.

Wilson, S., Goodmann, S., 2016. Icebreakers and Ice Breeding Seals. The Circle. WWF Magazine No. 3, pp. 16–21.

Wilson, S.C., Crawford, I., Trukhanova, I., Dmitrieva, L., Goodman, S.J., 2020. Estimating risk to ice-breeding pinnipeds from shipping in Arctic and sub-Arctic seas. Marine Policy 111. Available from: https://doi.org/10.1016/j.marpol.2019.103694.

Witting, L., Born, E.W., 2014. Population Dynamics of Walrus in Greenland, vol. 9. NAMMCO Scientific Publications, pp. 191–218. Available from: http://doi.org/10.7557/3.2612.

Wolkers, H., Van Bavel, B., Ericson, I., Skoglund, E., Kovacs, K.M., Lydersen, C., 2006. Congener-specific accumulation and patterns of chlorinated and brominated contaminants in adult male walruses from Svalbard, Norway: indications for individual-specific prey selection. Science of the Total Environment 370 (1), 70–79.

WWF (World Wildlife Fund), 2019. WWF Canada's Final Written Submission for the Nunavut Impact Review Board's Reconsideration of the Mary River Project Certificate for Baffinland's Mary River "Phase 2" Proposal. [submitted September 23, 2019 to Nunavut Impact Review Board (NIRB)], https://wwf.ca/wp-content/uploads/2020/03/wwf_final_written_submission_for_mary_river_phase_2_September-2019.pdf

Yesson, C., Fisher, J., Gorham, T., Turner, C., Hammeken Arboe, N., Blicher, M., et al., 2016. The impact of trawling on the epibenthic megafauna of the west Greenland shelf. ICES Journal of Marine Science. Available from: https://doi.org/10.1093/icesjms/fsw206.

Yurkowski, D.J., Auger-Méthé, M., Mallory, M.L., Wong, S.N.P., Grant Gilchrist, H., Derocher, A.E., et al., 2018. Abundance and species diversity hotspots of tracked marine predators across the North American Arctic. Diversity and Distributions 328–345. Available from: https://doi.org/10.1111/ddi.12860.

Zolotarev, P., 2002. Population density and size structure of sea stars on beds of Iceland scallop, *Chlamys islandica*, in the southeastern Barents Sea. Sarsia 87, 91–96. Available from: https://doi.org/10.1080/003648202753631776.

CHAPTER

The future of Atlantic walrus in a rapidly warming Arctic 13

Erik W. Born[1], Øystein Wiig[2] and Morten Tange Olsen[3]

[1]*Greenland Institute of Natural Resources, Nuuk, Greenland*

[2]*Natural History Museum, University of Oslo, Oslo, Norway*

[3]*Globe Institute, University of Copenhagen, Copenhagen, Denmark*

Chapter Outline

Introduction

Walruses (*Odobenus rosmarus*) are regarded as a focal ecosystem component (FEC) in assessments by the Circumpolar Biodiversity Monitoring Program (CAFF, 2017). An FEC is a biological element that is considered central to the functioning of an ecosystem, is of major importance to Arctic residents and/or is likely to be a good proxy for short- and long-term changes in the environment (ibid.). In certain areas of their range, walruses are of great value to Indigenous communities as a resource (e.g., Born et al., 2017; Krupnik, 2018). Hence, the status of the walrus is a key conservation question for national authorities throughout its range; both where it is hunted for subsistence

The Atlantic Walrus. DOI: https://doi.org/10.1016/B978-0-12-817430-2.00012-1

purposes (i.e., Alaska, Canada, Greenland and Russian Pacific) and where it is not (i.e., Norway and Russia Atlantic and Laptev).

The Arctic is undergoing large-scale environmental changes due to climate change (e.g., Post et al., 2019), including warmer temperatures and a marked reduction of sea ice in several areas inhabited by walruses (e.g., Laidre et al., 2015). Decreasing sea ice is projected to continue, and it has been forecasted that by 2030 the Arctic seas will experience ice-free summers (Overland and Wang, 2013; Fig. 13.1). The reduction of sea ice will likely have severe consequences for several ice-associated Arctic pinnipeds including the walrus, which uses sea ice for essential life-history events, such as giving birth and hauling out on drifting ice floes in the vicinity of foraging banks. The potential effects of a warming Arctic on walruses are multiple, cumulative and complex and still poorly understood. They may be both direct (e.g., decreasing availability of on-ice haul-out platforms) and indirect (e.g., changes in prey distribution and availability) (Boyd et al., 2019). Moreover, whereas some effects will be detrimental, others may actually be advantageous, and effects may certainly vary on temporal and

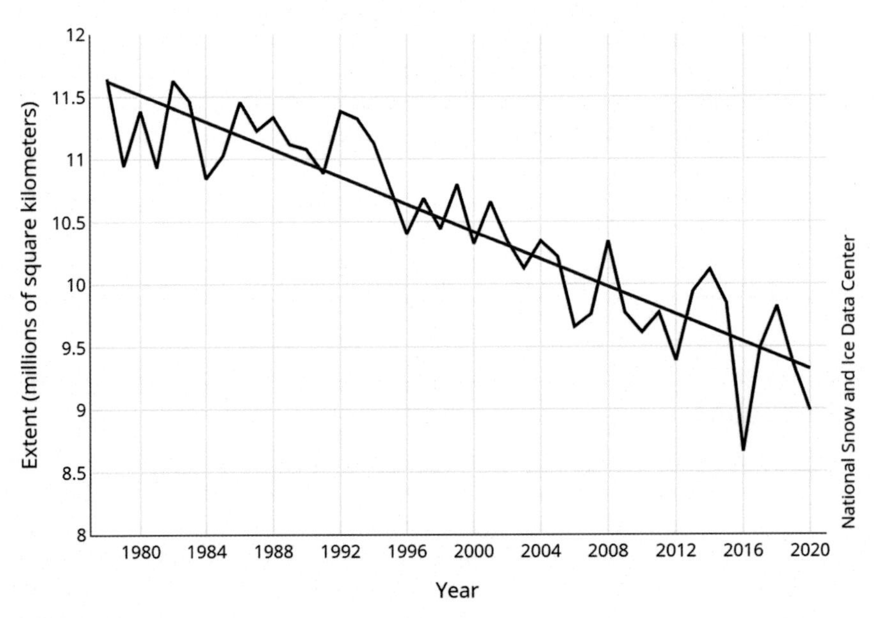

FIGURE 13.1

Decreases in Arctic sea ice in the period 1978–2020. The linear rate of decline for November sea ice extent is 5.1% per decade, corresponding to a loss of about 2.3 million km^2 of sea ice in November over the 42-year study period (NSIDC, 2020); an area larger than the entire area of Greenland.

geographical scales. For instance, Pacific walruses (*Odobenus rosmarus divergens*) in the Bering Strait–Chukchi Sea region are already showing signs of stress caused primarily by changes in their summer and winter sea-ice habitats (Garlich-Miller et al., 2011; MacCracken, 2012; Taylor et al., 2018; MacCracken et al., 2017; Higdon and Stewart, 2018). However, such changes have not yet been documented for the Atlantic walrus (*Odobenus rosmarus rosmarus*). Due to differences in sea ice habitats, movement patterns and area occupancy of Atlantic and Pacific walruses, it has been suggested that the former likely will be less negatively affected by the reduction in sea ice than the latter (Born, 2005, 2006; Laidre et al., 2008).

In this chapter, we review the current scientific literature to provide an updated assessment of the likely effects of climate change on walruses. We focus on the Atlantic walrus, but describe both the Pacific and Atlantic walrus scenarios to emphasise how the behavioural and ecological differences of the two subspecies likely alter the way they are exposed, and made vulnerable to, environmental changes. We suggest that walruses as a species generally show evolutionary, ecological and behavioural adaptations that render them less affected by the reduction of sea ice than other ice-associated Arctic pinnipeds. We also discuss other effects of warming, such as change in food resources, ocean acidification, increased predation, pathogens and UV radiation, and review the current conservation status of walruses. Finally, we provide recommendations for future research priorities. The potential negative impacts of human activities are discussed in Chapter 12 (this volume) and are not considered here.

Responses to declining sea ice: The contrasting Pacific and Atlantic scenarios

How important is sea ice for hauling out?

Walruses are flexible in behaviour and can use both land and sea ice (mainly drift ice) as platforms for hauling out (Salter, 1979, 1980; Born and Knutsen, 1997; Stewart et al., 2014; Laidre et al., 2008). According to Fay (1982, 1985), male walruses generally use beaches as substrate for resting, while females and young are more dependent of sea ice for hauling out. However, in several places within their range (e.g., on Southampton Island in northern Hudson Bay, the Canadian High Arctic, East Greenland, and Svalbard), Atlantic walruses of both sexes and all age classes haul out together on land (Salter, 1979; Born et al., 1995; Kovacs and Lydersen, 2008; Stewart et al., 2014; Øren et al., 2018). In situations where both sea-ice pans suitable for hauling out and a terrestrial haul-out are present in the same area, Atlantic walruses may be found on land only, on ice only or both on ice and land (e.g., Born and Knutsen, 1992; Stewart et al., 2014). At Bathurst Island in the Canadian High Arctic, the maximum number of walruses on land has been observed when ice cover near the haul-out ranges between 40%–60% (Salter, 1980). In contrast, Kovacs and Lydersen (2008) found that the occupancy pattern at a single haul-out site at Svalbard was most likely affected by the distribution of the subpopulation, and was independent of sea ice patterns. However, elsewhere on Svalbard, a photographic survey of walruses found a tendency for walruses to haul out mainly on land when sea ice was absent (i.e. within photographic distance, i.e., ≤ 100 m of the haul-out) (Øren et al., 2018). Hence, it is not entirely clear whether the presence of sea ice is a factor that greatly influences the haul-out behaviour of Atlantic walruses (Stewart et al., 2014).

Several factors such, as the presence of stranded sea ice on the beach blocking access to a terrestrial haul-outs, distance to foraging ground, the presence of predators and humans, age and sex class (e.g., juveniles and adult females vs adult males), group coherence and other social mechanisms likely determine the choice of haul-out platform (Fig. 13.2) (Hills, 1992; Born and Knutsen, 1992, 1997; Born et al., 1997a; Laidre et al., 2008). Thus while walruses may take advantage of ice floes for resting when given the opportunity, the ability to haul-out on land possibly makes them less sensitive to decreasing sea ice (Kovacs and Lydersen, 2008; Moore and Reeves, 2018). In fact, the use of the two seasonally-different haul-out substrates and the ability of walruses to make excursions approximately 100 km

FIGURE 13.2

A group of Atlantic walruses, consisting of adult males, an adult female and a calf at Svalbard. Walruses often use relatively small yet sturdy floes of sea ice for hauling out close to their foraging banks. *Photo: E.W. Born.*

away from their terrestrial haul-outs (Born and Knutsen, 1992; Wiig et al., 1996; Born and Acquarone, 2007; Freitas et al., 2009; Dietz et al., 2014) may potentially widen the feeding distribution, and thus theoretically positively affect the abundance of walruses in a changing climate (Kovacs et al., 2011).

The Pacific walrus scenario

Pacific walruses are distributed on the continental shelf of the Bering and Chukchi Seas (e.g., Higdon and Stewart, 2018) and are mainly found in shallow water areas providing access to suitable benthic prey at all seasons (Fay, 1982; Jay and Fischbach, 2008). This habitat is one of the largest shallow water areas in the Arctic and has also historically sustained a very large population of walruses (Fay, 1982; Taylor and Udevitz, 2015). The latest estimate of population size from 2014 was of 283,000 walruses (95% CI: 93,000–479,000; MacCracken et al., 2017). The potential foraging habitat (i.e., seabed $\leq$ 200 m depth) of the Pacific walrus is about 1.7 million km^2 (Garde et al., 2018), indicating an average density of approximately 0.17 animals/km^2.

Although some sub-structuring in the population is indicated (Jay et al., 2008; Sonsthagen et al., 2012), Pacific walruses basically constitute a single population. Most Pacific walruses make long-distance spring and autumn migrations from the Bering Sea through the Bering Strait to the Chukchi Sea (Fay, 1982; Speckman et al., 2011). During summer, the majority of adult males use terrestrial haul-outs along the mainland coast of Russia and the United States (Estes and Gol'tsev, 1984). However, thousands of male walruses do not migrate, but rather spend the summer in the Bristol Bay area in ice-free latitudes around 68°N (Fay, 1982; Estes and Gol'tsev, 1984). During this time, adult females with juveniles prefer offshore areas in the Chukchi Sea, where they haul out on drifting ice floes over foraging banks (Fay, 1982).

According to Laidre et al. (2015), the extent of sea ice in the Chukchi Sea has decreased in the period 1979–2013, whereas no such trend was detected in the Bering Sea. However, during the winters 2017–18 and 2018–19, the extent of the sea ice cover in the Bering Sea was exceptionally low (Rosner, 2019; Jones et al., 2020), indicating a general decrease in sea ice in this region as well. A decrease in summer sea ice may negatively affect the ability of Pacific walruses to obtain food (Kelly, 2001). For instance, in low-ice years (e.g., in 2007, 2011, 2012, 2014, 2016, 2017) (Krupnik, 2018) substantial portions of the sea ice edge in the Chukchi Sea (where females with juveniles and dependent calves occur during summer) receded north of the continental shelf and the females found themselves in waters which were too deep for foraging (Jay and Fischbach, 2008; Udevitz et al., 2013). Moreover, the general lack of summer sea ice is forcing Pacific walruses to use land-based haul-outs in large numbers of tens of thousands of animals (Monson et al., 2013). Such aggregations increase the risk of juveniles being trampled, and may also lead to increased rates of pathogen transmission (see below). Pacific walruses may experience an increase in energy expenditure when travelling from land haul-outs to offshore feeding areas (Rosen, 2021) as available haul-out sites may be situated far from offshore feeding areas, leading to a decline in reproductive or survival rates, and population size (Noren et al., 2012, 2014). When walruses are able to forage close to haul-outs their large numbers may cause the depletion of local benthic food resources (Jay et al., 2012, 2014).

Pacific walruses showed signs of having reached carrying capacity in the 1970's–80's at around 250,000–300,000 animals (Fay et al., 1997; MacCracken, 2012; Tempel and Atkinson, 2020; and references therein). During this time, the population likely became food limited and began to decline; and it is currently thought to be below its carrying capacity (MacCracken, 2012; USFWS, 2014). However, the population is expected to return to carrying capacity close to its current population size (approximately 283,000) given the future adverse effects such as sea-ice loss, declining prey availability and increased mortality at terrestrial haul-outs (MacCracken et al., 2017).

Overall, it is expected that the effects on Pacific walruses of a warming climate, sea ice retreat and associated decline in prey availability will lead to (1) a continuing northward shift in their distribution; (2) an increased use and associated trampling risk at coastal haul-outs in summer and autumn; (3) increased energy expenditure as travel time increases between resting and feeding areas and (4) a decrease in population size corresponding to the carrying capacity of near-shore foraging grounds (Jay et al., 2011; MacCracken, 2012; Taylor et al., 2018).

The Atlantic walrus scenario

Atlantic walruses are more coastal than Pacific walruses, with a distribution concentrated in coastal areas of the eastern Arctic (Born et al., 1995; NAMMCO, 2006) where the continental shelf is relatively narrow (Stein and MacDonald, 2004). The potential foraging habitat (i.e., seabed areas $\leq$ 200 m depth) of Atlantic walruses totals roughly 1.2 million km^2, or approximately 70% of what is available to Pacific walruses in the Bering Sea–Chukchi Sea region (Garde et al., 2018). With an estimated total number of approximately 30,000 individuals (Laidre et al., 2015; Higdon and Stewart, 2018), the density of Atlantic walruses is approximately 0.03 animals/km^2; five to six times less than that of Pacific walrus.

In contrast to the Pacific walrus, a decrease in sea ice with earlier spring break-up and later autumn ice formation (Laidre et al., 2015) may actually be beneficial to the Atlantic walrus in some areas by providing access to foraging areas for a longer period of the season (Born, 2005). Several Atlantic walrus subpopulations are distributed inshore (COSEWIC, 2006; Stewart, 2008), and nearly all potential and productive walrus feeding areas in the Atlantic region are closer to terrestrial haul-outs than in the Pacific region (Fig. 13.3).

In many areas, the inshore or near-shore habitat is characterised by the presence of thick solid winter ice, which seasonally exclude Atlantic walruses from many of their feeding areas—for example in Arctic Canada, North-West and

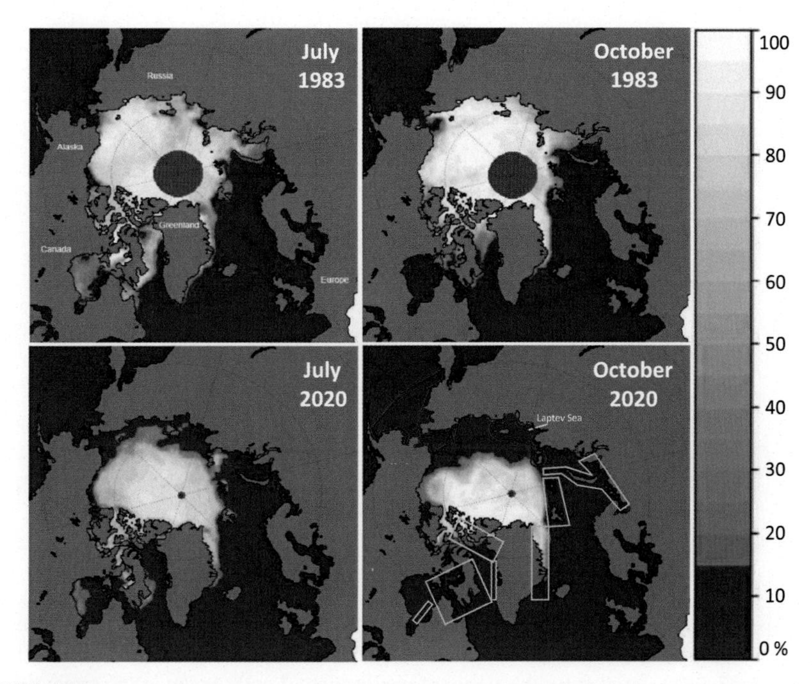

FIGURE 13.3

With the decrease in sea ice, the Atlantic walrus has increasingly gained access to inshore foraging banks for a longer period of the year. These maps show the percentage of Arctic sea ice cover in 1983 and 2020 for the months of July and October. In the 1970's–80's, the fast ice along the coasts and in fjords usually broke up in July, and formed again in October (see Born et al., 2004 and references therein); however, since the acceleration of sea ice loss in the 1990's, the ice breaks up much earlier and forms later in the season. Polygons in lower-right panel indicate the approximate distribution of Atlantic walruses (orange) and Pacific walruses (red). *Source for sea ice: NSIDC (2020). Distribution of walrus: Modified from Higdon and Stewart (2018).*

North-East Greenland and Svalbard (Vibe, 1950; Born, 2005; Born et al., 1997a; Stewart, 2008). As the open-water season increases in duration—as has been observed across the Atlantic Arctic for the last several decades (Rysgaard et al., 2007; Laidre et al., 2015; Yurkowski et al., 2020)—Atlantic walruses may increasingly gain access to their inshore and near-shore foraging grounds for a longer period of time. For instance, it has been predicted that by the end of the 21st century the duration of the open-water period will have increased from two and a half to about five months in North-East Greenland fjords (Rysgaard et al., 2003), thus greatly increasing the season during which walruses can forage in the area (Fig. 13.4) (Born et al., 2003; Born and Acquarone, 2007).

To summarise, although their biology is similar, the adaptive scenarios of Pacific and Atlantic walruses to reduction in sea ice are markedly different.

FIGURE 13.4

A group of Atlantic walrus females with young hauled out on the edge of the shore-fast sea ice in early June 1993 at 80° 57′ N in North-East Greenland. Hauling-out on such shore-fast ice is unusual given the frequent occurrence of polar bears in the region (Born et al., 1997a,b). *Photo: E.W. Born.*

Pacific walruses constitute more or less one coherent population that makes large-scale seasonal migrations. They are numerous and occur in high densities that may be close to the carrying capacity of their environment. Decreases in sea ice impede the ability of Pacific walruses to exploit their offshore feeding banks, and the population is showing signs of stress from climate change. In contrast, Atlantic walruses are distributed in relatively small, semi-isolated subpopulations across their range. They occur in low density in relation to the extent of their habitat and are assumed to be below carrying capacity. They also show encouraging signs of population increase in several areas. That is, decreases in sea ice may lead to increased access for Atlantic walruses to their coastal and inshore foraging banks, and hence an increase in population size.

How tolerant are walruses to rising temperatures?

Walruses are known to occur in areas with warm summer temperatures and seem to have a relatively good heat tolerance. Their present classification as an exclusively Arctic species may in large part be due to intensive exploitation and disturbance by humans displacing them from their former range (Bosscha Erdbrink and van Bree, 1986). For instance, some thousands of male Pacific walruses do not migrate north in spring but spend the summer in South-West Alaska around 68°N (e.g. Fay, 1982; Hills, 1992) where surface water temperatures are around 10°C during August (Fay and Ray, 1968; Overland et al., 1999). Miller (1997) inferred that between 12800 and 2900 CE, a large population of walruses likely lived in the Bay of Fundy (approximately 44°N; New Brunswick, Canada) when summer sea surface temperatures were about 12°C–15°C. Furthermore, a comparison of hypothetical sea temperatures with 17th century historical records of walrus distribution indicates that walruses in the Northumberland Strait area (approximately 46°N; Nova Scotia, Canada) may have inhabited waters as warm as 18°C during summer (Miller, 1997). The now-extinct Icelandic walrus would also have lived under boreal climate conditions until this lineage of walruses was extirpated by the Norse around the 10th century (Keighley et al., 2019). The survival until recent times of cold-temperate, non-ice breeding populations of walruses in Iceland and the Canadian Maritimes—as well as the apparent broad climatic tolerance of now-extinct Pliocene and Pleistocene walrus species (Chapter 2, this volume) suggest that overall, walruses may be less affected by warming than other ice-associated pinnipeds (Fig. 13.5).

FIGURE 13.5

A group of Atlantic walrus males at a traditionally used haul-out in East Greenland. It is expected that the continued decrease in sea ice will increasingly force walruses to use terrestrial haul-outs. *Photo: E.W. Born.*

Effects of climatic changes on marine productivity and benthos?

Walruses mainly forage on benthic invertebrates (e.g., Vibe, 1950; Fay, 1982). It is uncertain if, or to what extent, walruses are affected directly or indirectly by changes in their food resource due to climate change. Information on the potential effects of a decrease in sea ice on primary production, benthic productivity and biodiversity is inconclusive and effects likely differ on a regional basis (Josefson et al., 2013; Michel et al., 2013; Steiner et al., 2015; Jørgensen et al., 2017). However, the earlier onset of spring melt and sea ice retreat may result in longer growing seasons in marine systems and an overall increase in primary production and organic nutrients for shelf areas (Tremblay et al., 2012; Steiner et al., 2015; Wassmann et al., 2006; Sonsthagen et al., 2020). Within the range of Atlantic walruses, studies indicate that decreased sea ice and increased bottom water temperatures will lead to an increase in pelagic and benthic production (Rysgaard and Glud, 2007a,b). It has been suggested that an increase in average benthic macrofauna biomass in the Barents Sea between the 1960's and 2003 (Jørgensen et al., 2015) may be attributable to climate change and rising sea temperatures (Denisenko, 2013 in Jørgensen et al., 2017: 96). Moreover, considering the close coupling of pelagic and benthic productivity in Arctic marine systems (Piepenburg et al., 1997; Grebmeier and Dunton, 2000; Sejr et al., 2000, 2002; Darnis et al., 2012; Wassmann et al., 2006; 2011; Tremblay et al., 2011; Grebmeier et al., 2015) it is expected that a higher influx of pelagic organic matter to the seafloor during the spring bloom will benefit the nutrient-limited benthos (Sejr et al., 2004; Rysgaard and Glud, 2007b). This improved nutrient availability will likely stimulate bivalve growth and production (Sejr et al., 2009) and support greater biodiversity (Josefson et al., 2013; Michel et al., 2013), generally increasing the amount of prey available to Atlantic walruses.

However, the effects of decreased sea ice on the benthic fauna are complex and may differ between mid-Arctic and high-Arctic regions. A recent study involving carbon and nitrogen isotopes ($\delta^{13}C$, $\delta^{15}N$) indicated that a continuing decrease and thinning of sea-ice cover will influence benthic community biomass and structure differently in areas occupied by Atlantic walruses (Yurkowski et al., 2020). In association with decreased sea ice in the period 1982–2016, a decline in the contribution of sea ice-derived carbon to a shift to phytoplankton-derived carbon was observed at mid-Arctic latitudes (Foxe Basin, Canada) where less sea-ice algae are now incorporated into benthic fauna. A similar pattern was not observed at high-Arctic latitudes (Jones Sound, Canada). However, in both areas the $\delta^{15}N$ of the baseline and trophic position of Atlantic walruses did not change over time, indicating consistent nitrogen cycling and trophic roles of high bivalve consumption (ibid.).

Ocean acidification as a result of ongoing climate change may reduce the survival of benthic molluscs (Parker et al., 2013; Smaal et al., 2019) and thereby indirectly affect walruses. As the emission of greenhouse gases increases, carbon dioxide (CO_2) is absorbed from the atmosphere by seawater which lowers its calcium carbonate ($CaCO_3$) saturation state. Arctic marine waters are undergoing widespread, rapid ocean acidification (AMAP, 2013) and $CaCO_3$ undersaturation is already occurring on freshwater influenced shelves in the Arctic (Haug et al., 2017 and references therein). Lowered $CaCO_3$ saturation effects the survival of some marine molluscs by limiting their ability to form shells (Gazeau et al., 2013; AMAP, 2013, 2019). Heavily-calcified molluscs are likely to suffer the most significant effects (e.g., Hale et al., 2011; Fitzer et al., 2018). Renaud et al. (2019) modelled the potential future impact of ocean acidification and warming on the distribution of boreal, boreo-Arctic and Arctic benthic mollusc species. Generally, projections for bottom water temperature and the saturation state of aragonite (i.e., soluble form of $CaCO_3$ and an important mineral in mollusc shells) indicated substantial changes by the end of the 21st century. However, the assessment indicated that individual taxa of important walrus food items might experience considerably different changes in suitable habitat. Habitat loss was predicted for the bivalves *Mya truncata*, *Ciliatocardium* spp., *Hiatella arctica* and *Astarte borealis*, whereas an increase in habitat was predicted for *Serripes groenlandicus* (Renaud et al., 2019). Although walruses mainly forage on hard-shelled molluscs they also consume a variety of other taxa (Vibe, 1950; Fay, 1982; Gjertz and Wiig, 1992; Maniscalco et al., 2020), which makes it difficult to predict to what extent walruses will be affected by future ocean acidification (Fig. 13.6).

In contrast to the situation of the Atlantic walrus, it has been predicted that prey populations of the Pacific walrus will be negatively affected by warmer conditions (Grebmeier et al., 2006, 2015; Josefson et al., 2013; Michel et al., 2013; Jørgensen et al., 2017; MacCracken et al., 2017). Declines in bivalves and other benthic

FIGURE 13.6

Eighty-nine empty shells from three different important walrus bivalve prey species (*Hiatella arctica, Mya truncata, Serripes groenlandicus*), all consumed by a single adult male walrus during a six minute dive in North-East Greenland (Born et al., 2003. Photo: M. Sejr). Modelling of future impact of ocean acidification has predicted habitat loss for *Mya* sp. and *Hiatella* sp., whereas an increase in habitat was predicted for *Serripes* sp. (Renaud et al., 2019).

invertebrates may result in a trophic shift as walruses shift to predating on mammals and birds. Indeed, dietary studies suggest that Pacific walruses have increased their reliance on high trophic-level prey (such as pelagic fish, seals and sea birds) during the last 40 years (Mansfield, 1958; Lowry and Fay, 1984; Rausch et al., 2007; Seymour et al., 2014a,b; Metcalf, 2017; Scotter et al., 2019). Similarly, subsistence walrus hunters in Alaska have observed a decreasing volume of benthic prey, particularly clams, in walrus stomachs, and an increasing volume of pelagic fishes (Metcalf, 2017). The reliance of Pacific walruses on high-trophic level prey appears to fluctuate, and may be a gradual response to climate change-induced food web shifts (Seymour et al., 2014a, b). Changes in feeding patterns have not been observed in Atlantic walruses. In North-East Greenland, Atlantic walruses overwinter where waters are too deep for their preferred benthic prey (Born et al., 2005), and it has been suggested that where the distribution of seals and walruses overlap considerably seals may be an important food source for walruses (Born et al., 1997a; 2017). Thus, while decreasing sea ice and increasing ocean temperatures will likely affect walrus prey availability, effects will differ geographically and are not predicted to be universally negative.

Will exposure to pathogens increase?

Increased global temperature and warmer-water influxes to the Arctic have already resulted in a northward advance of some temperate-adapted marine mammals, such as humpback whales (*Megaptera novaeangliae*), fin whales (*Balaenoptera physalus*), minke whales (*Balaenoptera acuto rostrata*) (e.g., Laidre and Heide-Jørgensen, 2012), as well as several species of fish and invertebrates (e.g., Galkin, 1998; Haug et al., 2017). Changes in environmental conditions may also introduce pathogens (e.g., viruses and bacteria) novel to the Arctic, and potentially compromise the health of immunologically naïve wildlife (Jensen et al., 2010; Sonsthagen et al., 2014 and references therein; Parkinson et al., 2014; Scotter et al., 2019). For instance, it has recently been hypothesised that the spread of phocine distemper virus from North Atlantic to North Pacific marine mammals was associated with a recent reduction of Arctic sea ice (VanWormer et al., 2019). How walruses and other potential host species respond to these immunological challenges will depend in part on their adaptive potential (Scotter et al., 2019). The detection of limited genetic diversity in genes associated with adaptive immune response suggests walruses may have a reduced capacity to respond to novel immunological challenges associated with the shifts in ecological communities and environmental stressors predicted under changing climatic scenarios (Sonsthagen et al., 2014). Much remains to be learned about the future effects of pathogens in walruses and other Arctic pinnipeds.

Increased predation risks

Predation of walruses by polar bears (*Ursus maritimus*) may increase along with ongoing climate change as walruses are forced to spend more time at terrestrial

FIGURE 13.7

The decrease of sea ice will likely lead to an increase in walrus-bear encounters on land. Here, an adult male polar bear hesitates to attack a large Atlantic walrus bull, Svalbard. Typically, polar bears are only able to kill young walruses and calves (Stirling et al., 2021). *Photo: I. Gjertz.*

haul-outs (Ovsyanikov, 1995; Rode et al., 2015; Øren et al., 2018). However, polar bears and walruses have a long history of contact during periods of limited sea ice, and the two species have established various hunting and defence strategies during their evolutionary histories (see Chapter 3, this volume) (Fig. 13.7). Predation by killer whales (*Orcinus orca*) is a cause of walrus mortality in regions where the distribution of both species overlap on a seasonal basis (Lowry et al., 1987; Kryukova et al., 2012). Given the fact that a continued decrease in sea ice may lead to larger seasonal and geographical overlap between walruses and their predators, it is likely that the number of walruses killed by polar bears and killer whales will increase in the years to come. However, it is difficult to quantify the magnitude of mortalities from such predation.

Ultraviolet radiation

It has been suggested that solar ultraviolet radiation (UV) may pose a significant threat to walruses in the form of skin damage if the ozone layer above the Arctic continues to thin (Martinez-Levasseur et al., 2016). A range of skin abnormalities consistent with UV damage were detected at the microscopic level in a limited sample of Atlantic walruses from Hudson Strait, Canada. However, long-term data from local subsistence walrus hunters and Inuit Elders did not report a relation between the increased sun radiation secondary to ozone loss and walrus health (ibid.).

Perspectives and conclusion

National and international organisations have assessed the general status of walruses based on expert reviews. The IUCN SSC Pinniped Specialist Group has recommended that walruses should be listed as 'Vulnerable' (see IUCN Standards and Petitions Subcommittee, 2017) due to uncertainty surrounding the impact of a future decline in habitat quality as well as limited abundance data and trends (Lowry, 2016). The IUCN listed the Pacific walrus as 'Data Deficient' because of a lack of data on actual abundance and trends over recent decades. The Atlantic walrus was listed as 'Near Threatened' due to a relatively low number of mature individuals (about 12,500, which took into account a likely population size decrease of more than 10% over three generations due to loss of sea ice) (Kovacs, 2016). In the most recent status assessment of the Pacific walrus, the US Fish and Wildlife Service concluded that the population is currently stable, but that declining sea ice will ultimately lead to a population decline. However, the magnitude of the decline is unknown. Based on this assessment, the Pacific walrus was not listed as 'endangered' or 'threatened' under the US Endangered Species Act of 1973 (USFWS, 2017).

Our evaluation of putative climate change effects on walruses reveals the complex interplay of adverse and beneficial factors which can strengthen or oppose each other. Negative consequences for the abundant Pacific walrus as a result of a warming Arctic have already been observed. In contrast, similar signs have not been observed in Atlantic walrus, which are still increasing in population size following centuries of overexploitation across their range. It is likely that decreases in sea ice at coastal and inshore foraging grounds may even be beneficial to Atlantic walruses. Despite a marked decrease in sea ice across the Barents Sea during the last few decades (Lind et al., 2018; MOSJ, 2020), the number of walruses in Svalbard increased significantly at about 8% per year between 2006 and 2012 (Kovacs et al., 2014). Since the 2012 survey (Kovacs et al., 2014), this subpopulation has increased further by 48% (NAMMCO, 2019).

According to Born (2005, 2006) walruses exhibit some physical and behavioural traits that generally make them more adaptable than other Arctic ice-adapted pinnipeds to future warming of the Arctic. These include (1) the ability of some walruses to summer and forage at low latitudes, where sea and air temperatures are relatively high and there is no or little sea ice (e.g., Bristol Bay in Alaska); (2) walruses are the only species among Arctic pinnipeds with the habit of hauling-out on land in dense groups; (3) the size, tusks and gregariousness of walruses aid them in fending off and—in some cases—killing their main natural predator, the polar bear; (4) the fact that despite their negative buoyancy, the inflatable pharyngeal pouches of walruses allow both sexes to sleep at sea in the absence of terrestrial or ice haul-outs (Freuchen and Salomonsen, 1959; Fay, 1960); and finally (5) the fact that historically, walruses lived further south than they presently do, indicating tolerance for subarctic and temperate climates. Despite the overall uncertainty of Arctic warming and decreasing sea ice, we believe that the effects will not be catastrophic to the Atlantic walrus. Rather, the biggest threat to walruses—now and throughout their history—appears to be direct human disturbance (see Chapter 12, this volume). Future efforts should seek to mitigate these effects.

To determine the effects of climate change, we recommend that systematic studies be conducted on Atlantic walrus subpopulations that live under different environmental conditions (e.g., across their latitudinal range). Such studies should consider (1) abundance trends, (2) changes in movements and distribution related to changes in food resources and haul-outs platforms, (3) changes in mortality related to crowding at terrestrial haul-out, and associated changes in pathogens and predation.

Acknowledgements

We wish to thank Harry Stern (Polar Science Center, Applied Physics Laboratory, University of Washington, Seattle, USA) for information on sea ice.

References

AMAP (Arctic Monitoring and Assessment Programme), 2013. AMAP Assessment 2013: Arctic Ocean Acidification. Arctic Monitoring and Assessment Programme (AMAP), Oslo, Norway, 1–99.

AMAP (Arctic Monitoring and Assessment Programme), 2019. Arctic Ocean Acidification Assessment: 2018 Summary for Policymakers, 1–15.

Born, E.W., 2005. An Assessment of the Effects of Hunting and Climate on Walruses in Greenland. Greenland Institute of Natural Resources, Nuuk, and University of Oslo, pp. 1–346, ISBN 82-7970-006-4:.

Born, E.W., 2006. Robben und Eisbär in der Arktis: Auswirkung von Erderwärmung un Jagd (Seals and polar bears in the Arctic: effects of global warming and hunt). In: Lozan, J.L., Grassl, H., Hubberten, H.W., Hupfer, P., Karpe, L., Piepenburg, D. (Eds.), Warnsignale aus den Polarregionen (Warning signals from the polar regions). Wissenschaftliche Auswertungen und GEO, Hamburg, Germany, pp. 152–159.

Born, E.W., Knutsen, L.Ø., 1992. Satellite-linked radio tracking of Atlantic walruses (*Odobenus rosmarus rosmarus*) in northeastern Greenland, 1989–1991. Zeitschrift für Säugetierkunde 57, 275–287.

Born, E.W., Knutsen, L.Ø., 1997. Haul-out activity of male Atlantic walruses (*Odobenus rosmarus rosmarus*) in northeastern Greenland. Journal of Zoology 243, 381–396.

Born, E.W., Acquarone, M., 2007. An estimation of walrus predation on bivalves in the Young Sound area (NE Greenland). In: Rysgaard, S., Glud, R.N. (Eds.), Carbon cycling in Arctic marine ecosystems: case study Young Sound, 58. Monographs on Greenland, pp. 176–191.

Born, E.W., Gjertz, I., Reeves, R.R., 1995. Population assessment of Atlantic walrus, Norwegian Polar Institute 138, 1–100.

Born, E.W., Dietz, R., Heide-Jørgensen, M.P., Knutsen, L.Ø., 1997a. Historical and present status of the Atlantic walrus (*Odobenus rosmarus rosmarus*) in eastern Greenland. Monographs on Greenland 46, 1–73.

Born, E.W., Wiig, Ø., Thomassen, J., 1997b. Seasonal and annual movements of radio-collared polar bears (*Ursus maritimus*) in NE Greenland. Journal of Marine Systems 10, 67–77.

Born, E.W., Rysgaard, S., Ehlmé, G., Sejr, M., Acquarone, M., Levermann, N., 2003. Underwater observations of foraging free-living Atlantic walruses (*Odobenus rosmarus rosmarus*) and estimates of their food consumption. Polar Biology 26, 348–357.

Born, E.W., Teilmann, J., Acquarone, M., Riget, F., 2004. Habitat use of ringed seal (*Phoca hispida*) in the North Water area (North Baffin Bay). Arctic 57, 129–142.

Born, E.W., Acquarone, M., Knutsen, L.Ø., Toudal, L., 2005. Homing behaviour in an Atlantic walrus (*Odobenus rosmarus rosmarus*). Aquatic Mammals 31, 23–33.

Born, E.W., Heilmann, A., Kielsen Holm, L., Laidre, K.L., Iversen, M., 2017. Walruses in West and Northwest Greenland. An interview survey about the catch and the climate. Monographs on Greenland Vol. 355. Man and Society Vol. 44. Museum Tusculanum Press, Copenhagen, pp. 1–256.

Bosscha Erdbrink, D.P., van Bree, P.H.J., 1986. Fossil Odobenidae in some Dutch collections (Mammalia, Carnivora). Beaufortia 36, 13–33.

Boyd, I., Hanson, N., Tynan, C.T., 2019. Effects of climate change on marine mammals. Encyclopedia of Ocean Sciences 416–419. Available from: https://doi.org/10.1016/B978-0-12-409548-9.11627-6.

CAFF (Convention of Arctic Flora and Fauna) 2017. State of the Arctic Marine Biodiversity Report. Conservation of Arctic Flora and Fauna International Secretariat: Akureyri, Iceland, 1–197.

COSEWIC, 2006. COSEWIC Assessment and Update Status Report on the Atlantic Walrus *Odobenus rosmarus rosmarus* in Canada. Committee on the Status of Endangered Wildlife in Canada, Ottawa, pp. 1–65.

Darnis, G., Robert, D., Pomerleau, C., Link, H., Archambault, P., Nelson, R.J., et al., 2012. Current state and trends in Canadian Arctic marine ecosystems: II. Heterotrophic food web, pelagic-benthic coupling, and biodiversity. Climatic Change 115, 161–178. Available from: https://doi.org/10.1007/s10584-012-0483-8.

Denisenko, S.G., 2013. Biodiversity and Bioresources of Macrozoobenthos of the Barents Sea: Structure and Long-term Changes. Nauka, Saint-Petersburg (fide Jørgensen et al. 2017) (In Russian).

Dietz, R., Born, E.W., Stewart, R.E.A., Heide-Jørgensen, M.P., Stern, H., Rigét, F., et al., 2014. Movements of Walruses (*Odobenus rosmarus*) Between Central West Greenland and Southeast Baffin Island, 2005–2008, NAMMCO Sci Publ 9, 53–74. Available from: http://doi.org/10.7557/3.2605.

Estes, J.A., Gol'tsev, V.N., 1984. Abundance and distribution of the Pacific walrus *Odobenus rosmarus divergens*: results of the first joint Soviet-American aerial survey, August 1975. In: Fay, F.H., Fedoseev, G.A. (Eds.), Soviet–American Cooperative

Research on Marine Mammals. Department of Commerce, National Oceanic and Atmospheric Administration, National Marine Fisheries Service, Washington, DC, USA, pp. 67–76. Volume 1. Pinnipeds. Technical Report 12. U.S.

Fay, F.H., 1960. Structure and function of the pharyngeal pouches of the walrus (*Odobenus rosmarus* L.). Mammalia 2, 361–371.

Fay, F.H., 1982. Ecology and biology of the Pacific Walrus, Odobenus rosmarus divergens, Illiger. North American Fauna, United States Department of the Interior, Fish and Wildlife Service 1982, 74, 1–279.

Fay, F.H., 1985. *Odobenus rosmarus*. Mammalian Species 238, 1–7.

Fay, F.H., Ray, C., 1968. Influence on climate on the distribution of walruses, *Odobenus rosmarus* (Linnaeus). I. Evidence from thermoregulatory behavior. Zoologica - New York 53, 1–14. plates I-IV.

Fay, F.H., Eberhardt, L.L., Kelly, B.P., Burns, J.J., Quakenbush, L.T., 1997. Status of the Pacific walrus population, 1950–1989. Marine Mammal Science 13, 537–565.

Fitzer, S.C., Gabarda, S.T., Daly, L., Hughes, B., Dove, M., O'Connor, W., et al., 2018. Coastal acidification impacts on shell mineral structure of bivalve mollusks. Ecology and Evolution 8, 8973–8984. Available from: https://doi.org/10.1002/ece3.4416.

Freitas, C., Kovacs, K.M., Ims, R.A., Fedak, M.A., Lydersen, C., 2009. Deep into the ice: over-wintering and habitat selection in male Atlantic walruses. Marine Ecology Progress Series 375, 247–261.

Freuchen, P., Salomonsen, F., 1959. The Arctic Year, 371. Jonathan Cape. Thirty Bedford Square, London, pp. 1–440.

Galkin, Y.I., 1998. Long-term changes in the distribution of molluscs in the Barents Sea related to the climate. Berichte zur Polarforschung 287, 100–143.

Garde, E., Hansen, R.G., Zinglersen, K., Ditlevsen, S., Heide-Jørgensen, M.P., 2018. Diving and foraging characteristics of walrus in Smith Sound. The Journal of Experimental Biology 500, 89–99. Available from: https://doi.org/10.1016/j.jembe.2017.12.009.

Garlich-Miller, J.L., MacCracken, J.G., Snyder, J., Meehan, R., Myers, M., Wilder, J.M., et al., 2011. Status Review of the Pacific Walrus (*Odobenus rosmarus divergens*). U.S. Fish and Wildlife Service, Anchorage, AK, pp. 1–155.

Gazeau, F., Parker, L.M., Comeau, S., Gattuso, J.-P., O'Connor, W.A., Martin, S., et al., 2013. Impacts of ocean acidification on marine shelled molluscs. Marine Biology 160, 2207–2245. Available from: https://doi.org/10.1007/s00227-013-2219-3.

Gjertz, I., Wiig, Ø., 1992. Feeding of walrus *Odobenus rosmarus* in Svalbard. Polar Record 28, 57–59.

Grebmeier, J.M., Dunton, K.H., 2000. Benthic processes in the northern Bering/Chukchi seas: status and global change. In: Impacts of Changes in Sea Ice and other Environmental parameters in the Arctic. Report of the Marine Mammal Commission Workshop, 15–17 February 2000, Girdwood, Alaska. Available from the Marine Mammal Commission, Bethesda, Maryland, 61–71.

Grebmeier, J.M., Overland, J.E., Moore, S.E., Farley, E.V., Carmack, E.C., Cooper, L.W., et al., 2006. A major ecosystem shift in the northern Bering Sea. Science (New York, NY) 311, 1461–1464.

Grebmeier, J.M., Bluhm, B.A., Cooper, L.W., Danielson, S.L., Arrigo, K.R., Blanchard, A. L., et al., 2015. Ecosystem characteristics and processes facilitating persistent macrobenthic biomass hotspots and associated benthivory in the Pacific Arctic. Progress in Oceanography 136, 92–114.

Hale, R., Calosi, P., McNeill, L., Mieszkowska, N., Widdicombe, S., 2011. Predicted levels of future ocean acidification and temperature rise could alter community structure and biodiversity in marine benthic communities. Oikos 120, 661–674. Available from: https://doi.org/10.1111/j.1600-0706.2010.19469.x.

Haug, T., Bogstad, B., Chierici, M., Gjøsæter, H., Hallfredsson, E.H., Høines, Å.S., et al., 2017. Future harvest of living resources in the Arctic Ocean north of the Nordic and Barents Seas: a review of possibilities and constraints. Fisheries Research 188, 38–57. Available from: https://doi.org/10.1016/j.fishres.2016.12.002.

Higdon, J.W., Stewart, D.B., 2018. State of circumpolar walrus (Odobenus rosmarus) populations. Prepared by Higdon Wildlife Consulting and Arctic Biological Consultants, Winnipeg, MB for WWF Arctic Programme, Ottawa, ON, 1–100.

Hills, S. 1992. The effect of spatial and temporal variability on population assessment of Pacific walruses. PhD-thesis, University of Maine, Orono, December 1992, 1–120.

IUCN Standards and Petitions Subcommittee, 2017. Guidelines for Using the IUCN Red List Categories and Criteria. Version 13. Prepared by the Standards and Petitions Subcommittee. < http://www.iucnredlist.org/documents/RedListGuidelines.pdf >.

Jay, C.V., Fischbach, A.S., 2008. Pacific walrus response to Arctic sea ice losses. U.S. Geological Survey. Fact Sheet 2008–3041.

Jay, C.V., Outridge, P.M., Garlich-Miller, J.L., 2008. Indication of two Pacific walrus stocks from whole tooth elemental analysis. Polar Biology 31, 933–943. Available from: https://doi.org/10.1007/s00300-008-0432-1.

Jay, C.V., Marcot, B.G., Douglas, D.C., 2011. Projected status of the Pacific walrus (*Odobenus rosmarus divergens*) in the twenty-first century. Polar Biology 34, 1065–1084.

Jay, C.V., Fischbach, A.S., Kochnev, A.A., 2012. Walrus areas of use in the Chukchi Sea during sparse sea ice cover. Marine Ecology Progress Series 468, 1–13. Available from: https://doi.org/10.3354/meps10057.

Jay, C.V., Grebmeier, J.M., Fischbach, A.S., McDonald, T.L., Cooper, L.W., Hornby, F., 2014. Pacific Walrus (*Odobenus rosmarus divergens*) resource selection in the northern Bering Sea. PLoS One 9 (4), e93035. Available from: https://doi.org/10.1371/journal.pone.0093035.

Jensen, S.K., Aars, J., Lydersen, C., Kovacs, K.M., Åsbakk, K., 2010. The prevalence of *Toxoplasma gondii* in polar bears and their marine mammal prey: evidence for a marine transmission pathway? Polar Biology 33, 599–606.

Jones, M.C., Berkelhammer, M., Keller, K.J., Yoshimura, K., Wooller, M.J., 2020. High sensitivity of Bering Sea winter sea ice to winter insolation and carbon dioxide over the last 5500 years. Science Advances 6 (36), eaaz9588. Available from: https://doi.org/10.1126/sciadv.aaz9588.

Jørgensen, L.L., Ljubin, P., Skjoldal, H.R., Ingvaldsen, R.B., Anisimova, N., Manushin, I., 2015. Distribution of benthic megafauna in the Barents Sea: baseline for an ecosystem approach to management. ICES Journal of Marine Science 72, 595–613.

Jørgensen, L.L., Archambault, P., Blicher, M., Denisenko, N., Guðmundsson, G., Iken, K., et al., 2017. Chapter 3.3 Benthos. In: CAFF, 2017. State of the Arctic Marine Biodiversity Report. Conservation of Arctic Flora and Fauna International Secretariat, Akureyri, Iceland. 978-9935-431-63-9, 85–106.

Josefson, A.B., Mokievsky,V., Bergmann, M., Blicher, M.E., Bluhm, B., Cochrane,S., et al., 2013. Chapter 8. Marine invertebrates. In: CAFF, 2013. Arctic Biodiversity Assessment. Status and Trends in Arctic Biodiversity. Conservation of Arctic Flora and Fauna, Akureyri, Iceland, 276–309.

Keighley, X., Palsson, S., Einarsson, B.F., Petersen, A., Fernández-Coll, M., Jordan, P., et al., 2019. Disappearance of Icelandic walruses coincided with Norse settlement. Molecular Biology and Evolution 36, 2656–2667. Available from: https://doi.org/10.1093/molbev/msz196.

Kelly, B.P., 2001. Climate change and ice breeding pinnipeds. In: Walther, G.-R., Burga, C.A., Edwards, P.J. (Eds.), Fingerprints of climate change. Kluwer Academic/Plenum Publishers, New York, pp. 43–55.

Kovacs, K.M., 2016. *Odobenus rosmarus ssp. rosmarus*. The IUCN Red List of Threatened Species 2016: e.T15108A66992323. Available from: https://doi.org/10.2305/IUCN.UK.2016-1.RLTS.T15108A66992323.en. Downloaded on 01 June 2019.

Kovacs, K.M., Lydersen, C., 2008. Climate change impacts on seals and whales in the North Atlantic Arctic and adjacent shelf seas. Science Progress 92, 117–150. Available from: https://doi.org/10.3184/003685008X324010.

Kovacs, K.M., Lydersen, C., Overland, J.E., Moore, S.E., 2011. Impacts of changing sea-ice conditions on Arctic marine mammals. Marine Biodiversity 41, 181–194.

Kovacs, K.M., Aars, J., Lydersen, C., 2014. Walruses recovering after 60+ years of protection in Svalbard, Norway. Research Note. Polar Research 33, 26034. Available from: https://doi.org/10.3402/polar.v33.26034.

Krupnik, I., 2018. 'Arctic Crashes:' revisiting the human-animal disequilibrium model in a time of rapid change. Human Ecology 46, 685–700. Available from: https://doi.org/10.1007/s10745-018-9990-1.

Kryukova, N.V., Kruchenkova, E.P., Ivanov, D.I., 2012. Killer Whales (*Orcinus orca*) hunting for walruses (*Odobenus rosmarus divergens*) near Retkyn Spit, Chukotka. Biology Bulletin 39, 768–778.

Laidre, K.L., Heide-Jørgensen, M.-P., 2012. Spring partitioning of Disko Bay, West Greenland, by Arctic and Subarctic baleen whales. ICES Journal of Marine Science 69, 1226–1233. Available from: https://doi.org/10.1093/icesjms/fss095.

Laidre, K.L., Stirling, I., Lowry, L.F., Wiig, Ø., Heide-Jørgensen, M.P., Ferguson, S.H., 2008. Quantifying the sensitivity of Arctic marine mammals to climate-induced habitat change. Ecological Applications: A Publication of the Ecological Society of America 18, S97–S125.

Laidre, K.L., Stern, H., Kovacs, K.M., Lowry, L.F., Moore, S.E., Regehr, E.V., et al., 2015. Arctic marine mammal population status, sea ice habitat loss, and conservation recommendations for the 21st century. Conservation Biology: The Journal of the Society for Conservation Biology 29, 724–737.

Lind, S., Ingvaldsen, R.B., Furevik, T., 2018. Arctic warming hotspot in the northern Barents Sea linked to declining sea-ice import. Nature Climate Change 8, 634–639.

Lowry, L., 2016. *Odobenus rosmarus*. The IUCN Red List of Threatened Species 2016. Available from: https://doi.org/10.2305/IUCN.UK.2016-1.RLTS.T15106A45228501.ene.T15106A45228501.

Lowry, L.F., Fay, F.H., 1984. Seal eating by walruses in the Bering and Chukchi Seas. Polar Biology 3, 11–18.

Lowry, L.F., Nelson, R.R., Frost, K.J., 1987. Observations of killer whales, *Orcinus orca*, in western Alaska: sightings, strandings, and predation on other marine mammals. Canadian Field Naturalist 101, 6–12.

MacCracken, J.G., 2012. Pacific walrus and climate change: observations and predictions. Ecology and Evolution 2, 2072–2090.

MacCracken, J.G., Beatty, W.S., Garlich-Miller, J.L., Kissling, M.L., Snyder, J.A., 2017. Final Species Status Assessment for the Pacific Walrus (*Odobenus rosmarus divergens*), U.S. Fish and Wildlife Service, Marine Mammals Management, 2017, 1–297 (Version 1.0).

Maniscalco, J.M., Springer, A.M., Counihan, K.L., Hollmen, T., Aderman, H.M., Toyukak, M., 2020. Contemporary diets of walruses in Bristol Bay, Alaska suggest temporal variability in benthic community structure. PeerJ, 2020 (3), art. no. e8735. Available from: https://doi.org/10.7717/peerj.8735.

Mansfield, A.W., 1958. The biology of the Atlantic walrus *Odobenus rosmarus rosmarus* (Linnaeus) in the eastern Canadian Arctic. Fisheries Research Board of Canada Manuscript Report Series 653, 1–146. Report.

Martinez-Levasseur, L.M., Furgal, C.M., Hammill, M.O., Burness, G., 2016. Towards a better understanding of the effects of UV on Atlantic Walrus, *Odobenus rosmarus rosmarus*: a study combining histological data with local ecological knowledge. PLoS One 11 (4), e0152122. Available from: https://doi.org/10.1371/journal.pone.0152122.

Metcalf, V., 2017. Box 3.3.3 Indigenous knowledge of benthic species. In: CAFF, 2017. State of the Arctic Marine Biodiversity Report. Conservation of Arctic Flora and Fauna International Secretariat, Akureyri, Iceland. 978-9935-431-63-9, p. 99.

Michel, C., Bluhm, B., Ford, V., Gallucci, V., Gaston, A.J., Gordillo, F.J.L., et al., 2013. Chapter 14. Marine ecosystems. In: CAFF, 2013. Arctic Biodiversity Assessment. Status and Trends in Arctic Biodiversity. Conservation of Arctic Flora and Fauna, Akureyri, Iceland, 486–527.

Miller, R.F., 1997. New records and AMS radiocarbon dates on Quaternary walrus (*Odobenus rosmarus*) from New Brunswick. Géographie physique et Quaternaire 51, 1–5.

Monson, D.H., Udevitz, M.S., Jay, C.V., 2013. Estimating age ratios and size of Pacific walrus herds on coastal haulouts using video imaging. PLoS One 8 (7), e69806.

Moore, S.E., Reeves, R.R., 2018. Tracking arctic marine mammal resilience in an era of rapid ecosystem alteration. PLoS Biology 16 (10), e2006708. Available from: https://doi.org/10.1371/journal.pbio.2006708.

MOSJ, 2020. Environmental Monitoring of Svalbard and Jan Mayen. http://www.mosj.no/en/climate/ocean/sea-ice-extent-barents-sea-fram-strait.html (accessed 26.07.20.).

NAMMCO (North Atlantic Marine Mammal Commission), 2006. NAMMCO Annual Report 2005. North Atlantic Marine Mammal Commission, Tromsø, Norway, 1–381. <https://nammco.no/topics/sc-working-group-reports/>.

NAMMCO (North Atlantic Marine Mammal Commission), 2019. NAMMCO annual report 2019. Report of the Scientific Committee 26th Meeting, October 29–November 1, 2019. Tórshavn, Faroe Islands, 1–78. <https://nammco.no/topics/scientific-committee-reports/>.

Noren, S.R., Udevitz, M.S., Jay, C.V., 2012. Bioenergetics model for estimating food requirements of female Pacific walruses *Odobenus rosmarus divergens*. Marine Ecology Progress Series 460, 261–275.

Noren, S.R., Udevitz, M.S., Jay, C.V., 2014. Energy demands for maintenance, growth, pregnancy, and lactation of female Pacific walruses (*Odobenus rosmarus divergens*). Physiological and Biochemical Zoology: PBZ 87, 837–854.

NSIDC (National Snow and Ice Data Center), 2020. NSIDC, University of Colorado, Boulder, U.S.A. http://nsidc.org/arcticseaicenews/ and https://nsidc.org/arcticseaicenews/sea-ice-analysis-tool/ (accessed 27.12.20.).

Øren, K., Kovacs, K.M., Yoccoz, N.G., Lydersen, C., 2018. Assessing site-use and sources of disturbance at walrus haul-outs using monitoring cameras. *Polar Biology* 41, 1737–1750. Available from: https://doi.org/10.1007/s00300-018-2313-6.

Overland, J.E., Wang, M., 2013. When will the summer Arctic be nearly sea ice free? Geophysical Research Letters 40 (10), 2097–2101. Available from: https://doi.org/10.1002/grl.50316.

Overland, J.E., Salo, S.A., Kantha, L.H., Clayson, C.A., 1999. Thermal stratification and mixing on the Bering Sea shelf. In: Loughlin, T.R., Ohtani, K. (Eds.), Dynamics of the Bering Sea: A Summary of Physical, Chemical, and Biological Characteristics, and a Synopsis of Research on the Bering Sea. University of Alaska Sea Grant, AK-SG-99-03. North Pacific Marine Science Organization (PICES), pp. 129–146.

Ovsyanikov, N.G., 1995. Polar bear predation of walruses on Wrangell Island. Bulletin of the Moscow Association of Natural Scientists, Section of Biology 100, 1–13.

Parker, L.M., Ross, P.M., O'Connor, W.A., Pörtner, H.O., Scanes, E., Wright, J.M., 2013. Predicting the response of molluscs to the Impact of ocean acidification. Biology 2, 651–692. Available from: https://doi.org/10.3390/biology2020651.

Parkinson, A.J., Evengard, B., Semenza, J.C., Ogden, N., Børresen, M.L., Berner, J., et al., 2014. Climate change and infectious diseases in the Arctic: establishment of a circumpolar working group. International Journal of Circumpolar Health 73 (1), 25163. Available from: https://doi.org/10.3402/ijch.v73.25163.

Piepenburg, D., Ambrose, W.G., Brandt, A., Renaud, P.E., Ahrens, M.J., Jensen, P., 1997. Benthic community patterns reflect water column processes in the Northeast Water polynya (Greenland). Journal of Marine Systems 132, 467–482.

Post, E., Alley, R.B., Christensen, T.R., Macias-Fauri, M., Forbes, B.C., Gooseff, M., et al., 2019. The Polar Regions in a 2°C warmer world. *Science Advances* 5 (12), eaaw9883. Available from: https://doi.org/10.1126/sciadv.aaw9883.

Rausch, R.L., George, J.C., Brower, H.K., 2007. Effect of climatic warming on the Pacific walrus, and potential modification of its helminth fauna. The Journal of Parasitology 93, 1247–1251.

Renaud, P.E., Wallhead, P., Kotta, J., Włodarska-Kowalczuk, M., Bellerby, R.G.J., Rätsep, M., et al., 2019. Arctic sensitivity? Suitable habitat for Benthic Taxa is surprisingly robust to climate change. Frontiers in Marine Science 6, Available from: https://doi.org/10.3389/fmars.2019.0053; https://doi.org/10.3389/fmars.2019.00538.

Rode K.D., Wilson R.R., Regehr E.V., St. Martin M., Douglas D.C., Olson J. Increased land use by Chukchi Sea polar bears in relation to changing sea ice conditions, PLoS ONE 10(11), 2015, e0142213. Available from: https://doi.org/10.1371/journal.pone.0142213.

Rosen, D.A., 2021. Resting and swimming metabolic rates in juvenile walruses *(Odobenus rosmarus)*. Marine Mammal Science 37 (1), 162–172.

Rosner, C., 2019. Bering Sea ice conditions: winter 2019. University of Alaska Fairbanks (https://uaf-iarc.org/2019/04/11/bering-strait-sea-ice-conditions-winter-2019/) (accessed 25.07.20.).

Rysgaard, S., Glud, R.N., 2007a. The annual organic carbon budget of Young Sound, NE Greenland. In: Rysgaard, S., Glud, R.N. (Eds.), Carbon Cycling in Arctic Marine Ecosystems: Case Study Young Sound, 58. Monographs on Greenland, pp. 194–203.

Rysgaard, S., Glud, R.N., 2007b. Carbon cycling and climate change: predictions for a High Arctic marine system (Young Sound, NE Greenland). In: Rysgaard, S., Glud, R.N. (Eds.), Carbon Cycling in Arctic Marine Ecosystems: Case Study Young Sound, 58. *Monographs on Greenland*, pp. 206–214.

Rysgaard, S., Vang, T., Stjernholm, M., Rasmussen, B., Windelin, A., Kiilsholm, S., 2003. Physical conditions, carbon transport, and climate change impacts in a Northeast Greenland Fjord. Arctic, Antarctic, and Alpine Research 35, 301–312.

Rysgaard, Glud, R.N., Rysgaard, S., Kühl, M., Hansen, J.W., 2007. The sea ice in Young Sound: implications for carbon cycling. In: Rysgaard, S., Glud, R.N. (Eds.), Carbon cycling in Arctic Marine Ecosystems: Case Study - Young Sound, 58. *Monographs on Greenland*, pp. 62–85.

Salter, R.E., 1979. Site utilization, activity budgets, and disturbance responses of Atlantic walruses during terrestrial haul-out. Canadian Journal of Zoology 57, 1169–1180. Available from: https://doi.org/10.1139/z79-149.

Salter, R.E., 1980. Observations on social behaviour of Atlantic walruses (*Odobenus rosmarus* (L.)) during terrestrial haul-out. Canadian Journal of Zoology 58, 461–463.

Scotter, S.E., Tryland, M., Nymo, I.H., Hanssen, L., Harju, M., Lydersen, C., et al., 2019. Contaminants in Atlantic walruses Part 1: relationships between exposure, diet and pathogen prevalence. Environmental Pollution 244, 9–18. Available from: https://doi.org/10.1016/j.envpol.2018.10.001.

Sejr, M.K., Jensen, K.T., Rysgaard, S., 2000. Macrozoobenthic community structure in a high-arctic East Greenland fjord. Polar Biology 23, 792–801.

Sejr, M.K., Sand, M.K., Jensen, K.T., Petersen, J.K., Christensen, P.B., Rysgaard, S., 2002. Growth and production of *Hiatella arctica* (Bivalvia) in a high-Arctic fjord (Young Sound, Northeast Greenland). Marine Ecology Progress Series 244, 163–169.

Sejr, M.K., Petersen, J.K., Jensen, K.T., Rysgaard, S., 2004. Effect of food concentration on clearance rate and energy budget of the Artic bivalve *Hiatella arctica* (L) at subzero temperature. Journal of Experimental Marine Biology and Ecology 311, 171–183.

Sejr, M.K., Blicher, M.E., Rysgaard, S., 2009. Sea ice cover affects inter-annual and geographic variation in growth of the Arctic cockle *Clinocardium ciliatum* (Bivalvia) in Greenland. Marine Ecology Progress Series 389, 149–158. Available from: https://doi.org/10.3354/meps08200.

Seymour, J., Horstmann-Dehn, L., Wooller, M.J., 2014a. Proportion of higher trophic level prey in the diet of Pacific walruses (*Odobenus rosmarus divergens*). Polar Biology 37, 941–952.

Seymour, J., Horstmann-Dehn, L., Wooller, M.J., 2014b. Inter-annual variability in the proportional contribution of higher trophic levels to the diet of Pacific walruses. Polar Biology 37, 597–609. Available from: https://doi.org/10.1007/s00300-014-1460-7.

Smaal, A.C., Ferreira, J.G., Grant, J., Petersen, J.K., Strand, Ø. (Eds.), 2019. Goods and Services of Marine Bivalves. Springer Open, ISBN 978-3-319-96775-2 ISBN 978-3-319-96776-9 (eBook) doi: 10.1007/978-3-319-96776-9.

Sonsthagen, S.A., Jay, C.V., Fischbach, A.S., Sage, G.K., Talbot, S.L., 2012. Spatial genetic structure and asymmetrical gene flow within the Pacific walrus. Journal of Mammalogy 93, 1512–1524. Available from: https://doi.org/10.1644/11-MAMM-A-344.1.

Sonsthagen, S.A., Fales, K.R., Jay, C.V., Sage, G.K., Talbot, S.L., 2014. Spatial variation and low diversity in the major histocompatibility complex in walrus (*Odobenus rosmarus*). Polar Biology 2014, 497–506. Available from: https://doi.org/10.1007/s00300-014-1450-9.

Sonsthagen, S.A., Jay, C.V., Cornman, R.S., Fischbach, A.S., Grebmeier, J.M., Talbot, S. L., 2020. DNA metabarcoding of feces to infer summer diet of Pacific walruses. Marine Mammal Science 1–16. Available from: https://doi.org/10.1111/mms.12717.

Speckman, S.G., Chernook, V.I., Burn, D.M., Udevitz, M.S., Kochnev, A.A., Vasilev, A., et al., 2011. Results and evaluation of a survey to estimate Pacific walrus population size, 2006. Marine Mammal Science 27, 514–553.

Stein, R., MacDonald, R.W. (Eds.), 2004. The Organic Carbon Cycle in the Arctic Ocean. Springer Verlag, Berlin, Germany, pp. 1–363.

Steiner, N., Azetsu-Scott, K., Hamilton, J., Hedges, K., Hu, X., Janjua, M.Y., et al., 2015. Observed trends and climate projections affecting marine ecosystems in the Canadian Arctic. Environmental Review 23 (2), 191–239. Available from: https://doi.org/10.1139/er-2014-0066.

Stewart, R.E.A., 2008. Redefining walrus stocks in Canada. Arctic 61, 292–398. Available from: http://www.jstor.org/stable/40513028.

Stewart, R.E.A., Born, E.W., Dunn, J.B., Koski, W.R., Ryan, A.K., 2014. Use of multiple methods to estimate walrus *(Odobenus rosmarus rosmarus)* abundance in the Penny Strait-Lancaster Sound and West Jones Sound Stocks, Canada. NAMMCO Scientific Publications 9, 95–116. Available from: https://doi.org/10.7557/3.2608.

Stirling, I., Laidre, K.L., Born, E.W., 2021. Do wild polar bears (*Ursus maritimus*) use tools when hunting walruses (*Odobenus rosmarus*)? Arctic, 74, 17–187.

Taylor, R.L., Udevitz, M.S., 2015. Demography of the Pacific walrus (*Odobenus rosmarus divergens*): 1974–2006. Marine Mammal Science 31, 231–254. Available from: https://doi.org/10.1111/mms.12156.

Taylor, R.L., Udevitz, M.S., Jay, C.V., Citta, J.J., Quakenbush, L.T., Lemons, P.R., et al., 2018. Demography of the Pacific walrus (*Odobenus rosmarus divergens*) in a changing Arctic. Marine Mammal Science 34, 54–86. Available from: https://doi.org/10.1111/mms.12434.

Tempel, J.T.L., Atkinson, S., 2020. Pacific walrus (*Odobenus rosmarus divergens*) reproductive capacity changes in three time frames during 1975–2010. Polar Biology 43, 861–875. Available from: https://doi.org/10.1007/s00300-020-02693-4.

Tremblay, J.E., Bélanger, S., Barber, D.G., Asplin, M., Martin, J., Darnis, G., et al., 2011. Climate forcing multiplies biological productivity in the coastal Arctic Ocean. Geophysical Research Letters 38, L18604.

Tremblay, J.E., Robert, D., Varela, D.E., Lovejoy, C., Darnis, G., Nelson, R.J., et al., 2012. Current state and trends in Canadian Arctic marine ecosystems: I. Primary production. Climatic Change 115, 161–178. Available from: https://doi.org/10.1007/s10584-012-0496-3.

Udevitz, M.S., Taylor, R.L., Garlich-Miller, J.L., Quakenbush, L.T., Snyder, J.A., 2013. Potential population-level effects of increased haul-out related mortality of Pacific walrus calves. Polar Biology 36, 291–298.

USFWS (U.S. Fish and Wildlife Service), 2014. Stock Assessment Report - Pacific walrus (*Odobenus rosmarus divergens*): Alaska Stock. Revised: April 2014. U.S. Fish and Wildlife Service, Anchorage, AK. (FWS-R7-ES-2012-0019-0003), 1–30.

USFWS (U.S. Fish and Wildlife Service), 2017. Endangered and threatened wildlife and plants; 12-month findings on petitions to list 25 species as endangered or threatened species. Federal Register/Vol. 82, No. 192/Thursday, October 5, 2017/Proposed Rules, 46618–46645. < https://www.regulations.gov/document?D = FWS-R7-ES-2017-0069-0001 >.

VanWormer E., Mazet J.A.K., Hall A., Gill V.A., Boveng P.L., London J.M., et al., 2019. Viral emergence in marine mammals in the North Pacific may be linked to Arctic sea ice reduction, Scientific Reports 9(1), pp. 1–11.

Vibe, C., 1950. The marine mammals and the marine fauna in the Thule District (Northwest Greenland) with observations on the ice conditions in 1939–1941. Monographs on Greenland 150, 1–115.

Wassmann, P., Reigstad, M., Haug, T., Rudels, B., Carroll, M.L., Hop, H., et al., 2006. Food webs and carbon flux in the Barents Sea. Progress in Oceanography 71, 232–287.

Wassmann, P., Duarte, C.M., Agusti, S., Sejr, M.K., 2011. Footprints of climate change in the Arctic marine ecosystem. Global Change Biology 17, 1235–1249. Available from: https://doi.org/10.1111/j.1365-2486.2010.02311.x.

Wiig, Ø., Gjertz, I., Griffiths, D., 1996. Migration of walruses (*Odobenus rosmarus*) in the Svalbard and Franz Josef Land area. Journal of Zoology (Lond) 238, 769–784.

Yurkowski, D.J., Brown, T.A., Blanchfield, P.J., Ferguson, S.H., 2020. Atlantic walrus signal latitudinal differences in the long-term decline of sea ice-derived carbon to benthic fauna in the Canadian Arctic. Proceedings of the Royal Society B 287, 20202126. Available from: https://doi.org/10.1098/rspb.2020.2126.

Index

Note: Page numbers followed by "*f*," "*t*," and "*b*" refer to figures, tables, and boxes, respectively.

A

U

V

W

Printed in the United States
by Baker & Taylor Publisher Services